INTELLIGENT SYSTEMS

Technology and Applications

VOLUME I

Implementation Techniques

Edited by
Cornelius T. Leondes

INTELLIGENT SYSTEMS

Technology and Applications

VOLUME I

Implementation Techniques

CRC PRESS

Boca Raton   London   New York   Washington, D.C.

Library of Congress Cataloging-in-Publication Data

Intelligent systems : technology and applications / edited by Cornelius T. Leondes.
    p. cm.
    Includes bibliographical references and index.
    Contents: v. 1. Implementation techniques -- v. 2. Fuzzy systems, neural networks, and expert systems -- v. 3. Signal, image, and speech processing -- v. 4. Database and learning systems -- v. 5. Manufacturing, industrial, and management systems -- v. 6. Control and electric power systems.
    ISBN 0-8493-1121-7 (alk. paper)
    1. Intelligent control systems. I. Leondes, Cornelius T.

TJ217.5 .I5448 2002
629.8--dc21   2002017473

Visit the CRC Press Web site at www.crcpress.com

© 2003 by CRC Press LLC

No claim to original U.S. Government works
International Standard Book Number 0-8493-1121-7
Library of Congress Card Number 2002017473
Printed in the United States of America 1 2 3 4 5 6 7 8 9 0
Printed on acid-free paper

Foreword

Intelligent Systems: Technology and Applications is a significant contribution to the artificial intelligence (AI) field. Edited by Professor Cornelius Leondes, a leading contributor to intelligent systems, this set of six well-integrated volumes on the subject of intelligent systems techniques and applications provides a valuable reference for researchers and practitioners. This landmark work features contributions from more than 60 of the world's foremost AI authorities in industry, government, and academia.

Perhaps the most valuable feature of this work is the breadth of material covered. Volume I looks at the steps in implementing intelligent systems. Here the reader learns from some of the leading individuals in the field how to develop an intelligent system. Volume II covers the most important technologies in the field, including fuzzy systems, neural networks, and expert systems. In this volume the reader sees the steps taken to effectively develop each type of system, and also sees how these technologies have been successfully applied to practical real-world problems, such as intelligent signal processing, robotic control, and the operation of telecommunications systems. The final four volumes provide insight into developing and deploying intelligent systems in a wide range of application areas. For instance, Volume III discusses applications of signal, image, and speech processing; Volume IV looks at intelligent database management and learning systems; Volume V covers manufacturing, industrial, and business applications; and Volume VI considers applications in control and power systems. Collectively this material provides a tremendous resource for developing an intelligent system across a wide range of technologies and application areas.

Let us consider this work in the context of the history of artificial intelligence. AI has come a long way in a relatively short time. The early days were spent in somewhat of a probing fashion, where researchers looked for ways to develop a machine that captured human intelligence. After considerable struggle, they fortunately met with success. Armed with an understanding of how to design an intelligent system, they went on to develop useful applications to solve real-world problems. At this point AI took on a very meaningful role in the area of information technology.

Along the way there were a few individuals who saw the importance of publishing the accomplishments of AI providing guidance to advance the field. Among this small group I believe that Dr. Leondes has made the largest contribution to this effort. He has edited numerous books on intelligent systems that provide a wealth of information to individuals in the field. I believe his latest work discussed here is his most valuable contribution to date and should be in the possession of all individuals involved in the field of intelligent systems.

Jack Durkin

Preface

For most of our history the wealth of a nation was limited by the size and stamina of the work force. Today, national wealth is measured in intellectual capital. Nations possessing skillful people in such diverse areas as science, medicine, business, and engineering, produce innovations that drive the nations to a higher quality of life. To better utilize these valuable resources, intelligent systems technology has evolved at a rapid and significantly expanding rate to accomplish this purpose. Intelligent systems technology can be utilized by nations to improve their medical care, advance their engineering technology, and increase their manufacturing productivity, as well as play a significant role in a very wide variety of other areas of activity of substantive significance.

Intelligent systems technology almost defines itself as the replication to some effective degree of human intelligence by the utilization of computers, sensor systems, effective algorithms, software technology, and other technologies in the performance of useful or significant tasks. Widely publicized earlier examples include the defeat of Garry Kasparov, arguably the greatest chess champion in history, by IBM's intelligent system known as "Big Blue." Separately, the greatest stock market crash in history, which took place on Monday, October 19, 1987, occurred because of a poorly designed intelligent system known as computerized program trading. As was reported, the Wall Street stockbrokers watched in a state of shock as the computerized program trading system took complete control of the events of the day. Alternatively, a significant example where no intelligent system was in place and which could have, indeed no doubt would have, prevented a disaster is the Chernobyl disaster which occurred at 1:15 A.M. on April 26, 1987. In this case the system operators were no doubt in a rather tired state and an effectively designed class of intelligent system known as "backward chaining" EXPERT System would, in all likelihood, have averted this disaster.

The techniques which are utilized to implement Intelligent Systems Technology include, among others:

Knowledge-Based Systems Techniques
EXPERT Systems Techniques
Fuzzy Theory Systems
Neural Network Systems
Case-Based Reasoning Methods
Induction Methods
Frame-Based Techniques
Cognition System Techniques

These techniques and others may be utilized individually or in combination with others.

The breadth of the major application areas of intelligent systems technology is remarkable and very impressive. These include:

Agriculture	Law
Business	Manufacturing
Chemistry	Mathematics
Communications	Medicine
Computer Systems	Meteorology
Education	Military
Electronics	Mining
Engineering	Power Systems
Environment	Science
Geology	Space Technology
Image Processing	Transportation
Information Management	

It is difficult now to find an area that has not been touched by Intelligent Systems Technology. Indeed, a perusal of the tables of contents of these six volumes, *Intelligent Systems: Technology and Applications*, reveals that there are substantively significant treatments of applications in many of these areas.

Needless to say, the great breadth and expanding significance of this field on the international scene requires a multi-volume set for an adequately substantive treatment of the subject of intelligent systems technology. This set of volumes consists of six distinctly titled and well-integrated volumes. It is appropriate to mention that each of the six volumes can be utilized individually. In any event, the six volume titles are:

1. Implementation Techniques
2. Fuzzy Systems, Neural Networks, and Expert Systems
3. Signal, Image, and Speech Processing
4. Database and Learning Systems
5. Manufacturing, Industrial, and Management Systems
6. Control and Electric Power Systems

The contributors to these volumes clearly reveal the effectiveness and great significance of the techniques available and, with further development, the essential role that they will play in the future. I hope that practitioners, research workers, students, computer scientists, and others on the international scene will find this set of volumes to be a unique and significant reference source for years to come.

Cornelius T. Leondes
Editor

About the Editor

Cornelius T. Leondes, B.S., M.S., Ph.D., Emeritus Professor, School of Engineering and Applied Science, University of California, Los Angeles, has served as a member or consultant on numerous national technical and scientific advisory boards. Dr. Leondes served as a consultant for numerous Fortune 500 companies and international corporations. He has published over 200 technical journal articles and has edited and/or co-authored more than 120 books. Dr. Leondes is a Guggenheim Fellow, Fulbright Research Scholar, and IEEE Fellow, as well as a recipient of the IEEE Baker Prize award and the Barry Carlton Award of the IEEE.

Contributors

Chua Chee-Kai
Nanyang Technological
University
Singapore

Cheah Chi-Mun
Nanyang Technological
University
Singapore

Andrew W. Crapo
GE Corporate Research and
Development
Niskayuna, New York

He Du
Nanyang Technological
University
Singapore

John Durkin
Intelligent Computer
Systems, Inc.
Akron, Ohio

Robert Gay
Nanyang Technological
University
Singapore

Angela Goh
Nanyang Technological
University
Singapore

Wey-Shiuan Hwang
Michigan State University
East Lansing, Michigan

Hisao Ishibuchi
Osaka Prefecture University
Osaka, Japan

James Jiang
University of
Central Florida
Orlando, Florida

Gary Klein
University of Colorado
Boulder, Colorado

Samuel Y.E. Lim
Nanyang Technological
University
Singapore

Chunyan Miao
Simon Fraser University
Burnaby B.C., Canada

Yuan Miao
Nanyang Technological
University
Singapore

Guy W. Mineau
Université Laval
Quebec City, Quebec
Canada

Tomoharu Nakashima
Osaka Prefecture University
Osaka, Japan

Roger Pick
University of
Missouri
Columbia, Missouri

Ryoji Sakamoto
Osaka Prefecture University
Osaka, Japan

Svetla Vassileva
Bulgarian Academy of
Sciences
Sofia, Bulgaria

George-C. Vosniakos
National Technical
University of Athens
Athens, Greece

Laurie B. Waisel
Concurrent Technologies
Corporation
Johnstown, Pennsylvania

William A. Wallace
Rensselaer Polytechnic
Institute
Troy, New York

Xue Z. Wang
The University of Leeds
Leeds, U.K.

Juyang Weng
Michigan State University
East Lansing, Michigan

Thomas R. Willemain
Rensselaer Polytechnic
Institute
Troy, New York

Zhonghua Yang
Nanyang Technological
University
Singapore

Xuan F. Zha
Gintic Institute of
Manufacturing Technology
Singapore

Yilu Zhang
Michigan State University
East Lansing, Michigan

Contents

Volume I: Implementation Techniques

Contents

Volume II: Fuzzy Systems, Neural Networks, and Expert Systems

Contents

Contents

Contents

Volume V: Manufacturing, Industrial, and Management Systems

Contents

1

The Quest for the Intelligent Machine

John Durkin
Intelligent Computer Systems, Inc.

1.1 Introduction

Man has been searching for the intelligent machine for centuries. We are not there yet, but we're getting closer. With patience and a great deal of more work, it is inevitable that the day will arrive.

For most of our history the wealth of a nation was limited by the size and stamina of the work force. With the Industrial Revolution came a new opportunity for increasing national wealth. Machines fueled by steam and oil were developed to assist labor-intensive tasks, providing dramatic increases in productivity. Today, national wealth is measured in intellectual capital. Nations possessing skillful people in such diverse areas as science, medicine, business, and engineering, produce innovations that drive the nation toward a higher quality of life. To better use these valuable resources, new machines were sought that could capture the expertise of these talented people. This quest has placed us in search of machines powered not by steam but by knowledge.

Success in this quest could provide an enormous benefit to mankind. Nations lacking the expertise could use these machines to improve their medical care, advance their engineering technology, or increase their manufacturing productivity. Companies could use them to assist decision making or to proliferate throughout their organization the skills of a scarce number of experts. Researchers could also use them to better understand how humans reason.

The quest for the intelligent machine is top-heavy with moral and ethical questions. Can we replicate human consciousness in machines? If so, then what does that tell us about consciousness? What does it mean to be human? These questions have become the subject of serious debate among scientists and science fiction writers.

Most of us associate the intelligent machine with artificial intelligence (AI), a term first coined at the Dartmouth College conference in 1956. However, the quest for the intelligent machine has been going on for over 2000 years, motivated in part by the practical benefits offered with success and in part by the sheer fascination of the effort. Philosophers, mathematicians, and technicians have contributed to the quest. In most cases, new discoveries rely on previous labors by others, so that while those other achievements may be small, they are seminal. This chapter reviews these important contributions and shows how they shaped our present day intelligent machine.

1.2   Philosophy (470–322 BC)

Perhaps the genesis of the quest for an intelligent machine can be traced to Greek mythology. In the myth of Hephaestus, as a present from Zeus to Europa, Hephaestus creates Talos, a man of bronze whose duty is to defend the beaches of Crete. Talos thwarts invaders by hurling huge rocks at them, or by heating himself red-hot and caressing them in his warm embrace.

In the story of the Golem, a clay statue is brought to life by a Jewish Rabbi. The Golem, which is a perfect servant, is animated by placing in its mouth a piece of paper containing a sacred word or the name of the one of the Gods. The Rabbi was thrilled with his creation, until he realized that the only fault of the Golem was to take his master's orders too literally, with frightening results. This legend marks the first written account of man's fear of the artificially created person.

There is also a myth where an object of love is created and brought to life. Pygmalion, the king of Cyprus, fell in love with an ivory statue of his own making. He was fixated, obsessed, and in love as he lay with his statue on a soft bed. During the festival of the goddess Aphrodite, he prayed to the goddess for a wife as beautiful as his statue. Aphrodite answered his prayer by breathing life into the statue, and stood at the side of Pygmalion and his ivory maid as they wed.

The myth of Prometheus points out the danger in searching for knowledge. In an act of transgression against the gods of Olympus, Prometheus tricked Zeus into accepting the bones and fat of sacrifice instead of the meat. Zeus responded by hiding fire from man; where fire symbolized enlightenment or knowledge. Prometheus succeeded in stealing the fire back to enlighten humanity through intelligence or *nous*, the "arational mind." While his action freed humanity from ignorance, it earned Prometheus the wrath of Zeus. Zeus had him chained to a barren rock and sent an eagle to eat his immortal liver, which constantly replenished itself for eternity. The notion that human efforts to gain knowledge represent a transgression against the laws of God or nature, and ultimately leads to disaster, is deeply ingrained in human thought.

These myths point to man's early fascination in creating artificial beings and the human emotions involved in the effort. Over the centuries they influenced others who likewise conjured up man-made beings: Shelley's Frankenstein's monster, who occupied our nightmares, and Collodi's Pinocchio, who made us smile, are some examples. More recent ones, that are more in tune with AI, are Arthur C. Clarke's HAL, who we viewed as a ominous entity, and Steven Spielberg's young lad John in the movie A.I., who we cherished because he could love even though he was not real.

These fascinating stories form an early backdrop to the quest of the intelligent machine. They influenced early philosophers who developed theories surrounding human reasoning that guided the quest. Maybe a good place to start to gain insight into these efforts is ancient Greece, with a look at the work of Socrates, Plato, and Aristotle, each of whom helped to define the philosophies that shaped western culture.

Socrates (470–399 BC) held a strong belief in the doctrine that the "mind" is the source of all cosmic order. He was a man with strong ethical beliefs, extraordinary insightful rational thought, and, at times, a touch of arrogance—he believed himself charged by God with the mission to make his fellow men aware of their ignorance. It is in his attempts to obtain an understanding of human reasoning that we find his genius, and his contribution to present day AI.

Socrates's vision of the *universal definition* (what the Phaedo calls a form), for example, is an attempt to formulate the meaning of a significant predicate, that is, an important point of life that would be universally accepted through pure logic. To illustrate, Plato reported a dialogue in which Socrates asks Euthyphro, "I want to know what is characteristic of piety which makes all actions pious . . . that I may have it to turn to, and to use as a standard whereby to judge your actions and those of other men." Socrates's concept of a universal definition (or form) contains a pattern of a particular category of things in the world, such as man, woman, and stone, or even intangible things, such as beauty and piety. Using a piety form as sort of a template, e.g., he could use it to judge a man's moral character, or, more likely when considering Socrates's ego, to point out his imperfections. Viewing Socrates's universal definition today, it is easy to argue that it is the basis of a *class* in modern-day object-oriented programs and frame-based expert systems.

In another example, we can find his contribution to today's *rule-based* expert systems. Socrates was intrigued by logical thinking, particularly concerning what *deductions* one could make from available facts. In his studies, he would start with some reasonable hypothesis given the available information and then consider the consequences that naturally follow from it. If these consequences can be reasonably expected, he argued that the hypothesis might be regarded as provisionally confirmed. Aristotle would later refine this method with the introduction of a syllogism.

Socrates was the first of the great Greek philosophers to study human intelligence and to set forth concepts to explain human reasoning. Unfortunately, he was not a writer. He conveyed his ideas verbally to his students, who expanded them and, fortunately for the rest of us, put them in print. His most notable pupil was Plato.

Building on the life and thought of Socrates, Plato (428–348 BC) developed a profound and very broad study of philosophy that was based on the proposition that reason must be followed wherever its leads. We can see his brilliance in his writings across a wide range of subjects, which include mathematics, physics, astronomy, politics, and philosophy. Carrying forward Socrates's work on universal definitions (forms) is one of his major contributions.

Socrates was convinced that knowing rests in the identification of objects or ideas that could be described in inrefutable terms. Plato discovered that such predicate items do exist, e.g., a circle. To say, "This is a circle," is to attribute a certain property to the specific object, such as its radius. In this example, Plato is distinguishing between a specific circle that can be drawn knowing its radius, and the common property of the predicate object. Plato called the predicate object a *form* and the specific object a *particular*. Today, we simply have renamed these *class* and *instance*, respectively.

To the extent that humans have knowledge, Plato believed that they attained it through reasoning that leads to the discovery of forms. His theory of intelligence is thus composed of two parts: the *description of knowledge* contained in forms and the quest for knowledge by way of *reasoning*. These two parts represent the basic architecture of today's expert system where we find a *knowledge base* and *inference engine*.

Plato claimed that the search for definitions, and thereby the nature of forms, is a search for knowledge. He wrote that Socrates attempted to find a definition that would be immune to counterexamples. A definition is proposed and if any counterexample can be found, then it is rejected; otherwise it is accepted. Aristotle credits Socrates with another method based on "inductive arguments," which attempts to derive a definition through the consideration of concrete illustrations. For example, the consideration of "shoemakers and carpenters," which the fashionable speakers regarded as "vulgar laborers." Here, *induction* is viewed as a function which puts the meaning of a proposed definition before the mind. Today, we build *induction-based* expert systems using somewhat the same approach. Plato continued the work begun by the sophists and by Socrates. This effort was later raised to a high art by Plato's pupil Aristotle.

Aristotle (384–322 BC) inherited from Plato a deep appreciation of theories that might explain the human intellectual process and a vast body of problems to test these theories. Fortunately, Aristotle possessed a highly analytical mind and sought to develop logical methods to support his studies.

This rare combination of an understanding of both philosophy and logic enabled him to expand on the intellectual tradition bequeathed him.

Aristotle believed that the study of logical thought was the path to an understanding of human intelligence. In his *Logic*, he viewed knowledge in terms of necessary propositions that express causal relations that take the form of categorical *syllogisms*. A syllogism is a form of an argument that proposes that a given proposition can be said to be true because it is related to another proposition that is known to be true. According to Aristotle, a syllogism contains three categorical propositions (two premises and a conclusion). The form of a "universally affirmative" syllogism is: If every β is an α and every γ is a β, then every γ is an α, where α, β, and γ are variables. Any argument that fits this pattern is a valid syllogism. For example, if we know that "every person is mortal" and that "every Greek is a person," then we can conclude that "every Greek is mortal." Aristotle was the first logician to use variables, and his use of them in syllogisms later became a standard way of processing knowledge in the *predicate calculus*.

Aristotle argued for the application of his logical theory for obtaining a fundamental understanding of the sciences. In his *Posterior Analytics*, he wrote that each science must depend on a set of first principles, or axioms, that are necessarily true, and that the theorems that compose a science are deduced from its axioms. He also believed that all of these scientific deductions could be formed by way of syllogisms. Following this line of thought, he devoted much of the second book of the *Posterior Analytics* to Socrates's theory of the universal definition, because he felt that the most important axioms of any science would be the definitions that would constitute the field.

When we reflect on the contributions of philosophy to AI, our thoughts are naturally drawn to early Greece, where the study of *epistemology* was born; the name is derived from the Greek words *episteme* (knowledge) and *logos* (reason). Today it represents one of the four main fields of philosophy. During the past 2000 years, through the Renaissance, the Age of Enlightenment, the Industrial Revolution, and the Information Age, philosophers in the field have added to our understanding of the human intellect, but it will always be the three great Greek philosophers—Socrates, Plato, and Aristotle—to whom the field of AI is indebted. In our tour of the history of the intelligent machine, let us fast forward to the 1800s, where we find efforts to build such a machine.

1.3   Mechanics (1800s)

During the early 1800s an enterprising group of individuals toured Europe and America demonstrating a device they called the "Chess Automaton." This was a large mechanical box which they advertised as being capable of playing chess on a par with humans. Like other mechanical devices of the day, it contained gears and levers that the spectators were told would move in accordance with deciding the next chess move. However, somewhat unlike the other devices, it also included one of the better chess players of the day. This famous hoax illustrates man's fascination with intelligent machines even in those early times.

In 1834, Charles Babbage conceived the design of the first mechanical computer, which he called the "Analytical Engine" (McCorduck, 1977), that foreshadowed many of the architectural plans of the modern computer. His design was the first mechanical rendition of Plato's notion of the separation of knowledge (the engine's memory which Babbage called the "store") and reasoning (the engine's processor which he called the "mill"). His design also introduced the concept of a programmable machine that could execute a series of operations encoded on punched cards. Babbage's friend and collaborator, Ada Lovelace, daughter of the poet Lord Byron, wrote programs for the Analytical Engine and went on to suggest that the machine might someday be able to play chess. We might consider Lovelace to be the world's first programmer.

Beyond the technological innovations offered by the Analytical Engine, we can find a value that is closer to the core objective of artificial intelligence. Babbage's work put forth the idea that an intellectual activity could be represented in a pattern of steps that could be encoded in a machine, which could then be studied in order to better understand the activity. On his effort to build the engine he said, "If I should be successful ... it will thus call into action a permanent cause of advancement toward the truth, continually leading to the more accurate determination of established fact, and to the discovery and measurements of new ones." Unfortunately, the Analytical Engine was never built, for a reason that has become an all too often event within the history of the quest for the intelligent machine.

To obtain funding to build his machine, Babbage met with Prime Minister Lord Melbourne, who was responsible for government sponsored research in England. To reach a decision, Melbourne sought the advice of Reverend Richard Sheepshanks, a secretary of the Royal Astronomical Society (and also Babbage's archenemy). Sheepshanks wrote of Babbage in his book *The Exposition of 1851*, "He was ill-judged enough to press the consideration of this new machine upon the members of Government." Melbourne's decision was then obvious. Failing to obtain the necessary funding, Babbage gave up all expectations of building the Analytical Engine and later wrote of his disillusionment, "Thus bad names are coined by worse men to destroy honest people." Skepticism, personal bias, and frustration are some of the human emotions that have marred the quest.

Logic has been called the grammar of reason. With its exact syntax and logical rules, logicians have long argued that logic offers a tool for creating a formal language for thought. There are two mathematicians during this era who agreed with this argument and went on to make major contributions to the field, and whose work later would have a profound impact on AI.

The most important contributor to the field of logic during the early nineteenth century was the English mathematician George Boole, who established modern symbolic logic and the algebra of logic called Boolean Algebra. Although the importance of the role of his Boolean Algebra in designing digital logic circuits is well known, Boole's real motivation for developing his system of symbolic logic appears to be well aligned with the goals of contemporary AI researchers. In his 1854 writing, *An Investigation of the Laws of Thought, on Which Are Founded the Mathematical Theories of Logic and Probabilities*, he stated his goals as:

> ... to investigate the fundamental laws of those operations of the mind by which reasoning is performed: to give expression to them in the symbolical language of a Calculus, and upon this foundation to establish the science of logic and instruct its method; ... and finally to collect from the various elements of truth brought to view in the course of these inquiries some probable intimations concerning the nature and constitution of the human mind.

In 1847, Boole published another important work, *The Mathematical Analysis of Logic*, where he presented his general symbolic method of logical inference. Given a rule containing premises and conclusions represented as symbolic propositions, Boole described how one can draw logical conclusions through the symbolic treatment of the premises. He went further and discussed his ambitious goal to symbolize, systematize, and generalize the concepts of logic invented by the early Greek philosophers. He wrote, "I propose to establish the Calculus of Logic." In so doing, he laid the groundwork for the *propositional calculus* and the study of symbolic logic.

A proposition is a statement about the world that we consider as either true or false. For example, we might state, "Weather on weekends is lousy." In propositional calculus, symbols are associated with various propositions. By joining propositions together using *conjuncts* (ands), *disjuncts* (ors), and *implications* (if–thens), rules can be written to both capture and reason about knowledge. Propositional calculus was the first attempt to apply formal logic to the processing of symbols. However, it was limited in its ability to offer a practical tool for managing complex problems, because each symbol could represent a proposition of some complexity and there is no way to access the components of the proposition; a requirement in most real intelligent machine applications. Fortunately, this situation changed in only a few years.

In 1879, the German mathematician and logician Gottlob Frege published the book *Begriffsschrift*. In it he wrote of the use of a *predicate* that describes a relationship between components of a proposition. For example, instead of assigning a symbol to the entire statement, "Weather on weekends is lousy," we can create a predicate *weather* and write "weather (weekends, lousy)." This idea alone was important because it offered the opportunity to describe a given proposition at the level of granularity that would be needed in intelligent systems. However, even better, Frege went much further. He was as much a philosopher as he was a logician. His primary goal was to create a formal mathematical language for logical deductive reasoning. In addition to introducing the use of a predicate to describe a proposition, he introduced quantifiers, functions, and variables into his symbolic algebra, which offered a language for describing mathematics and its philosophical foundations. His language, now called the *predicate calculus*, later led to the popular AI programming language PROLOG in the twentieth century.

1.4   Computers and a First Glimpse at AI (1940s)

Prior to the twentieth century, much of the scientific and mathematic intellectual prerequisites needed to create an intelligent machine were in place. Next the proper vehicle was needed. Mechanical machines, with all the gears and levers in the world, were not the answer. The search for an intelligent machine was waiting for technology to catch up to it. It was waiting for the electronic computer.

The earliest rendition of the modern digital computer was developed independently by engineers in three countries embattled in World War II. In 1940, Alan Turing and his team of researchers in England built the first operational computer called the Heath Robinson for the exclusive purpose of deciphering German coded messages. The system was effective but slow, because its design relied heavily on the use of electromechanical relays. As the German codes became more sophisticated, making the Robinson obsolete, a new computer called the Colossus was built in 1943 using vacuum tubes, which provided the necessary speed to decode the messages.

The first general-purpose *programmable* computer, called the Z-3, was developed in Germany in 1941. It was built by Konrad Zuse, who also invented floating-point numbers for the computer. Later, in 1945, Zuse went on to invent the first high-level programming language called *Plankalkul*. The Z-3 project was partially funded by the Third Reich to assist in aircraft design. Beyond this application, however, the German military hierarchy did not place as much importance in computer technology as did their counterpart in England.

The credit for the first *electronic* computer should go to John Atanasoff and his graduate student, Clifford Berry, at Iowa State University. In 1939, they built a prototype computer using 300 vacuum tubes and capacitors for memory, that stored data in binary form. They called their invention the Atanasoff-Berry Computer, or ABC. The prototype was built in a couple of months and work continued for three years to build the complete system. However, with the onset of World War II, the project was abandoned when Atanasoff moved to Washington to work on a military research project.

Two other computer-related efforts in the United States were conducted during this period in support of military research. At Harvard University, Howard Aiken began work in 1939 on a machine that could perform the operations of addition, subtraction, multiplication, division, and reference to previous results. His machine, which he called the Mark I, was completed in 1944. The computer was used by the Navy for work in gunnery, ballistics, and design. At the University of Pennsylvania, John Mauchly and John Eckert were asked to devise ways to accelerate the recomputation of artillery firing tables for the U.S. Army. Their work led to the construction of a general-purpose digital computer called ENIAC, which contained more than 18,000 vacuum tubes and incorporated many of the features first introduced by Atanasoff in the ABC computer. Recognizing the potential commercial value of the technology, Mauchly later went on to develop UNIVAC I, which was specifically designed to handle business data.

The introduction of digital computers made AI a viable scientific discipline, because they provided the necessary memory and processing power needed to develop intelligent programs. Researchers were quick to seize the opportunity.

Most historic accounts of AI usually point to 1956 and the conference held at Dartmouth College as marking the birth of artificial intelligence. In the next section, we will take a look at this conference and its impact on the field. It can be argued, however, that the first work that is now generally recognized as AI (though it didn't come with the AI label) was done by Warren McCulloch and Walter Pitts (1943). Their work was biologically inspired where they drew upon knowledge of the basic physiology and function of neurons in the brain to propose the creation of a network of artificial neurons which could model human reasoning. They showed how propositions of logic and any computable function could be computed by some network of connected neurons. They also felt that future physiological studies of human reasoning could be aided by the design of various sorts of simple neural networks. This novel idea of modeling the structure and function of the brain offered a new hope for the quest of an intelligent machine.

The work of McCulloch and Pitts sparked a great amount of interest in neural computing. During the next 20 years, several developments occurred that seemed to indicate that researchers were on the right track using neural networks to replicate human reasoning in a machine. In 1949, Donald Hebb published the book *The Organization of Behavior* (Hebb, 1949) that proposed that the connectivity of neurons in the brain are continually changing as an organism learns differing functional tasks. He also proposed a learning law that showed, for the first time, that a network of neurons could exhibit learning behavior. In 1958, Frank Rosenblatt introduced the *perceptron*, and the *perceptron convergence theorem* (Rosenblatt, 1958). Following this revolutionary work, Rosenblatt built the first neurocomputer called the *Mark I Perceptron*.

In 1960, Bernie Widrow, working with his graduate student Ted Hoff (Hoff later went on to invent the microprocessor), introduced the *least means-square algorithm* and used it to formulate the *Adaline* (Widrow and Hoff, 1960). The Adaline was used for a variety of applications such as speech recognition, character recognition, weather prediction, and adaptive control. The success of the Adaline gave rise to the belief that artificial brains were right around the corner. This illusion was soon dispelled with the publication of the book *Perceptrons* by Minsky and Papert (1969). This book was mainly an attack on the perceptron. It showed the fundamental limits of a one-layer perceptron network in solving simple problems, such as the exclusive, or gate, function. The book also cast a shadow over the neural network field because it implied that essentially all neural networks suffer from the same flaw as the perceptron, and it went on to conclude that the technology will never make a major contribution to the AI field. Discouraged researchers left the field, funding dried up, and the study of neural networks lapsed into obscurity for nearly two decades. The damage was done. Rather than focusing on the data processing techniques of neural networks, efforts in the field centered on the symbolic processing methods favored by the attendees of the 1956 Dartmouth College conference.

1.5 Birth and Rise of AI (1950s and 1960s)

Most people in the field point to 1956 as marking the birth of artificial intelligence. During the summer of that year, ten of the leading computer scientists of the time attended a summer workshop at Dartmouth College sponsored by IBM. The main discussion was on their present research efforts in such areas as automatic theorem proving, neural networks, and new programming languages. They also discussed ways that their work might be directed to develop a computer that could simulate human reasoning. Two of the attendees, Allen Newell and Herbert Simon, both from Carnegie Tech (now Carnegie Mellon), had already developed a reasoning program for automatic theorem proving called the *Logic Theorist*, of which Simon claimed, "We have invented a computer program capable of thinking non-numerically." This statement served to convince the attendees that developing an intelligent machine was not only possible, it was probable. The Logic Theorist is considered by many to be the first AI program. No new breakthroughs came from this workshop, but it did put into motion the wheels that drove the research needed to create an intelligent computer. It also left us with one legacy which has both helped and haunted the field. The group of attendees decided to adopt John McCarthy's name for the field: **artificial intelligence**.

Artificial intelligence is an intimidating term. It brings forth images as threatening to us as the cotton gin was to farm laborers during the Industrial Revolution. The fear of displacement by automated machines with superior brute force has now been replaced by anxiety caused by the perceived risk of the loss to machines of our basic human trait: our intelligence. However, when we look beyond the alarming label to see what this field is about, a far less ominous picture is painted.

Most of the early work in AI was academic in nature, where programs were developed for playing games. A good example is the checkers-playing program developed by Arthur Samuel (1963). The program could beat most humans at the game and could improve its playing ability through automatic learning. Claude Shannon's (1955) chess-playing program is another good example. Many people outside the field were not impressed with these efforts; they considered the results only as interesting toys. Insiders, on the other hand, knew that ventures like these were exploratory in nature and were looking for ways to build intelligence into a computer.

Capturing the public's interest in a technology has always been important. Carl Sagan knew this. He recognized that communicating to the public the major events within the space program would draw the interest needed to further the program. His efforts had a profound effect on our imagination and NASA's projects. The AI field also has had its moments in the public's spotlight. Samuel's checkers-playing program was successfully demonstrated on television in February 1956, creating in the public's mind a vision of an intelligent machine. During the 1970s, the expert system PROSPECTOR generated enormous publicity by recommending exploratory drilling at a geological site that proved to contain a $100-million molybdenum deposit. More recently, the grand chess master Gary Kasparvor succumbed in a very well-publicized chess match to Deep Blue. A lesson for AI researchers from these events is that beyond their technical achievements, they should recognize the importance of communicating their successes to the public in order to capture the public's imagination.

In 1958, John McCarthy at MIT was looking for a high-level programming language which would enable him to perform symbolic processing. He thought that the ability to easily process symbols in a computer was needed in order to build an intelligent program. His search led him to develop the language LISP, which was first reported in MIT AI Lab Memo No. 1. LISP soon became the language of choice among AI researchers. It also played an important role in the expert systems field, where we can find many systems built using the language. LISP remains popular today and is the second oldest programming language, just one year younger than Fortran which was also developed at MIT.

In 1963, MIT received a two-million-dollar U.S. government contract to research machine-aided cognition. The contract by the Department of Defense's Advanced Research Projects Agency (ARPA) was granted to ensure that the U.S. would stay ahead of the Soviet Union in technological developments. The project served to increase the pace of development in AI research.

One of the more ambitious projects of this era involved the development of the General Problem Solver (GPS) (Newell and Simon 1972). GPS, which was built by the same pair who developed the Logic Theorist, was a general-purpose problem-solving technique, developed to solve a variety of problems ranging from games to symbolic integration. It was also designed to solve these problems using a human-like style of problem solving, by dissecting a given goal into subgoals and applying appropriate "operators" (such as legal chess moves) to accomplish them. In fact, having GPS solve a given problem in the same way that a human might was a major concern of Newell and Simon.

This theme is the basis of the interdisciplinary field of *cognitive science*, which brings together computer models from AI and experimental techniques from psychology to develop theories about the workings of the human mind. Also worthy of note is that GPS was the first attempt to separate the problem-solving methods from the knowledge of the problem—a key characteristic of expert systems. Though GPS demonstrated interesting results, it was soon recognized that its general-purpose problem-solving strategy was too weak for complex problems.

By 1970, the early euphoria surrounding AI had burned off and was replaced with the sobering realization that building intelligent programs to solve real-world problems was not as easy as first believed. To make matters worse, it was clear that the technology had not lived up to some rather bold predictions. For instance, in 1957 Herbert Simon said:

> *It is not my aim to surprise or shock you, but the simplest way I can summarize is to say that there are now in the world machines that think, that learn, and that create. Moreover, their ability to do these things is going to increase rapidly until—in a visible future—the range of the problems they can handle will be coextensive with the range to which human mind has been applied.*

Without losing steam, he later predicted (in 1958) that within 10 years a computer would be a chess champion. Similar optimistic predictions were being made by others in the field. Though it is important to spread the word about the potential of a new technology, it can be a dangerous practice. If more is promised than can be delivered, skepticism will rise and won-over converts may be lost. Worse, a general disinterest might set in, making it difficult to attract talented people and needed funding to the field—a scenario that can doom the field, as almost happened to AI following the Lighthill report (Lighthill, 1972).

1.6 Fall and Rebirth of AI (1970s)

In 1971, the Science Research Council of Britain commissioned Sir James Lighthill of Cambridge University to review the state of affairs in the AI field. The council was concerned. They had heard the many promises made and had provided considerable funding to support researchers in the field. They had also not seen much in return for their money and wanted to know if continuing to fund the effort was advisable. Lighthill reported, "In no part of the field have the discoveries made so far produced the major impact that was promised." He also noted that researchers at the time were predicting that " ... possibilities in the 1980s include an all-purpose intelligence on a human-scale knowledge base; that awe-inspiring possibilities suggest themselves based on machine intelligence exceeding human intelligence by the year 2000." Lighthill recognized that this was the same type of prediction researchers were making in the mid-fifties. In conclusion, he saw no need for a separate AI field and felt that if an intelligent computer were ever to be built, it would be a natural offspring of the combined efforts from the fields of automation and computer science.

The immediate impact of the Lighthill report was highly damaging to AI efforts in England and abroad. Funding started to evaporate and researchers began to look for greener pastures. Fortunately, mainly due to the efforts of a few researchers who remained and continued to labor within the shadow of this damning report, the field experienced a rebirth with a new direction.

On hindsight, the findings of the Lighthill report are highly debatable. We generally overestimate what a technology can do for us in a few years and underestimate what it can do in a few decades. When the report was written, AI was still a relatively new and emerging field with very few practical systems to show for the effort. Work to date was exploratory in nature. Most of the programs developed focused on games, such as checkers, chess, and robot control in a blocks world. To Lighthill, they probably appeared as toys with no real value for addressing practical problems. Researchers in the field, on the other hand, saw their programs as testbeds for studying the theories of intelligent machine design. Lighthill was right on one important point, however; no program had yet been written that could manage a real-world problem. Though his report was damaging, ironically, it led to events that would mark the rebirth of the field.

Following the report, researchers performed a critical assessment of their previous work, looking for reasons that might help explain why the technology was not living up to expectations. One point that became immediately clear was that most of the prior research efforts emphasized search techniques. The common thinking at the time was that intelligent human behavior was primarily based on smart reasoning techniques, and to model this in a computer required clever search algorithms. However, since earlier programs failed to produce meaningful results, researchers realized that search techniques alone were probably not enough to produce an intelligent program.

In 1965, there was an effort going on at Stanford University that caught the attention of the AI community. NASA was planning to send an unmanned spacecraft to Mars and wanted a computer program that could perform chemical analysis of the Martian soil. Given mass spectral data of the soil, NASA wanted the program to determine its molecular structure. In a chemistry laboratory, the traditional method of solving this problem is through a generate-and-test technique. Possible structures that could account for the mass spectrogram data are first generated and then each is tested to see if it matches the data. The basic difficulty with this technique is that there are millions of possible structures that might be generated. NASA turned to researchers at Stanford University to see if a program could be developed that could solve this problem. The Stanford team found in practice that there were knowledgeable chemists who first used *rules-of-thumb* (heuristics) to weed out structures that are unlikely to account for the data. They then decided to capture these heuristics within their program to constrain the number of structures generated. The result was a computer program called DENDRAL (Buchanan and Feigenbaum, 1978) that operated as well as an expert chemist in recognizing molecular structures of unknown compounds. More important, it was the first program whose success was attributed to what it knew about the problem, rather than relying on complex search techniques. The success of DENDRAL with its emphasis on knowledge led Ed Feigenbaum at Stanford to claim, "In the knowledge lies the power." It also led to the concept of a **knowledge-based system** or **expert system**. The expert system era had begun.

Revitalized by the new direction the field had taken, researchers quickly began to look for better ways of representing and searching knowledge in a computer program. Because DENDRAL was successful with its knowledge encoded in rule form, this seemed to be the way to go. In short order, programs were being built with a very simple architecture that contained the problem's knowledge in a set of rules which were placed in a module called the *knowledge base*, a processor of these rules by a second module called the *inference engine*, and another module called the *working memory* which contained the problem-specific facts and conclusions derived by the inference engine. Collectively, these three modules form what is commonly referred to as a *rule-based expert system*. Armed with this architecture, expert system developers during the 1970s knew it was now time to develop systems and learn from the experience. They also knew that it was time to produce systems that provided real benefits.

No other system contributed more to our understanding of building rule-based expert systems than MYCIN (Shortliffe, 1976). The objective of MYCIN was to diagnose infectious blood diseases. With about 500 rules, MYCIN was able to perform as well as some experts and considerably better than junior doctors. However, through the eyes of the expert system developer, the real success story of MYCIN rests in the insight it provided into rule-based expert system design.

During the 1970s, AI was moving at a faster pace, and further into the corporate sector. The first successful commercial expert system developed during this period was called XCON. XCON, originally called R1, was built at Carnegie Mellon and was designed to aid Digital Equipment Company (DEC) in the configuration of their VAX computer systems (McDermott, 1980). XCON was developed using OPS, a rule-based programming language, which remains today a popular language among expert system designers. XCON has proven to be a valuable tool for DEC. By 1986, it was saving the company an estimated $20 million dollars per year. The success of XCON led DEC to create a separate group within the company dedicated to AI efforts that by 1988 had developed over 40 other expert systems.

Another highly successful expert system developed during this decade was PROSPECTOR. It was developed at the Stanford Research Institute to aid geologists in the exploration of ore deposits (Duda et al., 1977). The system generated enormous publicity by recommending exploratory drilling at a site near Mount Tolman, located in eastern Washington, that proved to contain a large molybdenum deposit worth approximately $100 million.

In 1979, a group of individuals who were intimately involved with the development of earlier expert systems met at a workshop chaired by Don Waterman and Frederick Hayes-Roth to exchange ideas about the field. Viewing the developments of the last decade, and with an understanding of the capability and potential of the technology, they predicted;

Over time, the knowledge engineering field will have an impact on all areas of human activity where knowledge provides the power for solving important problems.

This was another rather bold prediction. A skeptic might suggest it would be nothing more than another one in a long line of predictions that failed. Things were different now, however. By the close of the 1970s, the field of expert systems was an established technology. The successes of systems such as MYCIN, XCON, and PROSPECTOR rekindled interest in the field and also provided a roadmap for others to follow for designing an expert system. The time was now ripe to build expert systems in large numbers across many application areas.

1.7 Proliferation of Expert Systems (1980s)

With expert systems now recognized as a practical tool for solving real-world problems, the 1980s marked the technology's coming-out party, and many attended. Most universities rapidly developed and offered expert system courses. Companies initiated expert system projects and often formed internal AI groups. DuPont, for instance, had their own group of AI specialists who by 1988 had built 100 expert systems that were saving the company an estimated $10 million per year, and had another 500 systems in development. Over two thirds of the Fortune 1000 companies became involved in applying the technology in daily business activities. Government agencies responded by dramatically increasing available funding for expert system research and development. There was also a surge of international interest in the technology. In 1981, the Japanese announced the Fifth Generation Project, a ten-year plan to build intelligent machines. In response to this plan in the United States, the Microelectronics and Computer Technology Corporation (MCC) was formed as a research consortium dedicated to furthering expert system development in the United States. In Britain, though still shaken from the Lighthill report, the government used the more upbeat Alvey report on the technology to reinstate funding in the area.

In the early 1980s, medical expert system applications dominated the scene. This is primarily due to the diagnostic nature of these applications and the relative ease of developing such systems. However, as we moved toward the mid-1980s and the low-hanging fruit was picked, it was time to reach for more difficult problems. It was also time to begin developing systems that produced benefits for the commercial sectors. Unfortunately, initial attempts frequently met with limited success. In hindsight, three primary reasons help to explain this result.

First, early applications of expert systems in industry often overchallenged the technology, leading to poor results. Fascinated by the notion of machines that could think, many designers attempted to build systems to solve problems that were beyond the reach of even the best experts. The thinking was, "Well, we can't solve this problem so let's try throwing AI at it." Second, other designers often took on a project whose scope was so broad that completing it within a reasonable time frame was impossible. Third, some designers developed remarkably intelligent systems; however, the effort was often divorced from an understanding of the client's need to integrate the system into existing hardware and software resources. The result was a powerful finished product that was left to collect dust on a shelf.

With the few successes being produced during this period, coupled with earlier glowing promises of the technology, critics crept out of the woodwork and quickly pounced on the situation. Journal and conference papers, newsletters, and the national media were swift to point to the shortcomings. Consider, for example, a report in Forbes Magazine (Simon, 1987). Forbes was asking, "What happened to those expert systems that were supposed to transform the world of business forever?" Expert system designers began to realize that finding a place for the technology could be as tedious as matching the glass slipper to Cinderella's foot.

The turning point came in the mid-1980s when designers began to focus on very narrow, well-defined, and even sometimes mundane tasks. They also took the time to look at where the technology would be embedded. Although the systems developed following this theme may have seemed unimpressive to the AI researcher viewing the scene from the ivory tower, they were well received by managers in industry because they produced commercially worthwhile results.

Given this new direction, the number of deployed systems began to accelerate rapidly. From only a handful of systems developed during the 1970s, the number rose into the thousands by the end of the 1980s. It is also interesting to view the widespread use of the technology. Systems have been built for applications ranging from aiding miners a mile below the earth's surface to helping astronauts aboard the space station. A survey of the literature conducted in 1992 in search of developed expert systems uncovered applications that naturally fell into 1 of 23 areas as shown in Table I.1.1 (Durkin, 1993). The breadth of the applications is remarkable. It is difficult now to find an area that hasn't been touched by the technology. In fact, we sometimes can find it in some strange places. Want assistance with your pig farm? Get a copy of TEP (The Electronic Pig), an expert system that assists with diagnosing the causes of problems with pig litter size. Having trouble with your love life? Fire up SEXPERT, which assesses and treats sexual dysfunction. Need help in getting rid of a problem employee? Check out CLARIFYING DISMISSAL, which assists employers in the dismissal of employees.

Up to the mid-1980s, the field was dominated by rule-based systems. However, during the later part of the decade, there was a shift toward object-oriented systems. In the world of expert systems, it was the frame-based system that began to take center stage. The word *frame* was coined by Minsky (1975), but its concept dates to Socrates's idea of the universal definition. Due to their ability to represent both descriptive and behavioral object information easily, and the powerful set of utilities they bring to the table, frame-based systems can address applications of greater complexity than

TABLE I.1.1   Major Application Areas of Expert Systems

Agriculture	Law
Business	Manufacturing
Chemistry	Mathematics
Communications	Medicine
Computer Systems	Meteorology
Education	Military
Electronics	Mining
Engineering	Power Systems
Environment	Science
Geology	Space Technology
Image Processing	Transportation
Information Management	

can be managed by rule-based systems. The areas where they are being applied is widespread and growing. Many of the early frame-based expert systems tackled simulation problems. This was a good choice since a simulation task inherently involves interacting objects. Today, we see frame-based systems built for problems that have been traditionally managed using a rule-based approach, such as diagnosis and design.

During the 1980s neural networks resurfaced after being buried in 1969 by the book *Perceptrons* by Minsky and Papert. Two major events occurred during the decade that led to the rebirth of interest in the technology. In 1982, John Hopfield wrote a paper describing a neural system called the *Hopfield Model*, which represents neuron activity as a threshold operation and memory as information stored in the interconnections between neurons (Hopfield, 1982). This model established, for the first time, the principle of storing information in a dynamically stable network. The next important event, particularly from an engineering viewpoint, was the development of the *backpropagation algorithm* for training multilayer perceptron networks (Rumelhart and McClelland, 1986). Backpropagation has emerged today as the most popular learning algorithm for the training of neural networks. It has been applied in a wide range of neural network applications, making the technology an effective tool for solving practical data classification problems.

In 1965, Lotfi Zadeh (Zadeh, 1965) introduced a new field called fuzzy logic, which provided, for the first time, the ability to perform common sense reasoning in a computer. Mamdani and Assilian during the 1970s established the framework of a fuzzy controller and used it to develop the first real-world fuzzy expert system to control a steam engine. The use of fuzzy logic as an effective expert system tool was set in the 1970s, but unfortunately, few took note of its potential value. In the United States the birth place of the technology, progress in the field proceeded very slowly during the 1980s, mainly because very few individuals in the United States were working in the field. The Japanese, fortunately, saw the potential commercial value of the technology and began to develop numerous systems. The success in Japan caught the attention of the expert system community. In the United States the number of fuzzy projects in both commercial and government sectors increased dramatically during the 1990s. Activity in Europe, South Korea, and Canada also has been on the rise recently.

1.8 State of the Field (1990s)

During the early 1990s, the expert systems field continued to build on the momentum established in the 1980s. The number of deployed systems continued to increase rapidly. Many organizations not involved earlier initiated expert system projects, and some began entire programs. New magazines, new special interest groups within professional organizations, and many more conferences dedicated to the technology emerged. To obtain a sense of the health of the field during this decade, we will consider a survey reported in the book *Expert Systems: Catalog of Applications* (Durkin, 1993) and another survey conducted in 1996 in preparation for a new edition of the book.

1.8.1 Application Areas

Recall the earlier prediction from the 1979 workshop chaired by Waterman and Hayes-Roth, " ... *field will have an impact on all areas of human activity*." The results from the surveys show that this prediction was right on the mark. Figure I.1.1 shows the number of systems developed in the different areas shown in Table I.1.1. We shouldn't be surprised by the range of applications. An expert system is inherently a tool to assist human decision making. We apply it to knowledge-intensive tasks that require expertise to accomplish. Therefore, wherever we find humans in such activities as diagnosing a system, designing a structure, or tutoring a student, we have also found a home for the technology.

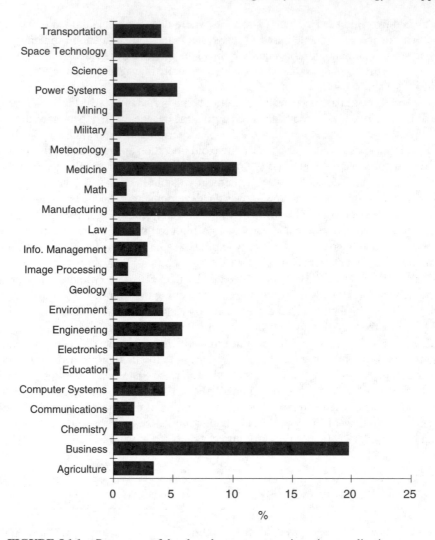

FIGURE I.1.1   Percentage of developed expert systems in various applications areas.

Figure I.1.1 also provides insight into the direction that the technology has taken. During the 1970s AI was a cult activity—almost a religion. Researchers centered on producing intelligent general-purpose reasoning machines. The fascination of achieving this academic challenge drove their efforts. By the 1980s, when fuel for the advancement of the technology came from sectors that demanded a return on their investment, researchers began to realize that AI is not a religious experience, but an economic one. A 1986 survey conducted by Waterman (1986) showed that the majority (30%) of the applications were in the field of medicine. Figure I.1.1 shows that this field remains attractive to expert system developers (10%); but percentagewise, activity in this area has decreased considerably. Even more revealing is the activity in the business and industrial areas over recent years. According to Waterman's survey, about 10% of the applications were in these areas in 1986. By 1996, these areas accounted for approximately 60% of the applications. In fact, it was found in the 1996 survey that one out of every five expert systems built to date has been applied in the area of business. A 1995 survey conducted in the United Kingdom alone found that 24% of the deployed expert systems were in the financial services sector (Coakes and Merchant, 1996). This is clearly a sign that expert systems are merging with the mainstream of information processing that

TABLE I.1.2 Types of Problems Solved by Expert Systems

Problem-Solving Type	Description
Control	Governing system behavior to meet specifications
Design	Configuring objects under constraint
Diagnosis	Inferring system malfunctions from observables
Instruction	Diagnosing, debugging, and repairing student behavior
Interpretation	Inferring situation description from data
Monitoring	Comparing observations to expectations
Planning	Designing actions
Prediction	Inferring likely consequences of given situations
Prescription	Recommending solution to system malfunction
Selection	Identifying best choice from a list of possibilities
Simulation	Modeling the interaction between system components

was previously dominated by conventional data processors. It also shows that the field has matured considerably over recent years, where we now find the technology well received by the commercial sectors.

1.8.2 Application Activity

Another way that we can sort expert system applications is by *problem-solving type*. Regardless of the application area, given the type of problem, experts collect and reason with information similarly. Likewise, expert systems are designed to accomplish generic tasks on the basis of the problem type (see Table I.1.2).

Figure I.1.2 shows the percentage of applications for each problem type in Table I.1.2. Be aware that many applications employ more than one activity. For example, a diagnostic system might first interpret the available data and later prescribe a remedy for the recognized fault.

As Figure I.1.2 illustrates, the predominant role of expert systems has been diagnosis. The survey showed that one out of every four expert systems built has employed a diagnostic activity. One reason for this result is that this is the role most experts play. Fields such as medicine, engineering, and manufacturing have many individuals who help diagnose problems. Another reason for the large percentage of diagnostic systems is due to their relative ease of development. Most diagnostic problems have a finite list of possible solutions and a limited amount of information needed to reach a solution. These bounds provide an environment that is conducive to effective system design.

Another explanation for the large percentage of diagnostic expert systems can be traced to the practical considerations of introducing a new technology into an organization. Most organizations prefer to take a low-risk position when considering a new technology. As such, they prefer projects that require the minimum resources and have the maximum likelihood of success. Because these systems are relatively easy to build, they are attractive to firms venturing into the field.

The drop-off from the large number of diagnostic applications to that of some other problem types is dramatic. Two reasons help explain this result. First, tasks such as design and planning are difficult to implement in an expert system because their steps vary greatly between application areas and it is often hard to precisely define these steps. Second, tasks such as instruction, control, and simulation, although they are excellent areas for expert system applications, are relatively new ventures.

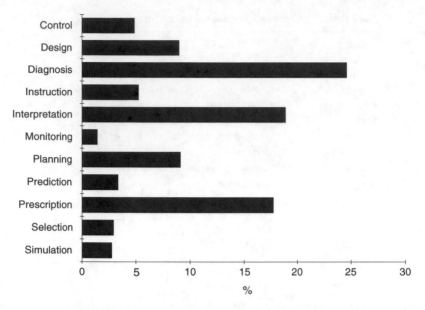

FIGURE I.1.2   Percentage of expert system applications by problem type.

1.8.3   Rate of Development

One of the better ways to measure the success of a new technology is to look at its rate of application. If we see an upward trend, we obviously interpret it as a positive sign. Conversely, if the trend is downward or stagnate, we should be concerned. Consider the expert systems field.

During the 1970s, when researchers were feeling their way with the new technology, only a handful of expert systems were developed. Work during this period was focused mainly on looking for ways to better represent and process knowledge in a computer. Most systems developed were therefore more of an academic exercise, where the researchers worked in relative isolation from the realities of business-oriented needs. During the 1980s, with a better understanding of how to build an expert system, the situation changed dramatically. To illustrate, consider a report by Paul Harmon.

In 1990, along with Brian Sawyer, Harmon published *Creating Expert Systems for Business and Industry* (Harmon and Sawyer, 1990). The book included estimates of the cumulative number of expert systems developed over the years 1985 and 1988 (Chapter 1). Figure I.1.3 shows these estimates along with results from the 1992 and 1996 surveys. The impressive growth rate of expert systems is an indicator of the acceptance of the technology by industry and a testament to the labors of earlier researchers.

We can also get a sense of the growth rate of AI by considering the number of U.S. patents filed that mentioned the name artificial intelligence and related terms such as knowledge based, fuzzy logic, and expert system. According to Robert Downs, the primary examiner for AI in the U.S. Patent Office, in 1988 about 100 patents mentioned AI specifically; while in 1998 around 1700 mentioned AI, and another 3900 or so mentioned related terms (Buchanan and Uthurusamy, 1999).

1.8.4   Software and Platforms

Another event that contributed to the large growth rate in developed systems was the introduction during the 1980s of new hardware and software technologies that provided support for expert system designers. The rise in the popularity of the technology during this decade led to the emergence of a new cottage industry that provided easy-to-use expert system development software called shells.

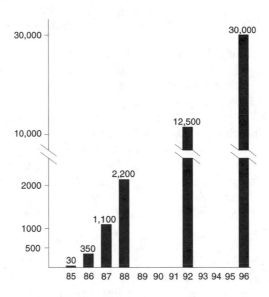

FIGURE I.1.3   Total number of expert systems developed by each year.

These shells were built to run on platforms ranging from personal computers (PCs) to mainframes. Most expert systems during the 1970s were developed on powerful workstations, using languages such as LISP, PROLOG, and OPS. This left the challenge of developing systems in the hands of the select few who could afford the platforms and had the patience to learn the complexities of the available languages. During the 1980s, we witnessed the proliferation of PCs. With these widely available platforms coupled with the shells, the opportunity to develop an expert system was now in the hands of many individuals from all disciplines.

The 1996 survey found that most expert systems were built on a PC using a shell. It showed that approximately 60% of the systems were developed on a PC and 20% on a workstation; the minicomputer and mainframe accounted for 10% each. Approximately 45% of the systems were built using a shell and 25% using LISP; PROLOG and OPS accounted for 10% each; and the balance using the languages C, Pascal, LOOPS, Fortran, Smalltalk, and Basic.

1.9   Epilogue

Aristotle opened his *Metaphysics* with: "All men by nature desire to know." It has always been a part of our nature to seek new truths and answer old questions. New discoveries evolved our civilization from one able to use fire to light our caves to one able to place a footprint on the moon. Along the way we faced some basic questions first proposed to us by Aristotle: What are we? What will we become? Perhaps no better opportunity exists for us to answer these questions, and continue our evolution, than the quest for the intelligent machine.

We cannot understand our world by intellect alone. We comprehend it as much by our feeling. Therefore our judgment of the intellect to understand is at best only half of the truth. Consciousness requires feeling. Our search for the intelligent machine is therefore an equal mix of technology and emotion. The technical side is analytical and often clear. We use and improve the technology of the day and, through trial-and-error, work toward developing the means to emulate human reasoning. Human emotions, on the other hand, are far more complex. We love and hate. We embrace and fear.

To "know" then, our machine must share our emotions. Can this be done? Or, maybe even a better question: Do we want it to be done?

When Garry Kasparov lost his chess match to the machine Deep Blue, we reacted with an equal share of fascination and fear. Here was a human, who many considered to be the greatest chess player ever, who lost to a machine. Moments like this naturally stir the emotions. Perhaps it stems from the notion that although we can easily accept being surpassed by machines in mechanical tasks, we consider the kind of creativity required by chess to be the province of humanity.

In the aftermath of the match many individuals were quick to voice their opinions and concerns about the consequences of the event. "New technology is threatening," said John C. Norcross, a psychology professor at the University of Scranton. He also remarked, "There's a hunk of emotional work in accepting that the world is moving and evolving. We fear change, particularly change that we did not initiate." Others were captured by the fascination of the moment. "Eventually, they may become more capable than we are and will no longer need us to build even better machines," said Edward Cornish, president of the World Future Society, in Bethesda, MD. Even the vanquished Kasparov was profoundly affected—"I think this moment could mark a revolution in computer science." He was also asked if he thought the machine was intelligent. "No," he said, quickly adding the caveat, "but you can definitely feel something stirring inside [Deep Blue], beneath the surface." The Kasparov vs. Deep Blue match was an event that brought forth the human emotions of fear, fascination, and hope—the backdrop of the history of the quest for the intelligent machine.

At the core of our concern is a simple question: If a machine can think, then where does that put us? We have always felt comfortable with theories that put us apart from and above the rest of our natural world. By tying our sense of self-importance to such theories, however, we feel less special when they turn out to be wrong. For instance, when Darwin put forth his version of our family tree, we felt somewhat diminished. To refortify our position, we argue to ourselves that it is our ability to reason that separates us from the beasts. Today, we feel threatened not by the beast but by our own creation—the intelligent machine.

We have come to accept computers as a part of our lives. They help us to do calculations, balance our checkbooks, and take us on a virtual tour of the world over the internet. Without them, planes would be grounded, cars would stall, phones would go dead, stock markets would stop, and man's footprint would never have appeared on the moon. We did not always embrace the technology, however. The introduction of the computer was met with open hostility by many individuals, who believed it was their birthright alone to partake in intellectual activities. After a while we became comfortable with the technology. Today, instead of seeing our silicon friend as a threat, we view it as a collaborator. Along the way we were provided a glimpse of tomorrow's machine.

Greek myths, as well as the more recent science fiction movies, have a rich history of being both the virtual stage on which we are presented an act to inspire our scientific discoveries, and the mirror where we come face-to-face with the potential consequences of our scientific endeavors. Art imitates life. For instance, consider HAL.

One of our most impressionable images of an intelligent machine was the computer HAL 9000 in the 1968 science-fiction classic *2001: A Space Odyssey*. HAL, the neurotic computer that director Stanley Kubrick and writer Arthur C. Clarke created for the movie, is the lethal embodiment of what happens when we give a machine consciousness, and then hurt its feelings. Speaking with the astronaut Dave, HAL says "I know that you and Frank were planning to disconnect me and I am afraid that is something I can't allow to happen." HAL, in a dispassionate way, responds by killing several members of the spaceship's crew—a simple calculated response to his obvious needs.

Kubrick was a scholar of the consequences of man's self pride. In most of the movies he directed, his persistent theme was that of man dreaming of being other, or more, than he is. His pessimistic view of humans was that of becoming as mechanical and unfeeling as the technology on which we depend. In the movie *2001*, the humans were unemotional, even machine-like. HAL, on the other hand, exhibited human-like feelings. For instance, as HAL's brain was being dismantled, it was his personality that disintegrated: Dave, stop ... Stop, will you? Stop, Dave ... Will you stop, Dave ... Stop, Dave. I'm afraid ... I'm afraid ... I'm afraid, Dave ... Dave ... my mind is going ... I can feel it ... I can feel it ... My mind is going ... There is no question about it. I can feel it ... I can feel it ... I can feel it ... I'm a ... fraid ... Good afternoon gentlemen, I am a HAL 9000 computer. I became operational on ...

With the simple use of a screwdriver, Dave changes the bloodred one-eye harbinger of the danger of technology into a simple, easily manageable, computer program. Will it be as easy tomorrow to control our technology?

Kubrick wanted to disturb the way we saw ourselves and show us things we don't want to face. In the end of the movie *2001*, an astronaut evolves into a Star Child and becomes not-man, better-than-man, by shrugging off its humanity. Which will come first, I wonder. Machines that are more and more like humans, or humans who are more and more like machines?

A more recent movie addressed the man–machine relationship, but in a much more friendly way. In the movie *A.I.*, we were introduced to David, a robot boy who loved but was not real. The movie was based on the 1969 short story "Supertoys Last All Summer Long," by British writer Brian Aldiss. The screenplay represented the collaboration of Stanley Kubrick, who spent 15 years on the project, and Steven Spielberg, who completed the project and directed the film after Kubrick's death in 1999.

A.I. is set in a future beset by environmental collapse where lifelike robots called "mechas" (mechanicals) live among but are subservient to humans called "orgas" (organics). Though possessing extremely advanced machine traits, the mechas lack one feature that Mr. Hobby, a noted scientist with the firm Cybertronics Manufacturing, wants to provide them. He tells his colleagues, "I propose that we build a robot who can love ... a robot that dreams." Mr. Hobby believes that love is the pathway to consciousness. After some debate on the proposed project with his colleagues, he asks, "Didn't God create Adam in his own image to love him?"

Two years later, Cybertronics assembles the perfect child, David, "always loving, never ill, never changing," and locates the ideal couple to adopt him: a couple who has sunk into deep depression because their real son, Martin, lay in a coma, leaving them to pray for the chance to love their son again and to receive love in return. But we know the danger of answered prayers. Love can break your heart. Even the heart of the robot David, who is abandoned by the one he loves most, and has to face a brutal world before he can gain her comforting touch. He asks his mother, Monica, who he believes to be his creator, "Mommy, if I become a real boy, can I come home?" David wants to prove that his heart beats, not ticks. And thus his difficult quest begins, with us hoping for success because we feel the innocent should not suffer like this. David's quest takes us on a journey that explores many of the basic human emotions, such as love, hate, betrayal, hope, prejudice, and fear.

Soon after being abandoned in the forest by his mother, David meets Joe, a street-wise mecha with some interesting human-like talents. David tells Joe that he wants to find a woman called the Blue Fairy. David knows of the fairy tale Pinocchio from his bedtime readings with his mother and believes that the Blue Fairy can make him real. The "toy boy" David dreams of fairy tales and happy endings, just as we did as children. Together David and Joe wander through garish landscapes that brilliantly evoke both fascination and dread.

One stop along the way is the Coliseum-like spectacle of the "Flesh Fair," where humans scream in delight as discarded (but still functioning) robots are crushed, melted, and pulled apart. This is where Spielberg increases our despair quotient by emphasizing the rift between humans and robots. Our relationship with machines has always been a tenuous one. We resent them when their physical superiority costs us our jobs and fear them when their intellectual potential threatens our vision of humanity. We worry, perhaps advisedly, that the intelligent machine imperils our unique position which we see as apart from and above the beast. But the humans at the Flesh Fair appear to be closer to the beast when they express the base emotions of destructive jealousy as they taunt and ultimately destroy the abandoned androids. Should we believe the human tormentors when they shout "They are becoming too smart, too many, too fast"? Should we believe them when they cry for mercy for David after perceiving him as a real boy? The master of ceremonies implores the audience to "not be fooled by the talent of this artistry." But it is easy to be fooled because AI blurs the line separating man and machine.

David's quest next takes him and Joe to meet Dr. Know, a hyperactive Einstein-like holographic projection who provides insight into finding the Blue Fairy. This knowledge, as is the case when we discover many new things, can come at a price: we learn more about ourselves and our limits. Shortly after leaving Dr. Know, Joe is hooked and reeled in by a scavenger police plane like a hapless fish. "I am, I was," Joe says to David as they part company, seemingly in response to Aristotle's questions: What are we? What will we become?

Eventually David arrives at a place told by Dr. Know where he can locate the Blue Fairy. On arriving, he first meets his inventor, Mr. Hobby, who introduces him to a replica of himself. David also sees dozens of them, hanging as finished products from an assembly line, ready for shipping. The moment is critical to David. He feels betrayed by the greed of Mr. Hobby, who he now knows created him for profit, and in so doing, robbed him of his individuality. David also now knows that he is, and will always be, a machine; never to be real, never to gain the love of his mother. This understanding brings out in him a very human-like response to a situation of total despair: He destroys the replica of himself in a violent fashion and, symbolically, destroys himself.

In the factory that produces the "David toys" is a mask that is used to produce the facial features of the android David, the machine. David walks up to the synthetic face, presses himself into the likeness, and peers through its eyes. We see the image as David, but once removed. The eyes capture our attention. Unlike the hope we saw in his eyes during his earlier journey to win back his mother's love, we now see eyes that appear cold and dead.

Had Stanley Kubrick lived to complete the script, the movie may have ended here. Kubrick wanted us to consider the consequences of our scientific discoveries and how these discoveries might shape our future. To accomplish this, he often presented his films in a way that made our blood run cold. The stage he set showed technology outpacing man's understanding of its potential. He also incorporated into his plays the arrogance of humans who believed that they could rise above and control the technology they developed. Perhaps he did this so that we would pause and think about what was needed for us to evolve into something better, rather than into something worse. At this point in our discussion of the movie *A.I.*, David is faced with the end of his quest, one of failure, a doomed Pinocchio yearning for human acceptance that will never be within his grasp. Kubrick would tell us that David has reached the point where he finally recognizes that he, like man, must learn the limits of hope, and then as a consequence often dies.

It has always been a difficult moment when scientists come face-to-face with the consequences of their discoveries. For instance, after witnessing the world's first nuclear explosion Robert Oppenheimer states, citing from the *Bhagavadgita*, "I am become death, shatterer of worlds." We are far from developing machines like HAL and David, but the day will come. At that time will their eyes appear cold and dead? At present we don't know. Perhaps we should prepare today to answer this inevitable question.

While Kubrick is likely to reveal the madness that lurks beneath our view of normalcy, Steven Spielberg is likely to show us that humanity survives human madness. Spielberg wants us to feel good about ourselves. He writes films with happy endings. He wants us to feel that our future is bright, as long as we continue to believe in ourselves.

Many scientists also like happy endings and only see the good in their discoveries: curing our diseases, increasing our life span and our quality of life, restoring us to Eden. On the other hand, some see both the bright and the dark sides of the technology they develop, and argue, often to convince themselves, that the potential good of the technology outweighs its potential evil. As an example, consider again Robert Oppenheimer. Several months after the atomic bombings in Japan, he was quoted as saying, "It is not possible to be a scientist unless you believe that the knowledge of the world, and the power which this gives, is a thing which is of intrinsic value to humanity, and that you are using it to help in the spread of knowledge and are willing to take the consequences." The field of artificial intelligence has produced many success stories; a history which should continue well into the future. But there is a potential dark side of AI that you hear about in the whispered conversation between AI researchers or in the shouts of sci-fi writers. Whether good or evil wins out is an issue that we presently still have the opportunity to address, before we lose control of the technology.

Since the movie *A.I.* is in the hands of Spielberg, and not Kubrick, it moves on to David, who in total despair following his encounter with his machine self, throws himself into deep waters to end his existence. Spielberg steps in and foils David's attempt at self-destruction, and allows him to discover his long-sought Blue Fairy at the bottom of the sea. A wonderful moment for David. But David's satisfaction is not immediate because he has not completed his quest. It takes some 2000 years, during which humans have disappeared from the planet, before his quest is renewed. His long-sought affection, his mother, has long ago expired. Here we see the machine David, driven by some of the most basic human emotions, wanting more. Through steps we won't go in to, he is given the opportunity to spend one and only one day with his mother.

David spends this day, the most important one of his existence, in the simplest way. He wants his mother to know the importance her role plays in his life, regardless of his species. He shows her drawings he created that depict his journey to win her love. He smiles at her, and together they play and laugh at the simplest things. His mother gives him a birthday party, his seventh, and says I love you. *"Mommy, if I become a real boy, can I come home?"*—David is home. In the final scene we see David peacefully lying in bed with his mother. He tucks her in for the final time, gently holding her hand as she vanishes forever. Even with her passing, David is at peace—his quest is done and he recognizes that the quest itself, not the end result, is of import.

Kubrick and Spielberg are master film makers. More important, they have produced films that not only entertain us, they have also created a stage that draws us into the act where we are forced to ask important questions about ourselves. Where is the line between machine and human? Where is the line that divides sentience from a programmed response? What human rights, if any, are owed to an intelligent machine? Can we throw away or destroy the intelligent machines we create, like we might do to a worn-out toy? If we can make a machine that thinks, can we, or should we, stop it from having feelings? If it is possible to replicate human consciousness in machines, then what does it mean to be human? Long after the images of the movies *2001* and *A.I.* have slipped from our memories, these questions will continue to haunt us.

Since the release of these films, we watched our computers get smarter—and smarter faster than us. Instead of just processing information, a role that took us some time to become comfortable with, we watched as our computers acquired the ability to reason about the information. Expert systems stood out in this new role. They diagnosed patients, made investment recommendations, and fixed our cars. Many applauded the new technology. Many others saw it as a threat to humans' rightful monopoly of rational thought. A few decades ago, the idea of a computer beating a human at chess seemed ridiculous. Today we take it for granted. We are even starting to accept that a computer can beat our best. We get used to things.

It is unlikely that computers will suddenly become as intelligent as people, in every way. The state of the art, the expert system, is nothing more than an idiot savant. It is only smart about a very narrow area. Given an IQ test, it would score less than a first grader. For now, we continue to hold the intellectual high ground. Machine intelligence will emerge gradually, improving decade by decade, giving us time to get accustomed to it. Computers will get more powerful. Software will get more clever. AI will creep up on us slowly, exhibiting along the way more and more human traits, and we won't even notice the change. David will not happen overnight, but his arrival is inevitable. He will be the product of a thousand little steps.

Today, most of us use word processors to compose our letters and spreadsheets to keep track of our accounts. We may even use a computer to find out how bad we are doing in the stock market. We have become accustomed to using computers in our everyday lives, to the point where we are, for the most part, dependent on the technology for information. Without the computer today, even with only some fifty-odd years under its belt, we would be in trouble. We have become dependent on the technology. If we were to sacrifice the technology, we would be lost from an informational perspective. But things could get worse. The computer is evolving from a processor of information to one that can make decisions with the information, and we are becoming more reliant on the computer to make these decisions.

We would not openly hand over all of our decision-making power to computers, but we might drift into a position of such dependence. As our problems become more complex and our machines more intelligent, we may let our machines make more and more of our decisions, simply because we recognize that machine-made decisions are better than the ones we might make. Eventually we will not be able to turn off our intelligent machines because we would rely too much on the decisions that they provide. At this point the machines will be in effective control.

Will we continue to fear intelligent machines? Of course. It is part of our nature to fear things that we don't completely understand. It is also our nature, as Aristotle pointed out long ago, to seek knowledge and use our new understanding to evolve into something better. Hopefully, when David arrives, we can coexist.

References

1. Buchanan, B. and Feigenbaum, E., DENDRAL and Meta-DENDRAL: Their applications dimension, *Artificial Intelligence*, vol. 11, pp. 5–24, 1978.
2. Buchanan, B. and Uthurusamy, S., The innovative applications of artificial intelligence conference, *AI Magazine*, vol. 20, no. 1, pp. 11–12, Spring 1999.
3. Coakes, E. and Merchant, K., Expert systems: A survey of their use in UK business, *Information and Management*, vol. 30, pp. 223–230, 1996.
4. Duda, R. et al., Development of a Computer-Based Consultant for Mineral Exploration, SRI Report, Stanford Research Institute, Menlo Park, CA, Oct. 1977.
5. Durkin, J., *Expert Systems: Catalog of Applications*, Intelligent Computer Systems, Inc., Akron, OH, 1993.
6. Harmon, P. and Sawyer, B., *Creating Expert Systems for Business and Industry*, John Wiley & Sons, New York, NY, 1990.
7. Hebb, D., *The Organization of Behavior: A Neuropsychological Theory*, Wiley, New York, NY, 1949.
8. Hopfield, J.J., Neurons with graded response have collected computational properties like those of two-state neurons, *Proc. Natl. Acad. Sci.*, vol 81, pp. 3088–3092, 1982.
9. Lighthill, J., Artificial Intelligence: A General Survey, Scientific Research Council of Britain, SRC Report 72–72, Mar. 1972.
10. McCorduck, P., History of artificial intelligence, in *Proc. IJCAI-77*, pp. 951–954, 1977.
11. McCulloch, W.S. and Pitts, W., A logical calculus of the ideas imminent in nervous activity, *Bull Math. Bio.*, vol. 5, pp. 115–133, 1943.
12. McDermott, J., R1: An expert system in the computer systems domain, *Proc. AAAI-80*, 1980.
13. Minsky, M.L., Frame System Theory, in *Thinking: Readings in Cognitive Science*, P.N. Johnson-Laird and P.C. Watson, Eds., Cambridge University Press, 1975.
14. Minsky, M. and Papert, S.A., *Perceptrons*, MIT Press, Cambridge, MA, 1969.

15. Newell, A. and Simon, H.A., *Human Problem Solving*, Prentice Hall, Englewood Cliffs, N.J., 1972.
16. Rosenblatt, F., The perceptron: A probabilistic model for information storage and organization in the brain, *Psychol. Rev.*, vol. 65, pp. 386–408, 1958.
17. Rumelhart, D.E. and McClelland, J.L., Eds., *Parallel Distributed Processing Explorations in the Microstructure of Cognition*, vol. 1, MIT Press, Cambridge, MA, 1986.
18. Samuel, A.L., Some studies in machine learning using the game of checkers, in *Computers and Thought*, E.A. Feigenbaum and J. Feldman, Eds., McGraw-Hill, New York, 1963.
19. Shannon, C.E., A chess playing machine, in *The World of Mathematics*, vol. 4, J.R. Newman, Ed., Simon and Schuster, New York, 1955.
20. Shortliffe, E.H., *Computer-Based Medical Consultation*, MYCIN, American Elsevier, New York, 1976.
21. Simon, R., The morning after, *Forbes*, vol. 140, no. 8, pp. 164–168, Oct. 19, 1987.
22. Waterman, D.A., *A Guide to Expert Systems*, Addison-Wesley, Reading, Mass., 1986.
23. Widrow, B. and Hoff, M.E., Adaptive Switching Circuits, IRE WESCON Convention Record, New York: IRE Part 4, pp. 96–104, 1960.
24. Zadeh, L.A., Fuzzy sets, *Information and Control*, vol. 8, pp. 338–353, 1965.

2

Soft Computing Framework for Intelligent Human-Machine System Design, Simulation, and Optimization

Xuan F. Zha
Gintic Institute of Manufacturing Technology

Samuel Y.E. Lim
Nanyang Technological University

Abstract. In this chapter, a new neuro-fuzzy hybrid approach to human workplace design and simulation is proposed. Problems related to human workplace design such as human-machine modeling, measurement and analysis, workplace layout design and planning, workplace evaluation and simulation are discussed in detail. The complex human-machine interactions in workplace design are described with human and workstation parameters within a comprehensive human-machine system model. Based on this model, procedures and algorithms for workplace design, ergonomic evaluation, and optimization are presented in an integrated framework. With a combination of individual neural and fuzzy techniques, the neuro-fuzzy hybrid scheme implements fuzzy if-then rules block for workplace design and evaluation by trainable neural network architectures. For training and test purposes, simulated assembly tasks are carried out on a self-built multi-adjustable laboratory workstation with a flexible motion measurement and analysis system. The trained fuzzy neural networks are capable of predicting the operator's posture and joint angles of motion associated with a range of workstation configurations. They can also be used for design/layout and adjustment of manual assembly workstations. The developed system provides a unified, intelligent computational framework for human-machine system design and simulation. In the end, case studies for workplace design and simulation are presented to validate and illustrate the developed neuro-fuzzy design scheme and system.

Key words: Human-machine system, soft computing, computational framework, neuro-fuzzy approach, neural networks, fuzzy logic, artificial intelligence, workplace design, layout planning.

2.1 Introduction

Although automated production systems are most widely used, there are still many jobs that prefer to be assigned to humans. Manual workstations tend to involve bulk or limp materials and relatively more unstructured requirements. Therefore, manual systems will still maintain their positions in workshop floors in the future, and with increasing competition, attention to the efficiency of manual work is growing (Braun et al., 1996). The design of industrial workplaces, as with most forms of design, is iterative and highly interactive. An important trend in industrial workstations is the focus on ergonomic design that considers the interaction between workers and their workplaces to increase safety and improve work efficiency. On the other hand, workstation modularity and flexibility are also becoming important features. The trend toward teamwork and flexible work groups further emphasizes the need for adjustability to individual workers. For instance, such adjustability may imply that:

- The work surface height should be adaptable to individual height.
- The work surface should be adjustable to improve variable workpiece position for increasing comfort and efficiency.
- An operator should be able to reposition the work frequently to accommodate changes in posture.

In the increasingly flexible work environment, a team should be able to quickly customize the workstation to the specific needs of variable tasks and alternating operators. Thus, the designer has to consider countless constraints with usually opposing goals. In order to assist the planning engineer doing this complex and time-consuming job, it is better to transfer such kinds of planning tasks to a computer program. However, the large number of conflicting constraints makes it impossible to find an optimal solution procedure (Braun et al., 1996). Generally, from the systematic point of view, the problems or tasks involved in workplace design can be classified into the following three categories (Zha, 2001; Zha and Lim, 2001):

- Type I: Given a designed workplace (e.g., workstation), the designer or the design system uses the analytic approach to analyze and check whether it is economically and ergonomically viable, i.e., the focus is on design evaluation, task time prediction, operator comfort, etc.
- Type II: Given the operators' anthropometric data, the designer or the design system uses the synthesis approach to lay out or design the workplace (workstation), e.g., work-surface height, bin arrangement, even human hand motion path, etc., from the working postures of an operator.
- Type III: Given a workplace layout and design, and operators' anthropometric data, the designer or the design system uses the analytic-synthesis approach for a given human-machine interaction or adjustment from operators to workplaces or vice versa.

The design and planning of a human-machine system (e.g., workplace) is a highly knowledge-intensive and ill-structured problem, which includes human-machine modeling, analysis, and prediction of the working posture of human operators in manual handling tasks and for design/layout of workstations and tasks. Most of the practical problems faced by designers are either too complex or too ill defined to analyze with conventional approaches. It is required to develop new intelligent methodologies that involve the integration of design, planning, analysis, and evaluation. In recent years, there have been many efforts made in applying individual intelligent techniques, such as knowledge-based (expert) system, neural network, and fuzzy logic, for operators' postural analysis and prediction, manual handling loads evaluation, and the design/layout of workstation and tasks.

These research efforts provide useful methods for workstation design/layout and adjustment, but they often take effect in a separated way and suffer many limitations or drawbacks such as simplicity, inaccuracy, inflexibility, and constraints for use. For example, Jung and Park's (1994) work only took three joints, i.e., shoulder, elbow, and wrist, into account when measuring and predicting reach posture. Similarly, due to the limitations of the flexible electrogonimeters in the measurement of shoulder and spine joint angles, the shoulder and spine joint parameters were not included in the model proposed by Lim et al. (1996), although they contribute and affect working postures. To overcome these shortcomings, new effective methods should be explored.

Every intelligent technique has particular computational properties (e.g., ability to learn, explanation of decisions) that make them suited for particular problems and not for others. For instance, the capabilities of rule-based (fuzzy) expert systems are inherently well suited to contribute to solutions to workplace design problems. The major drawback, however, is that the programmer is required to define the functions underlying the multi-valued, or ranked, possibility optimization. Furthermore, expert-type rules use a comprehensive language system that may have built-in biases, embedded goals, and hidden information structures, which may result in errors (Zurada et al., 1997). Neural networks use mathematical relationships and mappings to design and optimize human-machine systems. They are capable of statistical decision-making given incomplete and uncertain design information, and can be designed to adapt to the user/designer's requirements. Unlike rule-based (fuzzy) expert systems, they evaluate all the design conflict constraints simultaneously, and model or learn the knowledge base using black-box techniques. They do not use rules in the formal sense, so the design time can be greatly reduced from that of rule-based modeling. It must be conceded, however, that rule-based (fuzzy) expert systems are much easier for humans to error-check than an ensemble of continuous equations in neural networks (Zarefar and Goulding, 1992).

There is now a growing need in the intelligent systems community whereby complex problems require hybrid solutions. More research endeavors are necessary to develop general topologies for soft computing or hybrid intelligent models (e.g., neural fuzzy) and frameworks, learning algorithms, and approximation theory so that they are made applicable in system modeling and control of complex systems including workplace design and simulation.

This chapter proposes a new neuro-fuzzy hybrid scheme for human workplace design and simulation. In the proposed scheme, the motion analysis and measurement system and neural fuzzy hybrid system are used for measuring and predicting human reach posture and human workplace or workstation layout design and planning without any assumptions in theory. Such a scheme can build a unified framework to represent multi-view knowledge and to perform reasoning and learning for human workplace design and simulation in an integrated intelligent way. Accordingly, the following problems will be discussed: man-machine modeling; measurement and analysis, workplace layout design, ergonomic evaluation, and optimization; neuro-fuzzy hybrid scheme and application in workplace layout design and planning.

The structure of this chapter can be described as follows. Section 2.2 reviews the related works for workplace layout. Section 2.3 proposes a soft-computing framework for workplace design, simulation, and optimization. Section 2.4 discusses the human-machine system modeling, and introduces a method for measuring and analyzing the working postures and motions using the PEAK Motus system. Section 2.5 discusses the procedures and algorithms for workplace layout design, and ergonomic evaluation. Section 2.6 discusses adjustment and optimization of workplace layout. Section 2.7 details the neuro-fuzzy hybrid scheme and implementation for workplace layout design and planning; Section 2.8 studies several design case studies using the proposed scheme; Section 2.9 concludes with some discussions.

2.2   Related Work Review

From the literature review, we see that a number of design and planning procedures for the designs of human-machine systems (e.g., workplaces) have been already developed (Braun et al., 1996; Bullinger, 1986; Ben-Gal and Bukchin, 2001). One of the important steps in these procedures is the selection of part and tool bins, and their arrangement in the workplace with economic and ergonomic considerations. All these procedures are only general in nature, and they do not include rules or other detailed information about how to carry out the various steps (Braun et al., 1996). The design of industrial workplaces, as with most forms of design, is iterative and highly interactive. The designer has to consider numerous constraints and solutions for contradictory goals (Evans et al., 1986). Edwards (1970) developed the MTM-based task time prediction (TTP) model and a biokinematics human model for assembly workplace or assembly workstation design. A computer-aided tool can greatly assist the planning engineer in this complex and time-consuming job. However, a large number of conflicting constraints make it impossible to find an optimal solution procedure. Therefore, heuristic techniques have to be used in such a computer-based program (Braun et al., 1996).

There have been several computer-aided systems developed for the design and planning of human workplaces, which can be classified into three categories: systems for human modeling, systems for selection of furniture, and systems to select and place the bins and tools. The human models (Dooley, 1982; Beevis, 1989) can be used for corrective redesigning of the workplaces. The design process remains with the planning engineer, while the scrutiny of the solution with regard to anthropometry and biomechanics can be done with the help of the human model. Systems for selecting furniture have been reported by Abdel-Moty and Khalil (1986), Brennan et al. (1990), and Fernandez et al. (1990). The systems for choosing and placing of bins and tools (Ho and Lim, 1990; Haller, 1982) also consider the economic goals. However, most of these systems can only be used interactively because of missing heuristics and lacking of ability to heuristic reasoning. The planning engineer has to choose and place the bins and tools manually, while the computer-aided system visualizes and evaluates the solution. For example, in the ARPLA system, the selection of the bins has to be done interactively. After the requirements concerning the functions of each bin are defined, all the suitable bins are depicted on the screen for the planning engineer to select (Braun et al., 1996; Haller, 1982). ARPLA considers economic goals only. These goals can be achieved by positioning the bins next to the place from which the time necessary to bring the part to the place where it has to be assembled is minimal. The existing computer-aided planning systems for the design of assembly workplaces require many interactive processes, and consider economic goals only.

Braun et al. (1996) reported a computer-aided planning system called EMMA for design and planning of a manual assembly system. The described system was constructed from AutoCAD and consisted of a database of workstation elements (such as tools and bins) and anthropometric data combined with an MTM-1 analysis module. It was claimed that with EMMA, workplace layouts can be designed and evaluated subject to both ergonomic (human factors) and economic (assembly time and cost, MTM time analysis) considerations. Based on the 3D representation of the product to be assembled, bins and tools are chosen and placed on the workplace automatically. The system uses the following evaluation indices: work area coverage; coverage of the optimal visual area; use of left and right hand; balanced motion pattern; degree of control; and quota of sensomotorical motions. However, Braun et al. (1996) did not deal with the system structure, geometric design model, and system simulation. It is therefore not suitable for concurrent intelligent design, planning, and simulation of a manual assembly system.

The complexities of human-machine system design led to the application of many kinds of individual intelligent techniques, such as knowledge-based (expert) system, ERGO-SPEC (Brennan et al., 1990), neural network (Jung and Park, 1994; Lim et al., 1996), and fuzzy logic. For proper evaluation of the interactions between the user (operator) and the product (workplace) through human-machine models, accurate prediction of the human reach or handling loads is one of the essential functions of those models. In postural analysis and prediction of manual handling loads, expert or knowledge-based systems have also been used, such as in OWASCA (Vayrynen et al., 1990), LIFTAN (Jung and Freivalds, 1990), and ERGON-EXPERT (Rombach and Laurig, 1990). Nanthavanij (1994) developed an ergonomic assessment of VDT workstation in reference to the seated posture of the user by IntelAd, an intelligent computer workstation adjustment system. Merritt and Gopalakrishnan (1994) applied fuzzy set theory to predict the possibility of occurrence of the cumulative trauma disorders (CTDs) of the upper extremity. Further details can be found in Karwowski et al. (1990).

The back-propagation paradigm is a supervised learning network and was employed as a tool for predicting human movements, workplace layout design and planning, and adjustment of the design/layout of workstation and tasks (Jung and Park, 1994; Lim et al., 1996; Zha, 2001; Zha and Lim, 2001). Jung and Park (1994) examined the applicability of artificial neural networks to the prediction of human reach posture. They established three-dimensional motion trajectories of the joints of the upper limb (shoulder, elbow, and wrist) in the right arm from 5 percentile female to 95 percentile male through a motion analysis system that photographed the actual human reach. The data obtained were subsequently used to train and test a back-propagation network as a tool for predicting such human movements. Comparisons between prediction and real measurements were made using a pairwise t-test.

Lim et al. (1996) also examined the potential of neural network analysis to predict the range of anatomical joint motions for the design/layout of workstation and tasks. Simulated assembly tasks were carried out on a custom-built multi-adjustable workstation, and the posture and motion data recorded with a flexible electrogoniometric system. Prediction of joint motions was analyzed using a feed-forward back-propagation neural network. The trained neural network was capable of memorizing and predicting the maximum and minimum angles of joint motions associated with a range of workstation configurations. This work is very useful for workstation design/layout and adjustment.

To overcome these shortcomings, a new effective method was proposed for measuring and predicting the human working postures by using the motion analysis and measurement system and neural networks techniques (Zha and Lim, 2001; Zha, 2001). The network was trained by measurement data from the training set and the prediction capability of the trained network was tested for untrained test data. The structure of the employed neural network is dependent on the problem of design or planning, and can be categorized as: layout planning neural network, and configuration neural network (Zha, 2001). Zurada et al. (1997) developed an artificial neural network-based diagnostic system that can classify industrial jobs according to the potential risk for low back disorders due to workplace design. Such a system could be useful in hazard analysis and injury prevention due to manual handling of loads in industrial environments. The results show that the developed diagnostic system can successfully classify jobs into the low- and high-risk categories of LBDs based on lifting task characteristics.

2.3   Soft Computing Framework for Human-Machine Systems

The early proposed processes for human-machine systems design were similar to that of Singleton (1974), whereby human and hardware subsystems are developed in parallel, followed by subsequent integration both in terms of the interface and then the operational system itself. More recently, representations of the systems design process have handled the parallel hardware and human design needs by placing ergonomics considerations within all stages of the total system design process. The human-machine system development is mainly determined by the design of human model, product, and its operation process. Some different soft-computing schemes were proposed for human-machine system design for manual assembly (Zha, 1999).

In this chapter, a hybrid intelligent scheme and soft computing framework is proposed with integration of neural networks and fuzzy expert system as well as computer-aided workplace design, as shown in Figure I.2.1 Following this framework, the geometrical model from the CAD system constitutes the basis for layout and planning human-machine systems and operation processes. It has to be complemented to be a product model that contains all the information necessary for designing economic system. For example, the type of operational workstation or the dimension of the equipment to provide parts depends on the dimension and the weight of the operating parts. Furthermore, information like the class of fit, the depth of insertion, and the positioning direction can be taken from the product model.

To obtain all this information and make it available for the design, it is possible to simulate the whole operation (e.g., assembly) process within an extended CAD system. The instrument used in this case is built up on the CAD system and offers all the functions needed to simulate and record

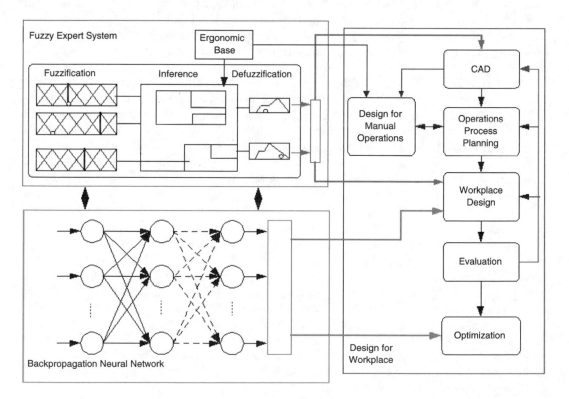

FIGURE I.2.1   Soft computing framework for human-machine system development.

the operation process (e.g., assembly) in addition to the CAD operations. During this simulation, the operations and the relations between the operating parts can be specified or completed with further information, e.g., relations among operation tasks or the combinations of parts to subassemblies for assembly tasks. Based on this information it is possible to derive the precedence diagram of the whole operation process. The precedence diagram is a directed graph that describes the logical and technological constraints among all tasks. This information and the additional description of the tasks are the basis for a computer-aided operation planning (Braun et al., 1996; Ho and Lim, 1990; Zha, 1999).

After obtaining the above information, the workplace design and its optimization can then be carried out. For the layout and design of a manual operation workstation, three areas need to be studied in detail: human modeling, selection of the furniture (tables, chairs, etc.), and selection and placement of bins and tools. The integrated design process for the workplace proposed in this chapter will generally be divided into two steps (Braun et al., 1996):

1. In the first step, the equipment (table, bins, tools, etc.) have to be selected and arranged on the workplace. Therefore, various ergonomic and economic goals, generally mutually opposing, have to be considered and optimized simultaneously. For example, short distances of motion are going to be realized to optimize the standard operation time and therefore the costs. This requires small bins placed next to the operation (e.g., assembly) device. On the other hand, the use of small bins implies a high frequency of replenishment, entailing increased cost for material distribution.
2. In the second step, the planning result that is already the workplace layout has to be evaluated according to this multi-criterion system of goals. As explained earlier, the planning and evaluation of manual workplaces are very complex tasks. Some heuristic methods for an approximate solving of these problems will be presented below.

Therefore, the proposed soft computing framework for the design of human-machine system requires the concurrent integration of the following procedures: design for manual operation (e.g., manual assembly), operation process planning, task assignment and line balancing, equipment selection, workplace design, evaluation, and simulation.

2.4 Man-Machine System Modeling, Measurement, and Analysis

In this section, we will discuss issues related to man-machine system modeling, measurement, and analysis of working posture and motion for human assembly workplace layout design and planning.

2.4.1 Man-Machine System Models

The application scenario of the man-machine system is that the operators seated or standing at a workstation in front of a conveyor belt or a processing line perform manual tasks such as part insertion operations. The operator's task is to pick up the required components from their respective bins that are positioned around the workstation and insert the components into the workpiece. Once the insertion has been finished, the completed workpiece is allowed to move with the conveyor belt for transfer to the next station. The task processes involve the interactions between human and machine that can be described by the human-machine system models.

2.4.1.1 Task Prediction Model

The task time prediction (TTP) model for assembly process is the component of a workplace design system that predicts the standard cycle time required for a specified method sequence and a specified layout (Edward, 1970). The best-known predetermined motion time systems are motion time analysis (MTA), work-factor (WF), basic motion time study (BMT), and methods time measurement (MTM). Among these, MTM has become very widely used. The basic motions in human workplace layout are based on the 39 time standards of the MTM-2 system. Process and resilience times are also included.

The designer first needs to define a layout of the component parts and the location of the man-model. The motions of the operator can thus be simulated by selecting the hand to be moved and associated MTM elements. The hand's destination must be digitized. The hand traveling distance will be generated automatically by the computer. Upon completion of the simulation, a process sequence chart is compiled to give the total cycle time and motion assigned indices (MAI), as

$$\text{MAI} = \frac{t_a - t_w}{t_c}$$

where t_a is the accumulated time of hand; t_w is the waiting time; and t_c is the cycle time. The ratio shows how efficiently the hand has been assigned work. If a low ratio is found when compared with a standard for the type of work, the analyst can try to improve the work methods by utilizing both hands more efficiently.

2.4.1.2 Man Model

The other important component of workplace design system is an ergonomic human or man model constructed to evaluate the effects of alternative workplace layouts and task sequences on the human. All models of this type analogize the human as a system of links, mechanical joints, masses, springs, dash pots, etc. The complexity of these models generally requires that they be analyzed on computers. Based on the human model, the human parameters are mainly the physical attributes of the human that would affect working posture as a result of the task being performed.

2.4.1.3 Man-Machine Model

Generally, a typical workstation system is comprised of a seat, a workbench table, part and product bins, and an operation center, as shown in Figure I.2.1. The workstation can affect the working posture of the human operator in the following four ways. The first is due to the location of the bins used to store the components. The locations of the bins influence the range that the wrist and elbow joints undergo. Bins placed in locations requiring excessive reach would result in the straining of the wrist and elbow joints. The second factor is the configuration of the table of the workstation. The improper table height and depth would result in the operator adopting awkward working postures to perform the task at hand. The other two factors are locations or positions of the seat and the operation center, respectively, which also influence the motion range of the wrist and elbow joints and the working posture. Similarly, improper seat and operation center locations would also result in the operator adopting awkward working postures to perform the task at hand.

The operation scenario described above involves interaction between the human (operator) and the machine (workstation). It is this interaction between the human and the machine that affects the working posture of the operator. Thus, the process parameters to be measured can be categorized into human parameters and machine parameters. The human parameters are mainly the physical attributes of the human, which would affect working posture as a result of the task being performed. In the case of the component insertion process with a seated operator, the following tasks can be identified:

- Stretch out the hand forward to reach the component bins
- Pick up the component in the bin
- Move the hand over to the workpiece
- Precise placement of the component into the workpiece

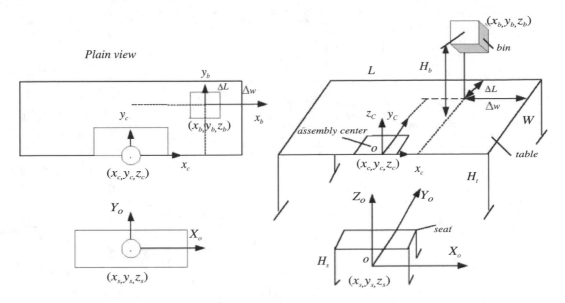

FIGURE I.2.2 Man-machine model of assembly workstation.

By analyzing the operations described above, it can be seen that the working posture concerned is mainly in the hand-wrist-arm-elbow-shoulder region, but sometimes in the head-neck-spine region, etc. The measurement of joint angles may include wrist, elbow, shoulder, head, neck, spine, and even knee joint, etc. Thus, by taking the upper limb and the seated nature of the task into consideration, besides human anthropometric data, the following human parameters, i.e., joint angles, denoted as $\theta_i (i = 1, 2, \ldots,)$ must be measured. Similarly, the following workstation parameters must be accordingly measured: seat height ($H_s = z_s$) and seat origin (x_s, y_s), distance of bin from origin in X-axis, Y-axis, and Z-axis (x_i, y_i, z_i), table work-surface height (H_t), operation center location (x_c, y_c, z_c), etc., where $x_b = L/2 - \Delta W$, $y_b = W/2 - \Delta L$, $H_b = H_t + z_b$, and $H_t = z_c$. In some cases, 1 and 0 can be used to represent the gender of the operator if necessary. For illustration of the modeling, a man-machine model is shown in Figure I.2.2. These notations are clearly described by the experimental man-machine model for PEAK motion analysis and measurement system in the next section.

2.4.2 Measurement and Analysis of Working Posture and Motion

The equipment for measuring and recording posture is the PEAK (1998) motion analysis and measurement system, which mainly consists of a Motus motion analyzer, two cameras, two video recorders, two video monitors, a video processor and human-machine workstation system. The video processor converted the images of body markers to digital outputs, which were further processed to determine postural coordinates and angles. The motion trajectory in the form of three-dimensional coordinates of each joint was measured via two cameras from the markers attached to the human joints. Figure I.2.3 shows the setup of PEAK motion analysis and measurement system.

Based on the PEAK system and the experimental task performed, the man-machine experimental model can be obtained from the simplification of models, as shown in Figure I.2.4. The data obtained are divided into two data sets—a training data set and a test data set for training and test or validation purpose of neural networks below. Test data were gathered from the arm movements to the targets, different from the ones used in order to obtain the training data set. More details about the measurement and analysis of working posture and motion can be found in Zha and Lim (2001).

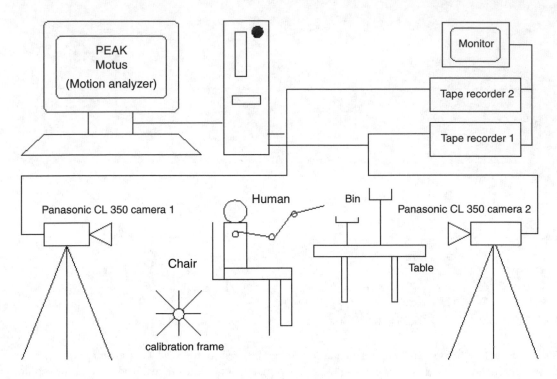

FIGURE I.2.3 PEAK Motus analysis and measurement system.

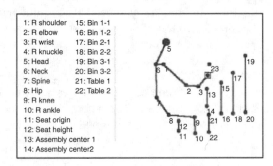

1: R shoulder	15: Bin 1-1
2: R elbow	16: Bin 1-2
3: R wrist	17: Bin 2-1
4: R knuckle	18: Bin 2-2
5: Head	19: Bin 3-1
6: Neck	20: Bin 3-2
7: Spine	21: Table 1
8: Hip	22: Table 2
9: R knee	
10: R ankle	
11: Seat origin	
12: Seat height	
13: Assembly center 1	
14: Assembly center2	

No.	Segment Label	Point 1	Point 2
1	R upper arm	6	5
2	R lower arm	6	1
3	R knuckle	1	2
4	R shoulder	2	3
5	Neck-head	3	4
6	Upper spine	6	7
7	Lower spine	7	8
8	R upper leg	8	9
9	R lower leg	9	10
10	Assembly center	13	14
11	Bin 1	15	16
12	Bin 2	17	18
13	Bin 3	19	20
14	Table	21	22
15	Seat	11	12

No.	Label	Angle type	Point 1	Point 2	Vertex
1	R shoulder angle (Q_1)	Anatomical-180	6	2	1
2	R elbow angle(Q_2)	Anatomical-180	1	3	2
3	R elbow angle(Q_3)	Anatomical-180	1	3	2
4	Head-vector angle(Q_4)	Vector	5	Y	6
5	Head-neck angle(Q_5)	Anatomical-180	5	1	6
6	Neck-upper spine angle(Q_6)	Anatomical-90	5	7	6
7	Upper spine-lower spine angle(Q_7)	Anatomical-180	6	8	7
8	Hip angle (Q_8)	Anatomical-180	7	9	8
9	Knee angle (Q_9)	Anatomical-180	8	10	9

FIGURE I.2.4 Experimental man-machine model.

However, due to inherently redundant degrees of freedom of human movement, there are a large number of possible postures that a person can naturally take; thus it is usually hard to achieve an accurate simulation of human postures in a traditional model. The prediction of human posture is essential for proper evaluation of an operator's usability for any workplace. Neural networks as an emerging technique have been used for this purpose in this research.

2.5 Human Workplace Layout Design and Evaluation

Based on the above discussion, three issues need to be specifically addressed in detail for the design of a workplace (Braun et al., 1996): human-machine modeling, workstation layout and planning including selection and placement of the furniture, tools (tables, chairs, etc.), and bins, and ergonomic and econo-technical evaluation. The first issue has been addressed in the previous section. In this section, we will only discuss the procedures and algorithms for human workplace layout design and evaluation.

2.5.1 Determining the Lot Size

In designing human workplaces, the first step is to determine the lot size Q, which is a quantity of parts provided at a workplace and dependent on the room space available to provide these parts (Braun et al., 1996). The maximum lot size Q_{max} can be determined by comparing the theoretical volume of workplace inside the area of reach V_T with the volume of the parts V_P, i.e.:

$$Q_{max} = \frac{V_T}{V_P} \qquad (I.2.1)$$

The calculation of V_T is described with an assumption of a simplified physiological area of reach, which is represented by a quarter of a sphere on each side of the shoulder and a cylinder between them.

2.5.2 Selection of Equipment

The selection of equipment, e.g., parts bin or product bin for providing parts at the manual workplace, is one of the most important and time-consuming tasks. The main decision variables in this context are the geometry and dimensions of the parts, their attributes (e.g., fragile), the way of providing parts, and the planned assembly lot size. As parts are provided in batches and bunkeredly, the planned lot size will determine the approximate dimensions of the bins. In this case, an average maximum packing depth for each part is ascertained. To determine the average maximum packing depth, the following assumptions are stated:

- The parts correspond to a ball with identical volume
- The balls are arranged in the most possible density
- The deviation of real and ideal volume of the parts can be considered with a correction factor

Thus, the ideal capacity C_{ideal} of a given bin providing balls can be obtained as (Braun et al., 1996):

$$C_{ideal} = \frac{b_b l_b}{d_b^2} \left[\frac{2}{\sqrt{3}} \left(\frac{h_b}{d_b} - 1 \right) + 1 \right] \qquad (I.2.2)$$

where b_b, h_b, and l_b are the breadth, height, and length of the bin. Following this formula, the capacity of all bins could be determined. To find the most suitable bin for each part according to its attributes, every part and bin is classified into a certain class, e.g., parts with a cylinder shape and a diameter less than 6 mm, bins of gravity-feed type with a tray. Then, each class of parts is associated with some classes of suitable bins. As it allows changing the bins to improve the workplace layout, this classification has proved to be advantageous especially for the equipment arrangement.

2.5.3   Planning of Workplace Layout

The planning task of workplace layout is to arrange the equipment and bins or to position the bins. Generally, the planning engineer has to take different ergonomic and economic goals into consideration simultaneously in this process. It is necessary to deal with such opposing goals with the balanced motion patterns (e.g., path, joint limits), or putting small parts into the visual area. Some further goals include minimizing the assembly time, minimizing the workload, or putting identical parts needed into one bin. In addition, there are a lot of constraints, such as minimum or maximum area of worker reach, existing supplies for the transportation of parts and assembled products, or supplies for electricity, existing tools, etc. (Braun et al., 1996).

Because of the complexity of the problem, it is not possible to formulate a complete decision model that guarantees optimal solutions to the layout. There are two bipartite heuristic ways that can be used. One is to find a good starting solution automatically by using several heuristics depending on the set goals. The other is to evaluate the starting solution that gives hints about possible improvements. At this stage, it is possible to redo to get better solutions with other strategies. Therefore, the basic adjustments or improvements should be considered for reach limitations of the area by defining upper and lower limits of the theoretical maximum range of the operator, or blocked zones. Different arrangement strategies, e.g., cyclical motion flow from the edge of the table to the middle or vice versa should be adopted.

Specifically, the basic adjustment may be overlapped with addition of some special arrangement or adjustment strategies. These strategies may include consideration of the visual field while positioning small or difficult-to-handle parts, gathering the same parts or similar parts together, realization of short motion distance for frequently used parts, and simultaneous work of both hands, one fixed sequence, starting inside with the left hand, etc. Based on these guidelines or regulations, the arrangement of 2D bins and tools can be carried out. The principal algorithms for 2D bins and tools arrangement are described in Braun et al. (1996). Figure I.2.5 illustrates some of the basic strategies for the arrangement of 2D bins and tools with 16 bins:

1. Bins in one row, starting outside, not alternating sequence
2. Bins upon/under its predecessor, starting outside, not alternating sequence
3. Bins in one row, starting inside, not alternating sequence
4. Bins upon/under its predecessor, starting inside, not alternating sequence
5. Bins in one row, starting inside, alternating sequence
6. Bins upon/under its predecessor, starting outside, alternating sequence
7. Bins in one row, starting outside, alternating sequence
8. Bins upon/under its predecessor, starting inside, alternating sequence

In the previous work (Zha and Lim, 2001), the algorithms for 2D bins and tools arrangement proposed by Braun et al. (1996) were extended for 3D cases. Figure I.2.6 gives a 3D layout of 6 bins created by the developed RAPID Assembly system (Zha, 1999). Therefore, the result of this planning step is a 2D or 3D layout of the assembly workplace and its computer-internal model containing all the information necessary for the evaluation of a planning solution. Further processing like interactive changing of the workplace layout or presenting the result in a CAD-system can be based on this model.

2.5.4   Evaluation of Workplace Layout

In order to evaluate human workplaces, e.g., manual operation (assembly) workstations, economically as well as ergonomically, effective variables have to be considered. Generally, the economical cost is labor cost determined by operation (assembly) time, whereas the ergonomic cost increases with the workload. Therefore, the evaluation of workplace layout should be based on a detailed analysis of the motions during the work (assembly) process. For this purpose, the

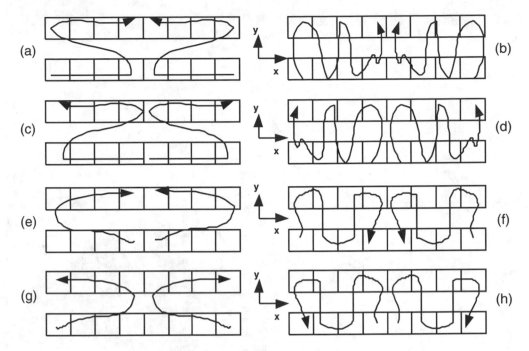

FIGURE I.2.5   Basic strategies for the arrangement of bins.

predetermined motion time standard like MTM or work-factor is considered most appropriate. In this research, the developed human workplace evaluation system includes the MTM-1 method so that the ergonomic evaluation can be done automatically. All information needed to carry out the MTM analysis is included in the computer internal model of the workplace and in a time standard database. The information includes: distances between bins or tools and the operation (assembly) area on the table, kind of motion, class of fit and mating, symmetry attributes, providing (feeding) ways, weight of parts, and orientation of parts.

The economic index can be found as the ratio between the calculated operation (assembly) time per unit and the lowest possible time, which represents the theoretical optimum. It can be calculated by replacing the real MTM codes with those time-optimal codes (Ho and Lim, 1990). The MTM analysis is the basis of the ergonomic evaluation of the assembly workplace layout. For evaluating workplace layout or configuration of tasks, different ergonomic indices are calculated (Braun et al., 1996; Ho and Lim, 1990):

1. Work area coverage. This index describes the proportion of the grasping motions to be executed within the reach area that the smallest person works at in this workplace to those which can be executed in a reduced reach area. The reduced area covers 90% of the maximum permissible area.
2. Coverage of the optimal visual area. This index, analogous to the work area coverage, represents the proportion of grasping motions within the visual field of the smallest person to all grasping motions.
3. Utilization of left and right hand. In order to achieve an optimal situation of stress and strain, a balanced use of both hands is necessary. This index indicates whether one hand is used more often than the other hand.
4. Balanced motion pattern. This index takes rhythmical motions into account while evaluating the planning solution. It describes the proportion of rhythmical motions of the reaching and

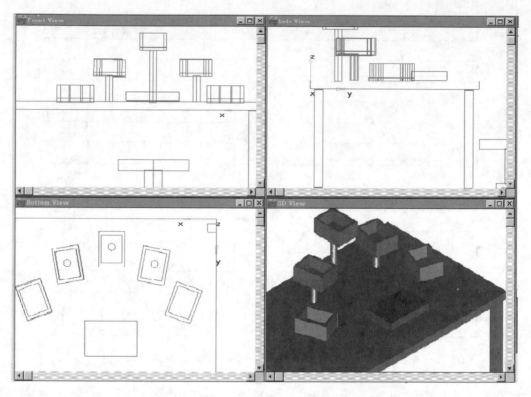

FIGURE I.2.6   3D arrangement of bins modeled in RAPID Assembly.

the moving kind to all motion of this kind. Motions are indicated as rhythmical when both hands carry out the same motion at the same time, or if they work out of phase, e.g., the right reaches to a bin.

5. Load-distance. This index uses the proximity criterion to locate the optimum placement of materials and tools. The load-distance analysis considers the number of inter-workplace movements and distance moved is:

$$C = \sum_{i=1}^{N} \sum_{j=1}^{N} L_{ij} D_{ij} \qquad (I.2.3)$$

where N = the number of work centers; L = the number of loads or movement of work element between work centers i and j, and D = the distance between centers i and j.

6. Motion range and frequency. This index is related to risk factors in body movements, task sequence, and motion speeds. Generally, the criteria such as the comfortable reach, optimal range of motions, and optimal motion paths (energy-saving) can be used for ergonomic evaluation of motion. Motion frequency is also useful for evaluating the ergonomic motion speed and human fatigue degree in motion in frequency-domain.

All six categories of values together amount to what is defined here as the ergonomic value of the workplace.

2.6   Adjustment and Optimization for Workplace Layout

The layout and planning result will be theoretically optimal if the standard time is equal to the lowest possible time and each of the six category evaluation indices discussed above are maximum or minimum. Relating to the index of a design and planning solution to the theoretical optimum, the quality of a planning result can be indicated as a percentage. In the following, we will discuss the model and approach for adjusting and optimizing workplace layout.

2.6.1   Adjustment of Workplace Layout

To help adjust a human-workstation system, the ergonomic assessment is required by suggesting a group of settings essential for obtaining a proper seated and standing posture as recommended by McAtamney and Corlett (1993). Two sets of information, personnel and workstation, are required in this assessment for estimating designed or suggested workstation settings. The personnel information includes the user's sex and anthropometric data. The workstation information includes adjustment ranges of the chair seat location (x_s, y_s, H_s), table-top height, and the bin locations (x_b, y_b, H_b). All components and parameters in the man-machine system as shown in Figure I.2.2 are assumed to be adjustable. In the case where any workstation component cannot be adjusted, input data for the lower and upper adjustment limits will be the same to indicate a fixed height feature. Traditionally, the adjustment system derives the recommended settings by using the following steps:

Step 1: Estimating human model parameters (anthropometric data) from a given operator
Step 2: Calculating (x,y,z) coordinates of key body joints when seated or standing properly
Step 3: Adjusting and calculating (x,y,z) coordinates of key workstation components
Step 4: Checking if adjustment constraints are satisfied

It is therefore noted that the intelligent adjustment system places a major emphasis on the adjustment of workstation and tasks. The main purpose is to find and adjust the settings that allow assembler to keep comfortable (such as all joint angles in their spine are straight) while operating. To achieve this, the angles at other key body joints, i.e., the working posture (shoulder, elbow, wrist, spine, head, knee, hip, etc.) might have to be different from that recommended by McAtamney and Corlett (1993).

2.6.2   Optimization of Workplace Layout

The theoretical optimum referring operation (assembly) time is defined as the corresponding MTM code that has the lowest motion time. A low motion time normally can be reached in two principal ways. One is to reduce the motion distance. The other is to change the way of providing the parts. On the basis of use of MTM codes, Braun et al. (1996) introduced a relative index to indicate the theoretical improvement feasibility of workplace layout. The advantage of this relative index is that it is not only possible to compare different solutions, but also to determine the inherent ergo-economical improvement feasibility. The improvement feasibility is split into changes concerning the design of the product and the layout of the workstation.

As mentioned above, the evaluation criteria, load-distance in Equation (I.2.3), can be chosen as an optimization objective function which considers both the number of movements and distance moved are minimized. For example, if there is a solution with low economical index because of the arrangement of the bins and tools, the planning engineer gets a hint to reduce the lot size. This step leads to smaller bins that can be arranged closer to the worker for getting shorter grasping distance as well as short motion times. On the other hand, a low ergonomic index because of the selection of bins indicates that improvements would be favorable by changing those bins which lead to low values in this characteristic into more suitable ones, e.g., magazines.

During the simulation and analysis, the relationship matrix based on the load-distance criterion is established. The total closeness rating for each tray is computed, and the components to be placed are ranked according to their total closeness ratings in a decreasing order. This establishes the sequence in which the components will enter the layout. The dimensions of the trays are converted to modular units specified by the user and the size of layout matrix established. The trays are placed one by one in modular units into the layout matrix according to the previously established sequence. Each time a new tray (bin) is placed, a sub-optimization problem is solved to determine the optimization position of this box to minimize the sum criterion of the load-distance between the new tray and those previously located. Constraints such as certain trays to be placed at fixed locations can be specified and taken into consideration during the location process.

2.7 Neuro-Fuzzy Hybrid Scheme for Workplace Design

As stated above, every intelligent technique has particular computational properties (e.g., ability to learn, explanation of decisions) that make them suited for particular problems and not for others. Thus, a neural fuzzy hybrid technique is needed to develop more general models for complex human-machine system modeling and workplace design and planning problems. In this section, we will first propose a neuro-fuzzy hybrid scheme for implementing fuzzy if-then rules block by trainable neural network architectures, and then discuss the application of the proposed neuro-fuzzy hybrid scheme and framework for human workplace design.

2.7.1 Neuro-Fuzzy Hybrid Scheme

The terminology related to neuro-fuzzy approach can generally be classified into three classes: neural fuzzy system, fuzzy neural network, and fuzzy neural hybrid system. The first two terms were widely used irrespective of their differences. The approach used in this chapter can be considered as neural fuzzy system or fuzzy neural network, which uses neural networks to adjust the membership function of the fuzzy system. Consider a block of fuzzy rules:

$$\text{IF } x \text{ is } A_i \text{ THEN } y \text{ is } B_i$$

where A_i and B_i are fuzzy numbers, $i = 1, \ldots, n$. Each of these rules can be interpreted as a training pattern for a multi-layer neural network, where the antecedent part of the rule is the input and the consequent part of the rule is the desired output of the neural network. The training set derived from these rules can be written in the form $\{(A_1, B_1), \ldots, (A_n, B_n)\}$.

If a form of two-input-one-output fuzzy systems is given as:

$$\text{IF } x \text{ is } A_i \text{ and } y \text{ is } B_i \text{ THEN } z \text{ is } C_i$$

where A_i, B_i, and C_i are fuzzy numbers, $i = 1, \ldots, n$, then the input/output training pairs for the neural network can be written in the following form: $\{(A_i, B_i), C_i\}, i = 1, 2, \ldots, n$. Similarly, if a form of two-input-two-output fuzzy systems is given as:

$$\text{IF } x \text{ is} A_i \text{ and} y \text{ is } B_i \text{ THEN } r \text{ is } C_i \text{ and } s \text{ is } D_i$$

where A_i, B_i, C_i, and D_i are fuzzy numbers, $i = 1, \ldots, n$, then the input/output training pairs for the neural network take the form as: $\{(A_i, B_i), (C_i, D_i)\}, i = 1, 2, \ldots, n$. There are two main approaches commonly used to implement the fuzzy if-then rule blocks above by standard error back propagation network (Medsker, 1995). One is to represent a fuzzy set by a finite number of its membership values (normally by linear functions). The other is to represent fuzzy numbers by a finite number of α-level sets. With simplicity but without loss of generality, the former approach is adopted in this research. Suppose that $[\alpha_1, \alpha_2]$ contain the support of all the A_i we might have as

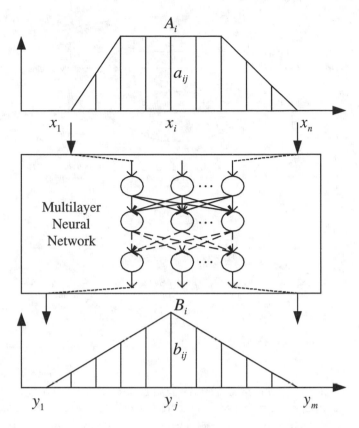

FIGURE I.2.7   A network trained on membership values for fuzzy numbers.

input to the system, and $[\beta_1, \beta_2]$ contain the support of all the B_i we can obtain as outputs from the system, $i = 1, 2, \ldots, n$. If $m \geq 2$ and $n \geq$ are positive integers, then:

$$x_j = \alpha_1 + (j - 1)(\alpha_2 - \alpha_1)/(n - 1)$$
$$y_i = \beta_1 + (i - 1)(\beta_2 - \beta_1)/(m - 1),$$

where $1 \leq i \leq m$ and $1 \leq j \leq n$. Thus, a discrete version of the continuous training set can be composed of the following input/output pairs: $(A_i(x_1), \ldots, A_i(x_n)), (B_i(y_1), \ldots, B_i(y_m)), i = 1, \ldots, n$. Using the notations $a_{ij} = A_i(x_j)$, $b_{ij} = B_i(y_j)$, the fuzzy neural network turns into an n inputs and m outputs crisp network, which can be trained by the generalized delta rule. Figure I.2.7 shows a network trained on membership values of fuzzy numbers. For more complex fuzzy systems, however, there are other more suitable approaches, such as ANFIS (adaptive-network-based fuzzy inference system), to be used for implementing the fuzzy system (Jang, 1993; Buckley and Hayashi, 1995). Thus, the neuro-fuzzy hybrid approach uses neural network to optimize certain parameters of an ordinary fuzzy system, or to preprocess data and extract fuzzy rules from data. In the next section, a framework will be proposed for the neuro-fuzzy hybrid workplace design.

2.7.2   Neuro-Fuzzy Hybrid Model and Framework for Workplace Layout

The application of the proposed neuro-fuzzy hybrid scheme to human workplace design can be illustrated in Figure I.2.8. In fuzzy logic, the fuzzy controller is generally reflected in three basic

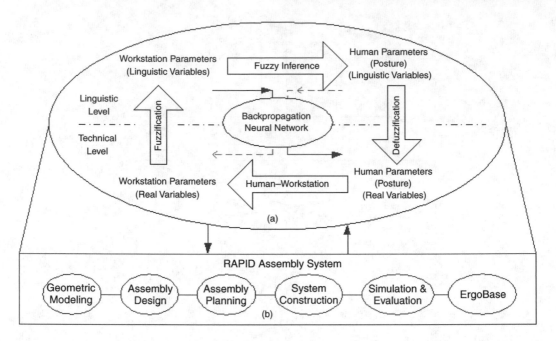

FIGURE I.2.8   The framework of neural fuzzy hybrid scheme for human workplace layout design and planning.

elements: fuzzification, fuzzy inference, and defuzzification. The fuzzification in the input interfaces translates analog inputs into fuzzy values. The fuzzy inference takes place in rule blocks containing the linguistic control rules. The outputs of these rule blocks are linguistic variables. The defuzzification in the output interfaces translates them back into analog variables. Accordingly, the complete loop of the neuro-fuzzy scheme for workplace layout has three steps, as shown in Figure I.2.8 (a). The variables for human/workstation parameters measured at the human-workstation system are translated into linguistic variables (fuzzification) and used to evaluate the condition of the fuzzy rule (fuzzy inference). The linguistic values of variables for human/workstation parameters are then translated back into numerical values (defuzzification). The backpropagation neural network is used to adjust fuzzy membership functions. Using this framework, the problems for workplace layout design and planning can be solved. In practice, however, to implement the integrated workplace design, it is important to design and simulate the whole operation (e.g., assembly) process within RAPID Assembly system (Zha, 1999), as shown in Figure I.2.8 (b). Thus, Figure I.2.8 can be considered as an overall framework for workplace layout design and planning in an integrated intelligent CAD environment.

As stated above, all components and parameters in the man-machine system are assumed to be adjustable. Figure I.2.9 shows the variation regions of the man-machine system to make the membership functions. The variations of bin locations, seating places, and operation (center) places are divided into 54 regions in space, respectively. But the variations of table-top surface height and the subjects are divided into 3 regions, respectively. These are inspired from the man-machine models established in Figure I.2.2 and observations of human assembly operation in practice. For output parameters of joint angles, they are quantified into linguistic variables or fuzzy terms such as S (small), M (medium), L (large), N (negative), Z (zero), and P (positive). These membership functions are determined by experiment and computer simulations of the human operation and neural networks models discussed in the above sections, as shown in Figure I.2.10. Controlled variables, i.e., joint angles, are reasoned out by a fuzzy rule base. Figure I.2.11 illustrates the neuro-fuzzy

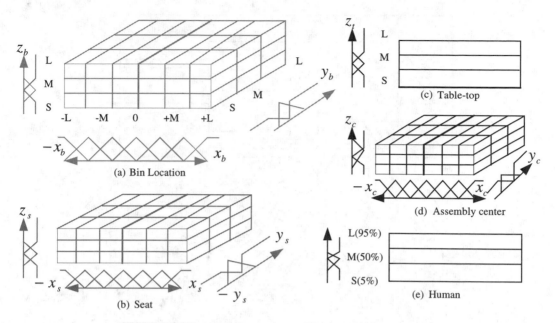

FIGURE I.2.9 Variation regions of man-machine parameters.

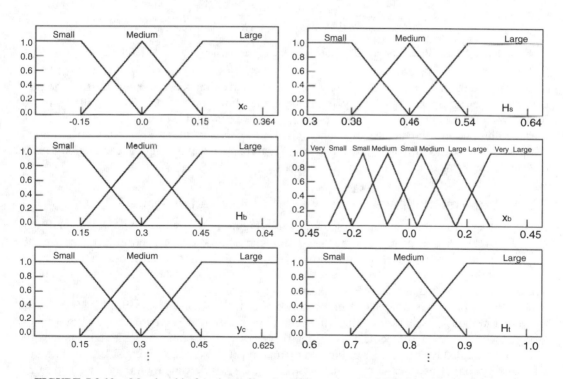

FIGURE I.2.10 Membership functions of man-machine system variables (H_b, H_t, H_s, x_b, x_c, . . .).

technologies mapping a back-propagation neural network to a fuzzy logic for workplace layout design and planning. If the neuro-fuzzy model is executed by a structured computer software package, e.g., fuzzyTECH (1996), which can update the structure and parameters by interactive debug, it allows

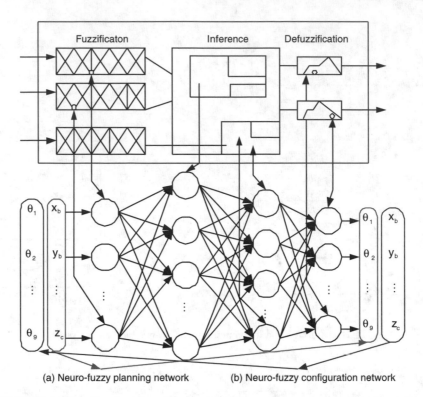

(a) Neuro-fuzzy planning network   (b) Neuro-fuzzy configuration network

FIGURE I.2.11   Illustration for neuro-fuzzy technologies mapping a neural network to a fuzzy logic.

inserting the human knowledge as previous experience, and during the training process, the designer can control the whole process directly. In this case, the neuro-fuzzy model can be trained using the information from multiple sources: designers' experience and sample data. It is worthwhile to note that the definition of the membership functions is an important but hard procedure, based on the designer's heuristic knowledge and training sets. Therefore, the membership functions should reflect different design cases, parameters, and constraints.

2.7.3   Implementation of Neuro-Fuzzy Hybrid Design System

In the real application of the technologies shown in Figure I.2.11, the following two neural fuzzy hybrid systems, layout planning system and configuration system, can be used for solving the workplace design problems discussed above. This is in line with the individual neural network scheme where the structures of the employed neural networks are categorized as layout planning neural network and configuration neural network, respectively.

The neuro-fuzzy hybrid layout planning network system, as shown in Figure I.2.11(a), is used to predict motions of shoulder, wrist, elbow, etc. joint angles associated with certain bin locations around the workstation, subject to anthropometric data and workstation configurations. In fact, this is pertinent to the layout planning of bin and operation path planning with consideration of human comfort. The input parameters are: bin location (x_b, y_b, z_b), percentile and gender (stature H_h), seat location and height (x_s, y_s, H_s), work-surface height (H_t), and operation (assembly) center location (x_c, y_c, z_c). The output parameters are joint angles $(\theta_i, i = 1, 2 \ldots, 9)$. The neuro-fuzzy hybrid configuration network system, as shown in Figure I.2.11(b), is used to predict bin locations on the workstation and optimal workstation parameters (e.g., work surface heights, seat heights, outputs) based on anthropometric data of the subjects and joint motion angles (shoulder, elbow, wrist, etc.). The main role is for the configuration of workstations with consideration of human parameters and anthropometric data. The input parameters are joint angles ($\theta_i, i = 1, 2 \ldots, 9$), while the output parameters are: bin location (x_b, y_b, z_b), percentile and gender (stature H_h), seat location and height (x_s, y_s, H_s), work-surface height (H_t), and even operation (assembly) center (fixture) location (x_c, y_c, z_c).

In the above two models, two hidden layers are chosen for the problem. From the literature, two hidden layers for back-propagation neural network are enough for most problems. All the rules are derived from the training of the neuro-fuzzy model based on the prior database. Once all the relevant measurement parameters are obtained, the training file is created (Zha and Lim, 2001). Each parameter is treated as a field (each field must be separated from the others by at least one space) and each example as a logical row. Thus, each logical row contains the inputs and desired outputs for one sample. The neuro-fuzzy hybrid design system is implemented through integrating a neuro-fuzzy system and RAPID Assembly system (Zha, 1999) which is used for virtual prototyping and planning with modules such as geometric modeling, assembly design, assembly sequence planning, assembly system construction, simulation and evaluation, ergonomic database (e.g., anthropometric data), etc.

2.7.4   Fuzzy Rule Acquisition through Machine Learning

In the developed system, the fuzzy rules in knowledge base are acquired using the following learning approach in artificial neural network area: fuzzy symbolic-connectionist network (FSC-Net), from Universal Problem Solvers, Inc. (1995). The fuzzy symbolic-connectionist network is a hybrid symbolic/connectionist network that utilizes fuzzy logic as its means to perform uncertainty management. The main purpose of FSC-Net is to act as a knowledge acquisition tool, which can be used by domain experts in the development of expert systems. The main learning algorithm is centered on a mechanism that supports the automatic construction of the network topology. The connectionist network encodes the learned knowledge. By incorporating symbolic structures into the network itself, it is possible to represent both structured and unstructured variables and rules. Rules can not only be added as domain specific knowledge, but also can be extracted after learning or refined during learning. Finally, FSC-Net supports the direct incorporation of fuzzy variable membership functions through the user, or by automatically learning the appropriate membership functions for any given input.

The FSC-Net supports two basic types of variables: structured and unstructured. Unstructured variables are of the simplest form and require the least amount of hidden unit representation. In general, an unstructured variable is a symbolic input that has an uncertainty measure associated with it, e.g., "X 0.8," where "X" designates an input with an uncertainty measure of 0.8. The structured variables, on the other hand, can come in two flavors: fuzzy and nominal. The fuzzy variables allow the user of the system to divide (i.e., partition) the numerical range of a variable into its fuzzy equivalent. In general, fuzzy variables are described by a set of membership functions, where each such function is associated with a linguistic hedge (i.e., high, small, large, etc.). The membership functions correlate a given numerical value with a degree of membership indicating the strength of the numerical value being a member of the predefined fuzzy sets. A linguistic hedge within the FSC-Net is described by four quantities:

Linguistic Hedge = ⟨Lower Bound, Upper Bound, Lower Plateau, Upper Plateau⟩

Whenever the value of a given fuzzy variable lies within the range defined by the two quantities "Lower bound" and "Upper bound," the membership value of the linguistic hedge is defined as 1 (i.e., complete presence). Besides allowing fuzzy variables, FSC-Net also supports variables with nominal inputs (i.e., inputs that can be assigned a finite number of symbolic values). At any one time only a single value can be assigned to a variable. Associated with every value assignment is a belief value indicating how certain it is that the assigned value is the right one.

TABLE I.2.1 Learning Example

Training File				Output Rule File
Test or training file formart:				Output rule file format:
NTrainingPatterns 2				
Fuzzy	Height	0.26	0.85	Rule 1: if and (Nominal(Size[Large]) = 0.95,
Nominal	Size	Large	0.70	Fuzzy(Height[Low]) = 1.0,
TopRaise	1.0			TopRaise = 0.85) then
Outputs				Successful (0.86);
Successful	0.80			Rule 2: if and (Nominal(Size[Small]) = 1.0,
End				TopRaise = 1.0) then
Fuzzy	Height	39.0	0.96	Successful (1.0);
Nominal	Size	Small	0.8	Rule 3: if or (Nominal(Size[Small]) = 1.0,
TopRaise	0.65			TopRaise = 0.9) then
Outputs				Successful (0.84);
Successful	0.59			
End				

The format for training files (extension*.trn) is quite simple, as shown in Table I.2.1. The first line of every file contains the string "NtrainingPatterns" and the number of training examples that should be read by the system. Next, each training pattern is listed. This pattern consists of two parts: the input list and the output list. Inputs can be preceded by the keywords "Fuzzy" or "Nominal," if they represent structured variables, or no keyword at all (unstructured formats). Next, the inputs identifier is stated followed by its value and the degree of certainty. Once all inputs have been listed, the keyword "Outputs" is used to separate the list of inputs from the list of outputs. Every output entry consists of the output identifier and the output value. The output value can either be a degree of certainty (if the data set contains classificatory data) or a scaled continuous value (for all continuous data). The user needs to scale all values onto the unit interval (i.e., [0, 1]).

2.8 Case Studies

To verify and validate the developed neuro-fuzzy hybrid design methodology and intelligent adjustment system in RAPID Assembly system, three typical cases corresponding to the above stated three categories of work in workplace design were studied within RAPID Assembly system. The experimental operation (assembly) tasks for a kind of micro-optic lens were carried out on a multi-purpose adjustable workstation (Mah, 1996; Zha and Lim, 2001), which can be used for simulating some real industrial assembly workstations. The ergonomic assessment and adjustment of the assembly workstation were performed.

The models for workstation design and simulation were constructed using standard feed-forward networks and trained with available data. The network model parameters were obtained by an iterative method (trained), using a modification of the standard back-propagation algorithm and moment equilibrium constraints. Training with a limited set of exemplars allowed accurate prediction of postures (generalization). Sensitivity analyses during training and testing phases showed that the choice of specific network parameters was not critical except at extreme values of those parameters.

TABLE I.2.2 Anthropometric Data of User Population Group

Subject	Users Category	Body Height
1	Male—5th Percentile	1.62
2	Male—50th Percentile	1.73
3	Male—95th Percentile	1.85
4	Female—5th Percentile	1.49
5	Female—50th Percentile	1.59
6	Female—95th Percentile	1.70

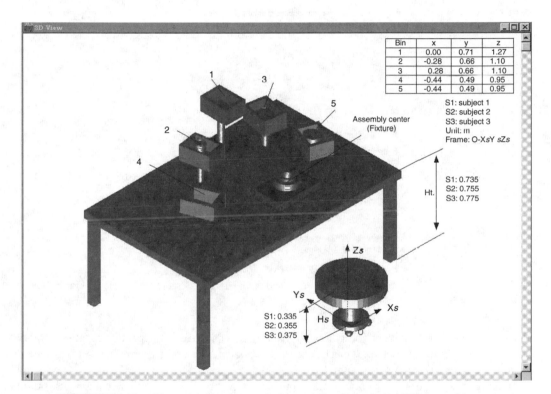

FIGURE I.2.12 3D workstation layout design and simulation in RAPID Assembly system.

The first case was for the layout of a manual assembly workstation. Given operators' anthropometric data, the system was used to design and plan workstation layout. The most often encountered case is to determine the work-surface height and the bin location or arrangement from working posture or joint angles of an operator. Table I.2.2 lists anthropometric data (body heights) of user population group. After using the system, the layout of laboratory assembly workstation can be determined. The results are partially illustrated in Figure I.2.12, in which a 3D solid simulation of a workplace for manual micro-optic-lens assembly is given.

The second case was for ergonomic assessment of manual assembly workstation layout. Given a workstation layout design, the system was used to check whether or not it is ergonomically viable for a human operator's comfort. This is actually a workstation design evaluation work. Figure I.2.13

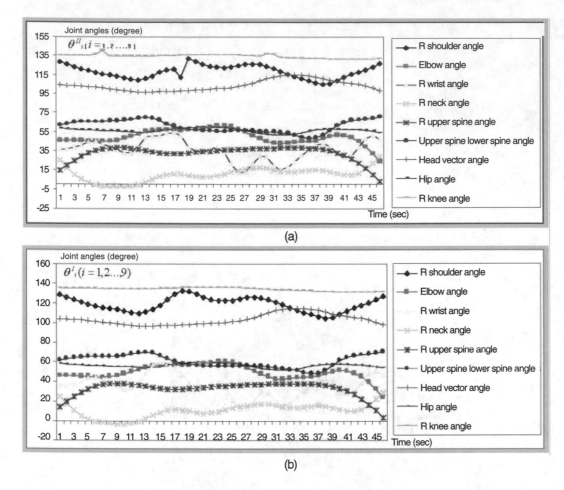

FIGURE I.2.13 Predicated and analyzed joint angles for a 95th percentile female operator during pick-place.

shows the predicted and analyzed joint angles for a 95th percentile female operator during assembly operation process (pick and place). It can be seen that model predictions are correlated and consistent with motion analysis from experimental data. The prediction error is within ±6 degrees. The partial assessment results of workstation layout are summarized in Table I.2.3. In practice, the workstation can be adjusted according to the suggested settings.

The third case was for adjustment of manual assembly workstation. Given a workstation layout and design, and operators' anthropometric data, the system was used to investigate the interaction between a given human (operator) and a machine (workstation) or to adjust operator-workstation system. The nature of this case is actually an incorporation of the above two cases' work. Figure I.2.14 shows a snapshot of neuro-fuzzy 3D-workplace design and simulation in RAPID Assembly system.

2.9   Discussions and Conclusions

In this chapter, the potentials of the neuro-fuzzy hybrid techniques have been examined to predict the range of anatomical joint motions, design/layout, and adjustment of workstation and tasks.

TABLE I.2.3 Ergonomic Assessment of Assembly Workstations

Assembly Workstation	User Population Group		Workstation Settings (m)				Proper Seated Posture?	Remarks
			Seat Height	Seat Distance	Assembly Center	Bin Height		
Fixed Height	M	5th	0.43	0.85	0.77	1.13	No	
		50th	0.43	0.85	0.77	2.50	No	Bin too high
		95th	0.43	0.85	0.77	2.50	No	
	F	5th						
		50th						
		95th						
Partially Adjustable	M	5th	0.43	0.85	0.77	1.13	No	
		50th	0.43	0.85	0.77	1.20	Yes	
		95th	0.43	0.85	0.77	1.50	Yes	
	F	5th						
		50th						
		95th						
Fully Adjustable	M	5th	0.33	0.50	0.30	0.80	Yes	
		50th	0.35	0.60	0.40	1.10	Yes	
		95th	0.37	0.70	0.60	1.30	Yes	
	F	5th						
		50th						
		95th						

Simulated assembly tasks were carried out on a custom-built multi-adjustable workstation, and the posture and motion data recorded with a flexible PEAK motion analysis system. The trained neural network was capable of predicting the angles of joint motions associated with a range of workstation configurations. This approach does not need any assumption usually required for working posture modeling, since the approach predicts human posture by directly emulating the human motion. The algorithm can generate the results more quickly and accurately than other existing heuristic or analytic algorithms. On the other hand, due to the use of learning functions, the accuracy of calculation can be improved continuously with the application of the system to more and more populations and cases.

The proposed neuro-fuzzy intelligent hybrid scheme can carry out design, simulation, and optimization of a human workplace in an integrated way. Such a scheme can provide a unified framework to represent multi-view knowledge and to perform reasoning and learning for intelligent layout design and simulation for a human workplace. The developed approach and system are best suited for human workplace prototyping in virtual design system. They are also potentially to be used for real industrial workstation design/layout and adjustment if the test and training data are collected in industrial *in situ* and the database is large enough. It should be noted, however, that the developed system in its present form in this research is only applied to laboratory simulation of industrial tasks and is not applied in real industrial environment.

In summary, the neuro-fuzzy approach showed a very promising prediction, design, and planning capability for human-machine modeling and analysis, and layout design of workstation. It is therefore

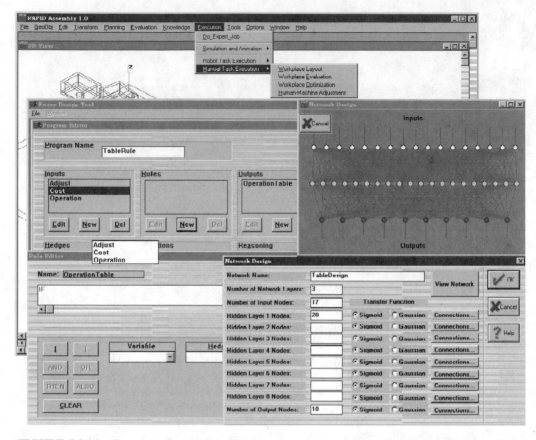

FIGURE I.2.14   Snapshot of neuro-fuzzy 3D workplace design and simulation in RAPID Assembly.

expected that this method be used to more accurately simulate human workplace than existing heuristic methods, as well as to improve a human-machine modeling capability in general. The future work is expected to enrich the data samples and knowledge base of the system to fit for more generic or real industrial applications.

References

Abdel-Moty, E. and Khalil, T.M.S., 1986. Computer aided design of the sitting workplace, *Comp. Ind. Engineer.*, 11(1–4), pp. 22–26.

Beevis, D., 1989. Workplace design-anthropometrical and biomechanical approaches, In *Applications of Human Performance Models to System Design*, (Mcmillan, R.G.) (Ed.), (New York: Plenum Press), pp. 327–339.

Ben-Gal, I. and Bukchin J., 2002. The ergonomic design of workstation using virtual manufacturing and response surface methodology, *IIE Trans. Design and Manufacturing*, 34, pp. 375–391.

Braun,W.J., Reboller, R., and Schiller, E.F., 1996. Computer-aided planning and design of manual assembly systems, *Int. J. Prod. Res.*, vol. 34, no. 8, pp. 2317–2333.

Brennan, L., Farrell, L., and McGlennon, D., 1990. ERGOSPEC: a prototype expert system for workstation design, *Computer-aided Ergonomics: A Researcher's Guide*, Karwowski, W., Genaidy, A.M., and Asfour, S.S. (Eds.), Taylor & Francis, Philadelphia, pp. 117–127.

Buckley, J.J. and Hayashi, Y., 1995. Neural nets for fuzzy systems, *Fuzzy Sets and Systems*, 7(1), pp. 265–276.

Bullinger, H.J. (Ed.), 1986. *Systemtische Montageplanung* (Munich: Hanser) (in German).

Chen, C.-L., Kaber, D.B., and Dempsey, P., 2000. A new approach to applying feedforward neural networks to the prediction of musculoskeletal disorder risk, *Applied Ergonomics*, 31(3), 269–282.

Dooley, M., 1982. Anthropometric modeling programs: a survey, In *Computer Graphics and Application*, New York: Institute of Electrical and Electronic Engineering (IEEE), 2(6), p. 17.

Edward, K.K., 1970. A model for the design of manual work stations, Ph.D. Dissertation, University of Michigan.

Evans, S.M. and Chan, D.B., 1986. Using interactive visual displays to present ergonomic information in workplace design, In *Trends in Ergonomics/Human Factors III*, W. Karwowoski (Ed.) (Amsterdam: Elsevier), pp. 17–24.

Fernandez, J.E., Marley, R.J., and Eyada, O.K., 1990. ErgoCAD: an ergonomic CAD system, *Comp. Ind. Engineer.*, 18(3), pp. 313–318.

fuzzyTECH 4.22, 1996. Reference Manual, 1996.

Haller, E., 1982. Rechnergetutzte Gestaltung ortsgebundener Montagearbeitsplatze, dargestellt am Beispiel kleinvolumiger Teile, IPA Forschung und Praxis, (Berlin: Springer-Verlag).

Ho, N.C. and Lim, T.E., 1990. Computer-aided workplace layout and line balancing, *Computer-Aided Ergonomics: A Researcher's Guide*, Taylor & Francis, Philadelphia.

Honavar, V. and Uhr, L. (Eds.), 1994. *Artificial Intelligence and Neural Networks: Steps Toward Principled Integration*, Boston: Academic Press, 653+xxxii pages.

Jang, J.-S., Roger, 1993. ANFIS: Adaptive-network-based fuzzy inference system, *IEEE Trans. Syst., Man, and Cybernetics*, 23, pp. 665–685.

Jung, E.S. and Freivalds, A., 1990. Development of an expert system for designing workplaces in manual material handling jobs, In *Computer-Aided Ergonomics: A Researcher's Guide*, Karwowski, W., Genaidy, A.M., and Asfour, S.S. (Eds.), Taylor & Francis, Philadelphia, pp. 279–298.

Jung, E.S. and Kang, D.S., 1995. An object-oriented anthropometric database for developing a man model, *Int. J. Ind. Ergonomics*, 15, pp. 103–110.

Jung, E.S. and Kee, D.Y., 1996. A man-machine interface model with improved visibility and reach functions, *Comp. Ind. Engineer.*, vol. 30, no. 3, pp. 475–486.

Jung, E.S. and Park, S.J., 1994. Prediction of human reach posture using a neural network for ergonomic man models, *Comp. Ind. Engineer.*, vol. 27, nos. 1–4, pp. 369–372.

Karwowski, W., Genaidy, A.M., and Asfour, S.S. (Eds.) 1990. *Computer-Aided Ergonomics: A Researcher's Guide*, Taylor & Francis, Philadelphia.

Kengskool, K., Goldman, J., and Leonard, M.S., 1987. An expert system for human operators workplace design, In *Trends in Ergonomics/Human factors IV*, S.S. Asfour, (Ed.) (Amsterdam: Elsevier).

Lim, S.Y.E., Fok, S.C., and Tan, I.T.Y., 1996. Neural network investigation of posture and motion, *Contemporary Ergonomics*, S.A. Robertson (Ed.), Taylor & Francis, Philadelphia, pp. 539–544.

Mah, S.C., 1996. Neural network analysis of working posture, Final Year Project Report, School of Mechanical and Production Engineering, Nanyang Technological University, Singapore.

McAtamney, L. and Corlett, E.N., 1993. RULA: a survey method for the investigation of work-related upper limb disorders, *Applied Ergonomics*, 24(2), pp. 91–99.

Medsker, L., 1995. *Hybrid Intelligent Systems*, Kluwer Academic Publishers Norwell, MA.

Merritt, T.W. and Gopalakrishanan, S., 1994. An application of fuzzy set theory to the prediction of cumulative trauma disorders of the upper extremity, *Int. J. Indust. Ergonomics*, 13, pp. 95–105.

Nanthavanij, S., 1994. An ergonomic assessment of fixed-height, partially adjustable, and fully adjustable VDT workstations by IntelAd, *Comp. Ind. Engineer.*, vol. 27, nos. 1–4, pp. 361–364.

PEAK Motion Analysis and Measurement System, Performance Inc., Englewood, CO, 1998.

Rombach, V. and Laurig, W., 1990. ERGON-EXPERT: A modular knowledge-based approach to reduce health and safety hazards in manual materials handling tasks, In *Computer-Aided Ergonomics: A Researcher's Guide*, Karwowski, W., Genaidy, A.M., and Asfour, S.S. (Eds.), Taylor & Francis, Philadelphia, pp. 299–309.

Singleton, W.T., 1974. *Man-Machine Systems*, Harmondsworth: Penguin Education.

Universal Problem Solvers, Inc., 1995–, http://www.upso.net/

Vayrynen, S. et al. 1990. OWASCA: computer-aided visualizing and training software for work posture analysis, In *Computer-Aided Ergonomics: A Researcher's Guide*, Karwowski et al. (Eds.), Taylor & Francis, Philadelphia, pp. 273–278.

Zarefar, H. and Goulding, J.R., 1992. Neural networks in design of products: a case study, *Intelligent Design and Manufacturing*, A. Kusiak (Ed.), John Wiley & Sons, Inc., Now York, pp. 179–201.

Zha, X.F., 1999. Knowledge Intensive Methodology for Intelligent Design and Planning of Assemblies, Chapter 7, Ph.D. Thesis, Nanyang Technological University, Singapore.

Zha, X.F., 2001. A neuro-fuzzy hybrid scheme for design and simulation of human machine systems, *Applied Artificial Intelligence*, vol. 15, No. 3, pp. 797–823.

Zha, X.F. and Lim, S.Y.E., 2001. Intelligent design and planning of manual assembly workstations: a neuro-fuzzy approach, *Comp. Ind. Engineer.*, accepted (in press).

Zha, X.F., Lim, S.Y.E., and Fok, S.C., 1998. Integrated intelligent design and assembly planning: a survey, *Int. J. Adv. Man. Technol.*, vol. 14, No. 9, pp. 664–685.

Zha, X.F. and Lim, S.Y.E., 2000. A hybrid neural network model and system for interactive and intelligent workplace design (revised), submitted to *Int. J. Ind. Engineer.*, Dec. 2000.

Zurada, J., Karwowski, W., and Marras, W.S., 1997. A neural network-based system for classification of industrial jobs with respect to risk of low back disorders due to workplace design, *Applied Ergonomics*, vol. 28, No. 1, pp. 49–58.

3

Visualization and Modeling for Intelligent Systems

Andrew W. Crapo
GE Corporate Research and Development

Laurie B. Waisel
Concurrent Technologies Corporation

William A. Wallace
Rensselaer Polytechnic Institute

Thomas R. Willemain
Rensselaer Polytechnic Institute

Abstract. Models are artifacts used to better understand our world. As such, they are embedded in intelligent systems as representations of knowledge. In the context of mining data to create knowledge, the modeler is often faced with discovering and understanding relationships in data that have no apparent analog in the laws of physical science. Sketches and diagrams as aids in problem solving and as a means of communication are as old as recorded history. The question now is: Can visualization help us not only to discover the patterns and relationships in these data but also to use newly discovered knowledge to build computational models?

The purpose of this chapter is to provide a description of the process of modeling and, based upon theories of cognition, to show how visualization can assist in developing and assessing computational models for intelligent systems.

3.1 Introduction

Every time we use a map, assemble a piece of furniture, or play a piece of music, we are using models. What do all of these activities have in common? They all use one thing to represent another: the map represents geography, the assembly instructions represent the relations of mechanical parts to one another, and musical notation represents sounds of specified pitch and duration. The map, the instructions, and the music are models.

Albert Einstein once wrote, "My particular ability does not lie in mathematical calculation, but rather in visualizing effects, possibilities, and consequences." (Pinker, 1997). As the domain of human endeavor moves increasingly beyond the three-dimensional space-time continuum of our visual experience into abstract problems with many variables and large quantities of data, the need for "visualizing effects, possibilities, and consequences" becomes greater than ever. The modeler's task is to explore the data, understand the relationships relevant to the problem at hand, and capture those relationships in a computational model that will meet a specific need.

Visualization is not new. Throughout recorded history, drawings and other visually oriented representations have been an integral part of human investigation of the world and of the historical record as well. From Da Vinci to Einstein to the present, visualization has contributed significantly to invention and discovery. What is new is the computational and visual-display capabilities of desktop computers. Data analysts have been quick to embrace computer-based visualization. Effective principles of graphical data representation have been elucidated by Tukey, Cleveland, Mosteller, and Tufte, among many others (Cleveland, 1995; Cleveland and McGill, 1988; Cleveland, 1988; Hoaglin, Mosteller, and Tukey, 1985; Tufte, 1983; Tufte, 1997). The modeling community, including intelligent-systems practitioners, has perhaps had less success in exploiting visualization to facilitate the development of and to improve the quality of models. The demands of building models in increasingly abstract, data-rich domains increase the potential benefits of successful model visualization.

Every one of the perhaps 30 million users of spreadsheet software can be considered a potential modeler (Savage, 1996). A surprisingly large number of these people have, in fact, constructed computational models. While it is easy to create various kinds of presentation graphics from these spreadsheet models, very limited support exists for the actual process of conceiving, building, validating, and communicating the models themselves. We believe that significant advances in modeling tools are possible using visualization.

What unique capabilities of our visual and cognitive systems can be leveraged to better help modelers exercise their craft or to place modeling within reach of more people? In attempting to answer this question, we will of necessity consider theories of model building, vision, and cognition and will examine the role of visualization in the modeling process. We will suggest areas of opportunity for advancing the state of the art in visualization tools to support modeling for intelligent systems.

3.2   Models and the Process of Modeling

A model is an abstraction of reality. Modeling, the act of building a model, can be described as the process of developing an analogical system of relations, with the resulting model comprised of entities and the relationships between them (Willemain, 1995). Modeling is "a fundamental way in which human beings understand the world" (Powell, 1995).

3.2.1   A Taxonomy of Models

Many methods of model, decision, and problem classification exist. Tables I.3.1, I.3.2, and I.3.3 depict the comprehensive model classification scheme of Tersine and Grasso (1979). Other researchers use similar classification schemes: Ackoff and Sasieni (1968) classify models according to their type of structure in the same fashion as Tersine and Grasso do. Eilon (1985) classifies decisions by *frequency* and *replication*. *Frequency* is a measure of repetitiveness, i.e., how frequently during a given time period a decision needs to be made. *Replication* refers to the uniformity of the problem: how much it varies from other problems of its type. Simon (1960) and Ackoff and Sasieni (as cited in

TABLE I.3.1   Model Classification

Dimension of Classification	Categories in Each Dimension	Examples	Comments
FUNCTION	Descriptive	Organizational structure	A descriptive model is empirical, the result of observation, and it describes a situation without making any predictions or recommendations.
	Predictive	Economic forecasting	A predictive model forecasts the state of affairs in the future, given a particular strategy.
	Normative	Procedural manual for a power plant	A normative model is instructive or advisory, making a recommendation on how to achieve an objective.
STRUCTURE	Iconic	Blueprints	Iconic models are physical representations of physical systems, such as scale models, photographs, drawings, and maps. Iconic models usually are on a smaller scale than the physical system and sometimes may reduce the dimensionality as well.
	Analog	Contour lines on a map	Analog models use one set of properties to represent another set of properties. A graph, for example, may be an analog model of physical magnitude.
	Symbolic	Any equation	Symbolic models use symbols such as letters or numbers to represent variables and the relationships between them.
DIMENSIONALITY	Two-dimensional	Photograph	
	Three-dimensional	Scale model	
	Multi-dimensional	Multiple regression	Dimensionality refers to the number of variables used to construct the model.

TABLE I.3.2 Model Classification

Dimension of Classification	Categories in Each Dimension	Examples	Comments
DEGREE OF CERTAINTY	Certainty	Linear programs	Certainty models present a single known state of nature.
	Conflict	Game theory	Have states of nature that are controlled by opponents or competitors.
	Risk	Statistical quality control, Actuarial tables	Have known states of nature that can be described probabilistically.
	Uncertainty	Long-term weather prediction	Lack good information on states of nature and their relative probabilities.
TEMPORAL REFERENCE	Static	Organization chart	Time is not a variable.
	Dynamic	Growth models, Forecasting techniques, Dynamic programming	Time is a variable.
DEGREE OF GENERALITY	Specialized	Computer simulation	Customized to one particular problem.
	General	Linear programming models	Can be used for many different problems.
DEGREE OF CLOSURE (degree to which model is influenced by its environment)	Closed	Anatomical drawings	All variables are controllable.
	Open	Economic forecasting	One or more variables from the external environment are uncontrollable.

MODELS 2

TABLE I.3.3 Model Classification

Dimension of Classification	Categories in Each Dimension		Examples	Comments
3 MODELS DEGREE OF QUANTIFICATION (How mathematical is the model?)	Qualitative	Mental	Mental image	A mental model is the internal conceptualization of a situation.
			Mental conception of a relationship in an equation	Qualitative models use no mathematical description or measurement. They tend to be "less precise, rational, and consistent than quantitative models; however, they are usually more flexible, robust, and reflective of reality." They "consider intangible, human, and behavioral factors that are ignored in quantitative models."[1]
		Verbal	Memorandum	A verbal model results when a mental model is anchored by verbalizing it or writing it down. "Verbal models attempt to communicate and improve on mental models but tend to be imprecise."[1]
	Quantitative	Inductive	Linear regression	Usually statistical, inductive models reason from the specific to the general, drawing inferences from data.
		Deductive	Math programming	Optimization models, which yield a best outcome within a constraining set of circumstances, are an example of deductive models, since they start with a general mathematical description in the form of constraints and use the optimization instruction (maximize or minimize) to arrive at an optimal solution for a given situation.
		Heuristic	Plant layout, Traveling salesman problem	Heuristic models aim for a workable solution, but not necessarily the best one. They often rely on expert knowledge for their construction and generally consist of rules and algorithms.
		Simulation	Monte Carlo simulation	Simulation models "attempt to recreate the essence of a system's behavior over time."[1] Like heuristic models, simulation models look for workable rather than optimal solutions. They are used for complex problems when other quantitative models are insufficient or inappropriate.

Source: [1] Tersine, R. and Grasso, E.T., Models: a structure for managerial decision making, *Ind. Management*, 21, 1979, 6–11.

Rivett, 1972) classify problems according to how structured they are. Rivett (1972) classifies models by content: *queueing, inventory, allocation, scheduling/routing, replacement/maintenance, search,* and *competition.*

Tersine and Grasso's classification system uses the term "variable." We define "variable" as a quantity that can assume any of a set of values. Note that the use of variables is not limited to quantitative models. Qualitative models may contain variables that are categorical in nature. For example, "interbreeding" (as described on p. 189 of Gruber and Barrett, 1974) is considered to be a categorical variable in the Theory of Evolution. The values it can assume are the degrees to which it may occur.

Although model classification schemes may not be particularly helpful in performing modeling tasks, model classification can nevertheless be a useful exercise in the context of studying visualization in the modeling process, since the process of creating different kinds of models may benefit from different kinds of visualization.

3.2.2   Stage Theories of Modeling

Traditional theories of problem solving and modeling tend to describe a linear sequence of stages (Bartee, 1973; Churchman, Ackoff, and Arnoff, 1957; Cowan, 1986; Eilon, 1985; Evans, 1991; Hadamard, 1945; Schwenk and Thomas, 1983; Simon, 1960; Smith, 1989; Urban, 1974; VanGundy, 1988; Wallas, 1926). The use of stages is a standard procedure, yet the reality is that seasoned modelers often arrive at their models by an intuitive process developed after years of trial and error. Why, then, do researchers insist on distilling the art of modeling into a set of rules or stages? The answer lies in the trade-off between portraying the process accurately and providing a tool that is helpful and easy to use. Stages provide a useful checklist and help organize an unstructured task (Evans, 1991). They "decompose a poorly understood process into components that can be analyzed and improved" (Smith, 1989).

The stages of these theories vary in number and in detail of description. At one end of the spectrum is the traditional operations research (OR) approach (Churchman, Ackoff, and Arnoff, 1957), which has six stages: *problem formulation, model construction, solution derivation, model and solution testing, establishing controls,* and *putting the solution to work.* At the other end is Simon's model (Simon, 1960), which has only three stages: *intelligence, design,* and *choice.*

Not all theories insist on linearity. Willemain (1995), for example, stresses that the five modeling topics identified in his research—*context, structuring, assessment, realization,* and *implementation*—do not necessarily occur in any particular order (or, more precisely, they are only stochastically ordered). Urban (1974), Eilon (1985), and Schwenk and Thomas (1983) all note that the order of their stage models is not rigid, that the stages don't necessarily occur in the order given, and that frequent looping back to an earlier stage may occur in what is essentially an iterative process. Others (Bartee, 1973; Hadamard, 1945; Smith, 1988; Wallas, 1926) focus more strongly on the cognitive processes taking place during each stage, such as perception, conceptualization, reasoning, and creativity.

The remainder of this section is devoted to a survey of stage theories of problem solving. A subset of these theories has been charted in Table I.3.4, with comparable stages on the same horizontal level. Table I.3.4 is organized around Johnson-Laird's theory of deduction, which will be discussed in Section 3.4.

One of the earliest stage theories of problem solving has been attributed to Dewey (Brightman, 1980; Evans, 1991): *definition, analysis,* and *solution.* During the *definition* phase, information is gathered, and the problem is conceptualized and formulated. During *analysis,* possible solutions are generated. During the *solution* phase, a solution is chosen. This theory is used as a template onto which all other theories are mapped. This theory's simplicity and universality, as well as its early date, make it a natural choice for a template.

TABLE I.3.4 Mapping of Theories of Problem Solving to Johnson-Laird's Mental Models Theory of Deduction

Johnson-Laird[1]	Dewey[2]	Simon[3]	Evans[4]	Wallas-Hadamard[4,5]	Traditional OR[6] (Hillier & Lieberman, 1974)	Smith[7]	Willemain[8]	Cowan[9]
Comprehension	Definition	Intelligence	Definition	Preparation	Problem Formulation	Problem Identification Problem Definition	Context	Gestation
Description	Analysis	Design	Analysis	Incubation	Model Construction	Problem Structuring Diagnosis Alternative Generation	Structuring	Categorization Diagnosis
Validation	Solution	Choice	Solution	Illumination	Solution Derivation Model and Solution Testing		Realization Assessment	
			Implementation	Verification	Establishing Controls Putting Solution to Work		Implementation	

Sources: [1] Johnson-Laird, P.N. and Byrne, R.M.J., *Deduction*, Lawrence Erlbaum Associates, Hove, U.K., 1991.
[2] Evans, J.R., *Creative Thinking in the Decision and Management Sciences*, South-Western Publ. Co., Cincinnati, OH, 1991.
[3] Simon, H.A., *The New Science of Management Decision*, Harper & Brothers, New York, 1960.
[4] Hadamard, J., *The Psychology of Invention in the Mathematical Field*, Princeton University Press, Princeton, NJ, 1945.
[5] Wallas, G., *The Art of Thought*, Harcourt, Brace and Company, New York, 1926.
[6] Hillier, F.S. and Lieberman, G.J., *Operations Research*, 2nd ed., Holden Day, San Francisco, CA, 1974.
[7] Smith, G.F., Defining managerial problems: a framework for prescriptive theorizing, *Management Sci.*, 35(8), 1989, 963–981.
[8] Willemain, T.R., Model formulation: what experts think about and when, *Operations Res.*, 43(6), 916–932, 1995.
[9] Cowan, D.A., Developing a process model of problem recognition, *Acad. Management Rev.*, 11(4), 763–776.

Simon (1960) proposed a similar three-phase theory of decision-making: *intelligence, design*, and *choice*. This theory is often cited in the literature as one of problem solving rather than decision making (Evans, 1991; McPherson, 1968; VanGundy, 1988), and it is treated as such since it describes a broader process than simply choosing among alternatives. *Intelligence* refers to searching the environment for information relevant to the problem, as in military intelligence. *Design* activity is "inventing, developing, and analyzing possible courses of action" (Simon, 1960, p. 2). The final phase is choosing a particular course of action from those available. It is worth noting that Simon's final phase, *choice*, ignores implementation.

Wallas (1926) and Hadamard (1945) expand problem solving to the four stages of *preparation, incubation, illumination*, and *verification*. During *preparation*, the problem solver studies the problem and—this is important—reaches a dead end. Wallas describes the preparation stage as "hard, conscious, systematic, and fruitless analysis of the problem" (1926, p. 81). During *incubation*, the problem solver turns to other tasks and allows the unconscious mind to work, constructing combinations of ideas (Hadamard, 1945). *Illumination* brings the well-known "aha" experience, which is followed by *verification*—testing and evaluating the proposed solution.

Evans (1991) also uses four phases: *definition, analysis, solution*, and *implementation*. Problem *definition* includes establishing controllable and uncontrollable variables, defining a relationship between those variables, setting up constraints, and determining the levels of knowledge available for both the present state and the desired state. During *analysis*, the problem solver sorts relevant from irrelevant information, classifies the problem, and builds an appropriate conceptual or mathematical model. Finally, the problem may be "solved" in the sense of optimization; "resolved," implying compromise and satisfying; or "dissolved" by changing the nature of the problem and/or the environment to remove the problem. Smooth implementation requires good planning and consideration of technical, organizational, and financial consequences.

Osborn and Parnes' Creative Problem Solving process (CPS) (Evans, 1991; Parnes, Noller, and Biondi, 1977; VanGundy, 1988) emphasizes "finding" as a central human activity in problem solving. The stages of CPS are *mess-finding, fact-finding, problem-finding, idea-finding, solution-finding*, and *acceptance-finding*.

The traditional *operations research* (OR) approach is presented by Churchman, Ackoff, and Arnoff (1957). *Problem formulation* involves analyzing the problem from both the consumer's and researcher's perspectives, defining objectives, and establishing possible courses of action. During *model construction*, the modeler finds mathematical relationships among the controllable and uncontrollable variables. In *solution derivation*, the modeler derives the optimal solution from the model. *Model and solution testing* determines how well the model represents reality by checking the model's predictive ability, verifying the data, and validating the model. In the step of *establishing controls*, the modeler develops acceptable ranges of fluctuation for the variables and makes rules for modifying the solution when the variables go "out of control." In the last step, *putting the solution to work*, the modeler translates the solution into operating procedures.

VanGundy (Evans, 1988, 1991) begins by *redefining and analyzing the problem*, which includes not only data collection and analysis but also an assumption that viewing the problem from a new perspective is a necessary part of the problem-solving process. The goal of *generating ideas* is to develop as many solutions as possible. Evaluating and selecting ideas is the *choice* phase, and finally, the ideas are *implemented*.

Urban's (1974) stage theory of modeling views the modeling process in the context of organizational change and portrays the modeler as an "interventionist." Although this research does not address organizational behavior, it reminds us that modeling activities in an organization often portend change and that human nature resists change. During *formulation of priors*, the modeler acknowledges his own biases in terms of problem approaches and solution techniques. *Entry* marks the entry of the modeling process into the organization. *Problem finding* means to identify the best problem to work on first. *Model development* and *model building* come next, followed by *estimation and fitting*, during which the data are fitted to the model. *Tracking* evaluates model validity.

Eilon's (1985) model of decision making begins with *information input* and moves through *analysis, performance measures, model strategies, prediction of outcomes, choice criteria*, and *resolution*. Bartee (1973) begins with *genesis*, the initial awareness that a problem exists, and moves on to *diagnosis*, where the problem is defined and described. During *analysis*, the problem is broken down into subproblems and studied, and finally, *synthesis* involves integrating the analyzed subproblems into a solution.

Several researchers have focused on analyzing the formulation stage of problem solving. Since problem formulation is a problem in and of itself, the stages of problem formulation also map into Simon's stages. Smith (1989) defines not only problem formulation phases but also the characteristic cognitive processes that take place during each phase, as well as delimiting events between phases. The *problem identification* phase is characterized by the cognitive process of perception and is triggered by some physical stimuli or evidence. A feeling of concern and a belief that a problem exists mark the transition to *problem definition*, whose main cognitive activity is conceptualization. Upon the formulation of an explicit or implicit representation of the problem, *problem structuring* ensues, characterized by instrumental reasoning. When a strategy for addressing the problem has been found, the process moves into the *diagnosis* phase, during which causal reasoning occurs. Once the problem's cause has been identified, *alternative generation* begins, characterized by creativity.

Willemain (1995) identifies five topics that arise during model formulation exercises; these topics do not necessarily occur in any particular order. Willemain's stages of model formulation mirror the entirety of the modeling or problem-solving process. *Context* refers to understanding features of the problem environment and the client. During *structuring*, the modeler identifies variables and relationships and builds the model. *Realization* refers to making the model fit the data and solving the model. *Assessment* involves "evaluating the model's correctness, feasibility, and acceptability to the client" (p. 921). *Implementation* involves "working with the client to derive value from the model" (p. 921).

Schwenk and Thomas (1983) use the following stage model of decision making: *gap identification* or *problem recognition, problem diagnosis or formulation, alternatives generation*, and *alternatives selection*. They emphasize that for different types of problems, different phases will assume greater prominence. For example, in problems that are primarily technical in nature, most of the activity takes place in the latter two stages. Problems that are primarily organizational in nature will require spending a great deal of time on the first two stages.

Cowan (1986) delineates three stages of problem formulation. *Gestation* is the period of time that precedes recognition of the problem. The situation is building but not yet recognized. *Categorization* is the process by which a problem reaches conscious awareness. *Diagnosis* yields an adequate description of the problem.

3.2.3 Attributes of Stage Theories of Modeling

3.2.3.1 Looping

The chart of theories given in Table I.3.4 implies that each stage is visited once and in order, yet the reality is that this linear portrayal of a sequence of stages represents a compromise between accuracy and ease of use. Some researchers have chosen to give more weight to the accuracy side of the equation by explicitly including descriptions of the nonlinearity of the process. Theories that address this issue are distinguished by us as having the attribute of *looping* (as in looping back through a flow chart). Urban (1974), Willemain (1995), Eilon (1985), and Schwenk and Thomas (1983) all note that their stage models are not rigid, that the stages don't necessarily occur in the order given, and that there may be frequent looping back to an earlier stage in what is essentially an iterative process. Morris (1967) discusses the alternation between modifying the model and testing the model with data.

TABLE I.3.5   Cognitive Activities in Cowan's (1986) Stage Theory of Problem Recognition

Stages	Process Variables	Descriptions of Process Variables
Gestation / Latency	Scanning	Attending to situational stimuli in one's surroundings
Categorization	Arousal	Ready to respond: a motivation to clarify an uncertainty
	Clarification	Attempting to understand/verifying
	Classification	Attaching to a discrepancy the label "problem" or "no problem"
Diagnosis	Information Search	Gathering additional evidence about a problem
	Inference	Drawing conclusions from information gathered
	Problem Description	Classifying the specific type and nature of problem at hand

Newell and Simon (1972) have used "generate and test" methods extensively in their work on problem solving. This iterative process may also be conceptualized as recursive. Each step contains the entire problem-solving process (Evans, 1991). Each stage in the problem-solving process is a subproblem to be solved and, as such, requires a journey through the stages of problem solving. Each subproblem also is a problem to be solved, and so on. Willemain (1995) found the entire problem-solving process embedded in the problem-formulation stage.

3.2.3.2   Divergent and Convergent Thinking

Journeying through the problem-solving stages, whether at the large scale of a complete problem or at the small scale of a subproblem, requires both divergent and convergent thinking at each stage (Couger, 1995; Evans, 1991). *Divergent thinking* follows many different paths and generates a multitude of answers to a question. *Convergent thinking* narrows to a single path, which leads to a single answer. Problem formulation, for example, requires considering different alternatives for structuring the problem and then selecting one. While every phase contains both types of thinking, some lean more toward one type or the other. Simon's *intelligence phase*, which includes problem definition and formulation, is in large part an exploratory process, characterized by divergent activities such as brainstorming. The *choice phase*, on the other hand, has as its goal a solution to the problem and, as such, focuses more on convergent thinking. Guilford's structure of the intellect theory (Couger, 1995; Guilford, 1957) identifies divergent and convergent thinking as the two most important of five types of mental operations.

3.2.3.3   Task vs. Process

The distinction between the decision-making paradigm and the problem-solving paradigm suggests that some stage theories of problem solving tend to be task oriented, while others tend to be process oriented. Process-oriented theories are distinguished by their inclusion of or emphasis on types of mental activity. Smith (1989) provides a description, discussed in Section 3.2.2, of the cognitive activities involved in problem formulation. Similarly, Cowan (1986) describes the cognitive processes involved in problem recognition (Table I.3.5). Wallas' and Hadamard's steps of *preparation, incubation, illumination*, and *verification* clearly highlight human information-processing activity. The traditional (OR) approach of Eilon and Urban, with their emphasis on *derivation, establishing controls, performance measures*, and *estimation*, are examples of theories that are more task oriented.

3.2.4   Summarizing Stage Theories of Modeling

This survey of the stage theories of modeling lays the groundwork for the derivation of the theory of visualization during model formulation. It establishes the overall pattern that stage theories of modeling follow and allows for an extraction of a generally applicable definition of model formulation: the process of deciding what kind of model to use. Table I.3.4, which maps the corresponding steps of various theories of modeling, allows a researcher working with any of these theories to determine the location of the theory of visualization and modeling (to be defined in Section 3.3.2) within the given theory. It also clarifies usage of terms that seem to have different meanings within different theories. For example, Willemain's use of the term "model formulation" (which corresponds to his topic of structuring) maps to the traditional OR approach's *model-construction* step, not its *problem-formulation* step. The extraction of the more process-oriented attributes common to a number of theories of modeling focuses attention on the aspects of modeling that are especially important for visualization and modeling for intelligent systems.

3.3   Visualization and Cognition

When the human brain is compared to that of other species, several differences stand out. One difference is the expanded prefrontal lobes, where deliberate thought and planning take place. Without this capacity, modeling would not be possible. A second difference is the human brain's greatly expanded areas for language, which make possible the communication of abstract ideas through speaking and writing. A third is the brain's expanded region for the processing of complex visual information. Last is the brain's expanded temporo-parietal areas, which carry this complex visual information to the language and conceptual regions of the brain (Pinker, 1997).

Sensory perception is closely related to two of these distinguishing features of the brain. Our auditory sense is closely associated with language, for which speaking and hearing are the primary expression. For example, the hearing impaired often experience difficulty in language development. The other associated sense, vision, is by far the predominant mode by which humans perceive the world in which we live (Pinker, 1984; Wade and Swanston, 1991).

3.3.1   Visual Perception and Mental Images

To better understand and appreciate our visual capabilities, consider language. Language is inherently one-dimensional, sequential. This is necessarily so because sound waves arrive at the ear of the listener over time. The cognitive processing that interprets this audiosensory information is such that even if multiple sources of sound (e.g., multiple people speaking) are picked up by the mechanisms of the ear, we are generally unable to comprehend multiple threads of conversations simultaneously. Nor does written language free us of this single-dimensionality. (Actually, it makes it worse. Even disregarding accompanying cues such as facial expression, eye movement, touch, etc., one can argue that intonation adds significant dimensionality to spoken language that is lost in the written word.) The symbols of written sounds or sequences of sounds form strings of symbols that have meaning and that can easily be converted back into meaningful audio sequences (read aloud). What was sequenced with respect to time in spoken language becomes sequenced with respect to space when the language is written.

Vision, by contrast, is not one-dimensional. The retina of the eye, which is an extension of the brain, is a two-dimensional surface with topographical mapping to visual portions of the brain (Kosslyn, Sukel, and Bly, 1999; Pinker, 1997; Wade and Swanston, 1991). A variety of mechanisms including ocular convergence, vertical disparity, occlusion, and motion parallax provide information about the distance from the eye to an object in the visual field, the third spatial dimension. The result is that we usually interpret the stimulus reaching our eyes in a three-dimensional manner. Because the brain appears to handle the depth information differently from the other two spatial dimensions, Marr (1982) dubbed our internal representation of the three-dimensional world as a "2-D sketch." In addition, neurological structures provide time-dependent processing, which adds a fourth dimension, time, to visual perception. This permits detection of motion, for example.

When text, i.e., symbols representing natural language, is placed on a page, which is inherently two-dimensional and can be traversed by the eyes in either direction, a person is rarely able to comprehend the meaning in other than a sequential fashion—line-by-line in the case of English. These sequential representations are sometimes referred to as *sentential* (Larkin and Simon, 1987). For example, we can define *sentential* as information that can be represented digitally using only ASCII. *Diagrammatic* representations may be distinguished from sentential representations by observing that information in a diagrammatic representation is indexed spatially, by location in two or three dimensions, and is processed by the visual portions of the brain (Lark and Simon, 1987). Natural language, whether heard by the ears, seen by the eyes, or felt by the fingers, is primarily processed by the language areas of the brain. To illustrate this difference, consider the two-dimensional array of text blocks that we commonly refer to as a table. While the content of each cell is clearly sentential, the arrangement of the cells is spatial, and cells can be, and usually are, indexed in a nonsequential manner.

Visual processing is largely parallel in nature. Each eye contains around 130 million photoreceptors that converge to about 1 million fibers (communication pathways) in each optic nerve. The integrity of each of these two million channels is maintained from the eye through most of the visual regions of the cortex, providing many layers of neural processing of visual information (Wade and Swanston, 1991). As a result, a number of visual features, including motion, color, intensity, size, intersection, closure, orientation, lighting direction, and distance from the observer, may be extracted "preattentively," i.e., without conscious effort and within 100–200 milliseconds. Such features seem to "pop out" of the visual (Healey, Booth, and Enns, 1996). Card, Mackinlay, and Shneiderman (1999) refer to this as "automatic" processing. Bertin referred to those graphical attributes that behave in this way as "retinal properties" (Card, Mackinlay, and Shneiderman, 1999, p. 29). Parallelism also exists in higher levels of the brain, where processing of spatial information appears to occur separately from processing of texture or motion (Kossyln, Sukel, and Bly, 1999).

Our ability to process and think about information relating to the three-dimensional world is not limited to what we see, any more than our ability to understand linguistic expression is limited to what we hear. For example, if someone asks you, "Which is darker, a Christmas tree or a frozen pea?" there is a good chance you will create a mental image of each and compare them (Pinker, 1997). Many of the same portions of the brain that are active when looking at a physical scene are also active when a detailed mental image is created and examined (Kossyl, Sukel, and Bly, 1999).

Some important differences exist, however. Experiments suggest that constructing and maintaining detailed images in memory is difficult (Finke, 1990; Kossyl, Sukel, and Bly, 1999). As this complexity of the images increases, our ability to examine or manipulate the images to solve problems becomes increasingly inferior to our ability to use an external visualization to solve the same problem (Reisberg and Logie, 1993). Furthermore, a mental image is apparently stored in a particular context, probably incident to the indexing mechanisms used to retrieve it, and it is difficult to "see" things in a mental image that are not normally a part of that context (Finke, 1990; Pinker, 1997). This is less true of visualizations, which are perceived more like the real world and are more amenable to a change of context. In this chapter, we use the words *image* and *imagery* to refer to mental images and the words *visual* and *visualizations* to refer to external representations—things that are actually perceived by the eyes. Nonsentential representations are referred to as *spatial, diagrammatic*, or *graphical*.

A number of researchers report that creative insights and problem-solving performance can be improved with appropriate visualizations (Hong and O'Neil, 1992; Roskos-Ewoldsen, Intons-Peterson, and Anderson, 1993; Pinker, 1997; Waisel, 1998). The literature offers at least two explanations for this improvement. The more common explanation is that visualization extends working memory by using the massively parallel architecture of the visual system to make an external representation function as an effective part of working memory (Larkin and Simon, 1987). Using concepts of expressiveness and effectiveness, Stenning and Gurr (1997) alternatively suggest that the strength of diagrammatic representations may, in fact, lie in their limited expressiveness. By reducing the degrees of freedom of expression, interpretation becomes easier, thus making diagrams more effective. In other words, sentential representations allow ambiguities that cannot exist in spatial visualizations. To illustrate, the statement "The lemon is next to the banana" tells us nothing about which fruit is on the left and which is on the right. Now try to imagine a lemon next to a banana. Ambiguity is not possible; the lemon must be on the right or the left when represented as an image (Pinker, 1997).

3.3.2 Representation

Representation, used repeatedly in the preceding paragraphs, is a critical concept. Gardner (1985) defines representation as the level of explanation, including symbols, rules, and images, between neuroscience and behavior. Marr (1982) wrote, "A representation is a formal system for making explicit certain entities or types of information, together with a specification of how the system does this." It follows that any particular representation will make explicit certain information at the expense of other information, which is pushed into the background and made more difficult to recover. Jones (1996) observes that representations suitable for a computer algorithm are often impenetrable to humans and vice versa, and that experts in modeling and optimization may prefer different representations than do experts in the problem domain. Representation will be central to any attempt to enhance the creation and use of models.

Reference has already been made to *sentential* vs. *diagrammatic* and *mental* vs. *external* representations. Until recently, most external representations, whether created by painting on a cave wall, depositing pigment on paper, or exciting phosphors on a computer screen, have been on flat (two-dimensional) surfaces. Advances in computational power and display technology are making possible representations that can be perceived as both dynamic and three-dimensional. Virtual reality allows the perceiver to interact with these representations. Interest in visual modeling is heightened by the expectation that technological advances will continue to open up new opportunities to leverage human visual perception.

3.3.2.1 The Effectiveness of a Representation

The effectiveness of a representation must always be evaluated in the context of the objective of using the representation. In the scientific domain, goodness is usually approximately equivalent to the speed and/or accuracy of perception of information that is needed to accomplish some task or to understand some concept or relationship. In education, representations may be evaluated based on the speed of comprehension, the depth of understanding, and the length of recall that occurs in the student. In marketing, the purpose of a representation is usually to have an emotional or (occasionally) rational reaction to the information, which will result in the perceiver buying or continuing to buy a product or service.

Recent interest in diagrammatic reasoning has led to a more rigorous understanding of what makes a representation useful and how reasoning with diagrammatic representations is similar to and different from reasoning with sentential representations (Gurr, Lee, and Stenning, 1998). Ideally, every relevant relationship between entities in the represented world should hold between corresponding entities in a model. Such a model is said to be *homomorphic* (Gurr, 1997) and represents a minimal requirement for a model to be considered tractable and well behaved. A more stringent requirement, leading to even stronger models, is that the homomorphic mapping from represented world to model must have an inverse mapping from the model back to the represented world, which is also homomorphic. Such a model is said to be *isomorphic* (Gurr, 1997). As a simple example, imagine a number line with negative numbers to the left and positive numbers to the right. Such a number line constitutes an isomorphic representation of the abstract concept of real numbers. The relationships "to the left of" and "to the right of" in the model (the number line) map exactly to

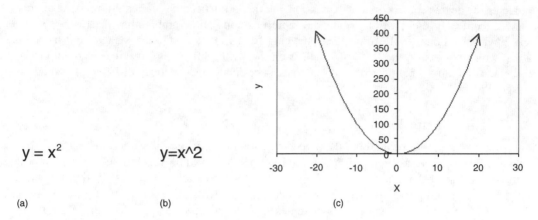

$y = x^2$ y=x^2

(a)   (b)   (c)

FIGURE I.3.1   (a) Sentential model in regular form; (b) Sentential model as input to computer; (c) Graphical form as generated by computer.

the relationships "is less than" and is "greater than," respectively, in the system of real numbers. The inverse mapping is equally robust.

As an example of sentential and nonsentential representations of a simple model, and the ways that representation can affect access to information, consider the high school algebra problem of solving a nonlinear equation of the form $y = f(x)$ for some set of conditions. The equation is a sentential representation of a model, and the graph, if properly drawn, is a diagrammatic visualization of the same model. Students usually need considerable practice before they can easily translate from one representation to the other. Even more skill is required to initiate creation of either representation as an aid in solving a "real-world" word problem. However, access to and understanding of the graph can clearly enhance overall understanding and insight and can sometimes provide immediate answers to questions such as, "Can y ever be negative?" On the other hand, the graphical representation hides other information, such as, "What is the exponent of x?" The use of graphing calculators in recent years has arguably facilitated the development of certain modeling skills by making the actual generation of the graph relatively quick and easy, thereby lowering the cost of creating graphs. Figure I.3.1 shows two sentential and a graphical representation of a model of this type.

3.3.2.2 The Basics of Graphical Representation

A number of authors and researchers have suggested guiding principles for effective graphical representation. Among these are the Gestalt principles of organization that describe ways in which people tend to perceive visual stimuli (Card et al., 1999; Ware, 2000). A number of these principles can be explained in terms of the neurological hierarchy of the retina.

1. *Simplicity:* Every stimulus pattern is seen so as to simplify the resulting structure.
2. *Proximity:* Objects near one another tend to be grouped together into a perceptual unit. This applies to concentrations of randomly arranged stimuli as well, allowing clusters of visual stimuli to be readily perceived as a cluster. In addition, contours very close to each other tend to be seen as one contour.
3. *Similarity:* Several stimuli presented together tend to be seen so that similar items are grouped together.
4. *Continuity:* Visual entities that are smooth and continuous are easier to perceive. In case of ambiguity, we are more likely to perceive entities that are smooth and continuous.
5. *Connectedness:* Neighboring elements are grouped together when straight or smoothly curved lines potentially connect them.

6. *Motion:* Elements moving in the same direction tend to be grouped together.
7. *Familiarity:* Elements are more likely to be perceived as a group if the group is familiar or meaningful.
8. *Symmetry:* Elements are perceived much more strongly as a visual whole if there is symmetry in the whole.
9. *Closure:* A closed contour tends to be seen as an object. The strong sense of closure offered by the rectangular frame of a "window" is an important enabling property for a windowing user interface such as Microsoft Windows.
10. *Relative size:* Elements in a pattern that are smaller tend to be seen as objects.

3.3.2.3   Representing Data Graphically

Since the accuracy of interpreting a representation is one of our proposed measures of effectiveness, let us categorize the kinds of information that we might wish to represent. The range of values that a particular attribute or property can assume is sometimes referred to as a *domain set*. Domain sets are usually classified as *nominal, ordinal,* or *quantitative*. A *nominal domain set* is a collection of unordered items, such as {blue, yellow, green, and red}. An *ordinal domain set* is an ordered set, such as {Sunday, Monday, Tuesday, Wednesday, Thursday, Friday, Saturday}. A *quantitative domain set* is a range of numerical values, such as [0,100 (zero to 100)] (Mackinlay, 1986; Card et al., 1999). Quantitative data can be classified as *interval* or *ratio*. *Ratio data* has an implied zero-value reference, making both ratios and differences in value meaningful. *Interval data* lacks an implicit zero-value reference, so only differences have clear meaning, although ratios of differences can be used. For example, the price of a commodity is a ratio value, permitting the Consumer Price Index to be defined as the ratio of the cost of a set of items today to the cost of the same set at a particular time in the past. Date is an interval value; it has been 226 years since the signing of the Declaration of Independence (a difference), but the ratio (2002/1776) is not inherently meaningful.

Cleveland and McGill empirically investigated the accuracy of various means of graphically displaying quantitative information (Cleveland, 1985). Ranked from most accurate to least accurate, their ordered list is shown in Table I.3.6. The concept of representational effectiveness is key to the empirical ranking of graphical attributes in this table. While Cleveland and McGill used accuracy alone as the measure of effectiveness, most modeling applications would include the cost in time (applied or elapsed), cognitive effort, or both as being of significance as well.

Figure I.3.2 shows a more elaborate ranking proposed by Mackinlay (1986), which adds nominal and ordinal data display. This model is based on various empirical psychophysical results but has not been empirically verified as a whole. The changes in relative effectiveness between different data types are shown by the lines connecting columns. Mackinlay proposed the principle of importance ordering: "Encode more important information more effectively." This is in agreement with the claim

TABLE I.3.6   Accuracy Ranking for Graphical Attributes

1	Position along a common scale
2	Position along identical, nonaligned scales
3	Length
4	Angle and Slope
5	Area
6	Volume
7	Color hue, Color saturation, and Density (amount of black)

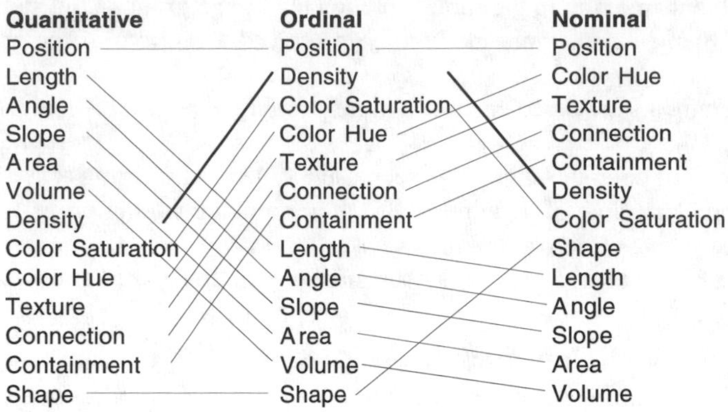

Quantitative	Ordinal	Nominal
Position	Position	Position
Length	Density	Color Hue
Angle	Color Saturation	Texture
Slope	Color Hue	Connection
Area	Texture	Containment
Volume	Connection	Density
Density	Containment	Color Saturation
Color Saturation	Length	Shape
Color Hue	Angle	Length
Texture	Slope	Angle
Connection	Area	Slope
Containment	Volume	Area
Shape	Shape	Volume

FIGURE I.3.2 Accuracy of information encoding by attribute for different kinds of data. *Source*: Mackinlay, J.D., Automating the design of graphical presentations of relational information, *ACM Trans. Graphics*, 5:2, 1986, 111–141.

by Card et al. (1999) that the most fundamental aspect of any visualization is the way in which it uses the 2-D or 3-D space it occupies, i.e., position. Therefore, for static displays, the most important information should be spatially encoded. For dynamic displays, the use of space and time may be combined as preattentive motion.

The possible ways in which information may be displayed by mapping data to a 2-D or pseudo-3-D display of graphic elements is less complex than one might expect. Current display technology can use a variety of techniques to give the illusion of three dimensions, but they are accomplished by projecting the information onto a 2-D surface. We will therefore limit our discussion to 2-D. Card et al. (1999) point out that most graphical presentations can be characterized by their use of spatial substrate (position), marks, and the graphical properties of the marks. In other words, creating a static visualization consists of deciding where in the 2-D space to place each of a set of graphic entities and what each entity's properties should be: color, size, shape, etc. Of course, this simplistic view of a graphical display does not take into account the powerful visual effects that can be achieved, for better or for worse, by the Gestalt principles. Violating these principles will result in slow comprehension and confusion (although humans have proven to be very adaptable, and it may be possible to train people to understand a very unintuitive display).

In the case of dynamic graphical displays, a series of displays are sequenced with respect to time. However, human perception places considerable constraint on the transitions between displays. If the intent is to introduce the sensation of motion, the visualization cannot introduce perceptions too disjointed from the other senses, such as from the inner ear, or nausea will be a likely result. If the changes with respect to time are more discrete, i.e., are changes of focus not intended to be perceived as motion, then the user may have difficulty maintaining a meaningful context in which to interpret the perceptions of the changing scenes. Even when the user initiates the transitions interactively, the maintenance of context is an important concern (Parker et al., 1998).

3.3.2.4 Applications of Graphical Representation Principles

Exploratory data analysis (EDA) provides some useful examples of applying the principles of effective graphical representation. Tukey (1977) is credited with inventing several types of graphical representations and techniques and thus significantly advancing the use of visualization in data exploration. A number of researchers have adopted and extended these graphical concepts (Hoaglin, Mosteller, and Tukey, 1983; 1985), with the result that several types of visual representation have become a standard part of statistical analysis. These include the following (from a poster supplement to the *RCA Engineer*, 30:3, May/June 1985).

1. ***Continuous curves*** plot the value of one variable vs. another on the 2-D spatial substrate. This type of graph uses the two spatial dimensions to represent the values of the two variables at each point on the line. These values are implicitly identified as the most important information in the graph by being assigned to position. Other information can be coded as different colors or patterns of lines, symbol types for discrete points, etc. The use of a smooth curve is particularly powerful because of the eyes' proficiency at seeing continuous curves. Descriptions of special cases of this type of graph follow.

   - An *X-Y curve plot* uses two available spatial dimensions as perpendicular axes (X and Y). The axes are labeled with the names of the two variables and tick marks and associated values to show quantitative domain. A symbol often indicates actual measurements, while a "smoothed" curve which "fits" the data is shown as a line. Usually the domain of the independent variable is shown on the horizontal axis and the domain of the dependent (correlated) variable is shown on the vertical axis.
   - A *Coded X-Y plot* is an $X - Y$ plot that shows the values of more than one dependent variable against the independent variable. Different symbols and line styles are used to allow differentiation of the two variables that share an axis.
   - An *R-Theta plot* is used when one of the variables is periodic or circular. The spatial substrate is assigned to polar coordinates with an angle ($0 - 360°$) as one dimension and the distance from the origin or center as the other.
   - A *Quantile-quantile (Q-Q) plot* is a special type of coded $X - Y$ plot that plots the individual quantiles of an unknown distribution vs. the expected quantiles for that distribution. This allows a person to visually determine how well the unknown distribution matches a certain known distribution and is normally used to check model assumptions.

2. ***Control Charts*** are a special case of $X - Y$ plots that show the upper and lower bounds of a controlled variable. It is easy to see if the value of the controlled variable deviates from the accepted range.

3. ***Scatterplots*** are also a special case of $X - Y$ plots that represent observations of two variables as points in the spatial substrate. Patterns in the data indicate a relationship between the variables. If more than two variables are associated with each observation, scatterplots can be arranged in a matrix formation so that each variable is plotted vs. every other variable, or some subset thereof. In the case of the scatterplot, there is not a continuous line for the eye to perceive unless the data is arranged so as to suggest one. That is precisely the power of such a graph—the eyes perceive patterns in the data.

4. ***Boxplots***, also called *box-and-whisker plots*, are used to provide information about the distribution of measured values of a variable. The horizontal line within the box is the median value. The top and bottom of the box represent the third and first quartiles, respectively. The vertical line, or whisker, shows the range of values that are within the interval $[Q1 - 1.5(Q3 - Q1), Q3 + 1.5(Q3 - Q1)]$. Any data values outside of this range are shown with "*" symbols. Boxplots use one of the spatial dimensions to indicate the values of the various attributes mentioned and the other dimension to identify different values of other variables. Length of lines and height of boxes are important but do not normally have the same base, so comparison is not easy or accurate. Considerable effort is usually required to learn to read boxplots and interpret the information in a meaningful manner.

5. ***Histograms*** use the two spatial dimensions to show a distribution. One axis shows the values as quantitative intervals or categorical values. The other axis uses any of several methods to

show how many values in the data set fall in the interval or category. Histograms provide a visual representation of the distribution of the values. One spatial dimension is dedicated to showing the ordered bins, with the width of the bar representing the size of the range of each bin. The other spatial dimension indicates the average value of all measurements falling within the bin. If the height of the bars varies in any kind of an increasing or decreasing manner, it is relatively easy for the eye to compare the curve passing through the top of each bar with a standard distribution and see how closely the data matches that distribution. Stem-and-leaf plots and inside-out plots are special types of histograms.

6. ***Multivariate Techniques*** are often needed when more than two dimensions are present in the data. Several approaches are briefly described.

- ***3-D plots*** are used when there are three dimensions to be shown. Many of the graphs described above can be rendered as 3-D, showing an additional variable, by using a point of view that does not fall on any of the three axes. It is often difficult to read a 3-D graph as accurately as a 2-D graph, and their effectiveness is questionable in many circumstances.

- ***Contour plots*** are like $X - Y$ plots in that two dimensions are represented in an $X - Y$ plane. Values of a third variable are shown by drawing curves for constant values of this third variable. For example, a contour map shows longitude and latitude as the two spatial dimensions and constant values of altitude as smooth curves. Contour lines on a good map used for hiking in the mountains will normally be 20 feet apart. Contour plots can also fill the space between contour lines with color. By ranging from a lighter shade to a darker shade as contour interval ranges change, a more visually powerful effect is achieved. However, most color gradations are not universally perceived as having any particular order, so many color contour plots are of questionable value.

- ***Glyph plots*** draw some type of figure, such as lines radiating from a point, on an $X - Y$ plane. The length of the various features represent values of a third, fourth, fifth, etc. variable. The most important variables normally are assigned to the X and Y axes, i.e., occupy position. The desire is for the eye to recognize shifting patterns in the glyphs.

- ***Chernoff-faces*** are a special kind of glyph in which the various attributes of a face, i.e., degree of "smile," aspect ratio, amount of hair, are keyed to the values of the third, fourth, ... variables. The idea is that human vision and cognition show special abilities to differentiate faces, and therefore the patterns of facial attributes will be more easily perceived. At least one study found object-like glyphs, including Chernoff-faces, to be more accurately and quickly perceived than less object-like displays (Ware, 2000).

- ***Parallel coordinates*** use one of the two spatial dimensions, usually the vertical axis, to show quantitative, normalized value. The other axis is used to locate all the variables in a nominal manner. In essence, a vertical line is drawn for each variable, resulting in a set of n parallel vertical lines for n dimensions. For each set of values, a point is placed on the vertical line assigned to that variable, and the points are connected with a series of straight lines. A simple example is shown in Figure I.3.3. Color coding or other techniques can be used to differentiate between lines, but large numbers of data values eventually make the lines indistinguishable. The intent is to create patterns that the eye will perceive and that higher cognitive processes can understand with sufficient reflection.

- ***Pixel maps*** use the pixels of a computer display to represent information. A block of n pixels for n dimensions is defined so that each variable has an assigned position in the block. The value of each variable is then mapped to a color map. The result is a block with a particular color pattern for each observation in the data. The blocks are then arranged on the screen in any of a variety of patterns—spiral from center out, left-to-right, top-to-bottom like text, etc. The result is a large pattern of small patterns. A criticism of the technique is that it uses one of the most important visual dimensions, position, in an arbitrary manner to

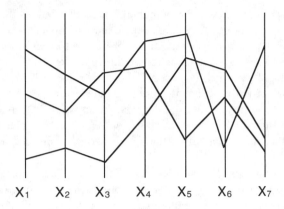

FIGURE I.3.3 Parallel coordinates.

show subsequent observations. Unless the data are dependent on the variable that separates the samples (time dependent for observations made over time), the most important visual attributes are "wasted."

7. Contrary to popular usage, some types of graphs are *not recommended*.

- *Pie charts* represent a percentage as an area. People are known to perceive area inaccurately. To use Tufte's terms, a lot of ink is used to occupy a lot of space with very little information represented in a manner that can be accurately perceived.
- *Bar charts* share some of the criticisms of pie charts. Most data found in a bar chart can be represented more perceptually accurately in an $X - Y$ line plot. An exception might be the histogram, which makes effective use of the bar width to represent bin size.
- *Cumulative line charts* are not accurately perceived for at least two reasons. First, to accurately perceive length comparisons, people need the two lengths to start at the same location on some scale. Cumulative line charts demand that the person compare the vertical distance between lines by making the mental shift to a common base, which is inaccurate. Cumulative line charts often color the region between the lines in a different color. This creates a set of colored regions, with the result that the observer tends to compare the areas of the colored regions. What the area signifies is unclear, and using the area for perceiving values is inaccurate.

3.4 Mental Models, Views, and Visualization

3.4.1 Visualization in Modeling from a Mental Models Perspective

The cognitive process that goes on during modeling is considered to consist of the creation and use of a mental model. Kenneth Craik (1952) first proposed the idea of using a "small-scale model of external reality" to analyze problematic situations and react effectively to them. Johnson-Laird developed Craik's idea into the theory of mental models, which Johnson-Laird presents as a method for reasoning without rules of inference (1983; 1988):

> *One way in which a valid inference can therefore be made is to imagine the situation described by the premises, then to formulate an informative conclusion that is true in that situation, and finally to consider if there is any way in which the conclusion could be false. To imagine a situation is, I have argued, to construct a "mental model" . . . (Johnson-Laird, 1988, p. 227).*

Johnson-Laird defines a mental model as a mental system of relations that has a structure similar to some other system of relations. Applying mental models theory to the process of deduction, Johnson-Laird describes a three-stage operation: *comprehension, description*, and *validation*. During the *comprehension* phase, one uses available knowledge to construct an "internal model of the state of affairs that the premises describe" (Johnson-Laird and Byrne, 1991, p. 35), i.e., a preliminary mental model. *Description* involves drawing a putative conclusion, i.e., developing a revised mental model that not only takes into account the original information but also makes assertions that had not been explicitly stated previously. During *validation*, the modeler searches for alternative models of the original knowledge that falsify the putative conclusions reached during the description phase.

In addition to mental models, Johnson-Laird (1983) argues that there are two other kinds of mental representation: *propositional representations* and *images*. A *propositional representation* is comprised of a set of natural language formulations, such as a paragraph of text. *Images* (when used in the context of mental models theory) "correspond to views of mental models" (Johnson-Laird, 1983, p. 157); as such, they are mental images of mental models.

Larkin and Simon (1987) distinguish sentential representations, which are sequential, such as natural language text; and diagrammatic representations, which are indexed by location in a plane.

The theory of visualization in model formulation will be presented in the context of Johnson-Laird's description phase, which corresponds to Dewey's analysis phase of modeling. During the comprehension phase, the information necessary for forming a mental model of the problem is gathered. During the description phase, a mental model is formed based on the preliminary knowledge gathered during the comprehension phase. It is helpful to conceive of the mental model as analogous to a Structured Query Language (SQL) database. The database exists, frequently in many dimensions, but its contents are inaccessible unless the database is queried and a view is extracted. One cannot go "inside" the database; one can pull out data only by extracting one view at a time. Each view of the database is limited to the two-dimensional tables that can be displayed on a computer monitor, much as each view of the mental model is restricted by the limitations of short-term memory and mental imagery. The mental model is an internal representation of the problem situation. So a mental model, like an SQL database, is inaccessible and incommunicable; the internal representation must be translated into "views" in order to be assessed and communicated. "Views" may be either *imagistic* or *propositional*. *Imagistic views* correspond to Johnson-Laird's image representations, in that they are mental images of the mental model. *Propositional views* are equivalent to Johnson-Laird's propositional representations; they are representations of the mental model that are expressed in natural language formation, i.e., in sentences or equations. During the description phase, the modeler manipulates these views in his or her mind. See Figure I.3.4 for a diagram of the relationships among mental models, views, and visualizations.

The manipulation of mental images is a complex mental task that stretches the limits of working memory. It can be enhanced through the use of external memory aids such as visualization (Reisberg and Logie, 1993). Accordingly, the modeler translates the views into visualizations, either by drawing and/or writing or by using visualization software. The visualizations may contain both text (representing propositional views) and pictures (representing imagistic views). Visualizations comprised mainly of text are considered "sentential," and visualizations comprised mainly of pictures are considered "diagrammatic," after Larkin and Simon (1987). Diagrammatic visualizations are those whose meaning depends on the location of marks on a two-dimensional plane. The use of the term "sentential" differs slightly from Larkin and Simon's in that it refers to visualizations that are in natural language format and can be translated into ASCII.

The views and visualizations are used for both within-model testing during description and between-model testing during validation. During within-model testing, the modeler tests the mental model for internal consistency and consistency with what is known about the problem situation. To do this, the modeler generates a series of views from the mental model. The modeler then anchors

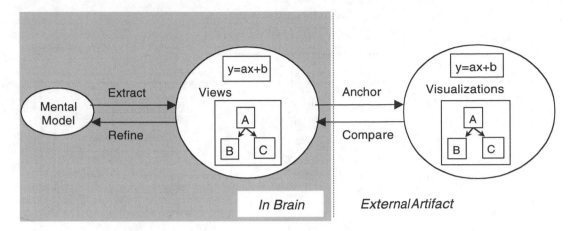

FIGURE I.3.4 Mental models, views, and visualizations. *Source:* Waisel, L.B., Modeler's Assistant V.1, Mac 058 and Win 95/98, Troy, NY, 1998. With permission.

these views by visualizing them on paper or with computer software. The next step is to compare the visualizations with the views—do they correspond? Do any of the views contradict each other? Do the views and visualizations correspond with what is known about the problem? If a series of views generated from the mental model of the problem are noncontradictory and correspond to their visualizations, then the modeler concludes that the model is satisfactory in comparison with itself, and the description phase ends. If a contradiction or noncorrespondence occurs, then the modeler checks whether (1) the visualization has been inaccurately transcribed, (2) the view has been incorrectly extracted from the mental model, or (3) the mental model is incorrect.

The modeler responds accordingly, modifying visualization, view, or mental model as needed. The validation phase involves the generation of alternative mental models with their attendant views and visualizations; if the alternative mental models can't explain the data any better than the original mental model, and if none of the mental models contradict known facts, then the original model is satisfactory in comparison with other models. See Figure I.3.5 for a diagram of this process.

Both the within-model testing, which occurs during Johnson-Laird's description phase, and the between-model testing, which takes place during Johnson-Laird's validation phase, are iterative processes. The modeler continues them until the modeler is satisfied that the model has been sufficiently validated. This iterative quality of modeling, addressed in the earlier section on model attributes, is further supported by Morris (1967), who describes the process of model building as including "alternation between modification of the model and confrontation by the data" (p. B-709). Confrontation by the data, in the visualization in modeling theory, takes place when the modeler checks that the views and visualizations are in accordance with what is known about the problems.

3.4.2 Imagery in Science

How does the model fit in with the case studies? Scientists have an "appetite for pattern" (as do people in general), yet they also appreciate "wildness," and "it is in their lively interplay that understanding moves forward" (Gruber, 1978, p. 122). The French mathematician Hadamard (1945) distinguished among three types of mental imagery: "free thought," which is completely undirected, such as day-dreaming; "controlled thought," which is purposeful and goal-directed, but not particularly intense; and "tense thought," which occurs during problem solving or any concentrated effort. He compared

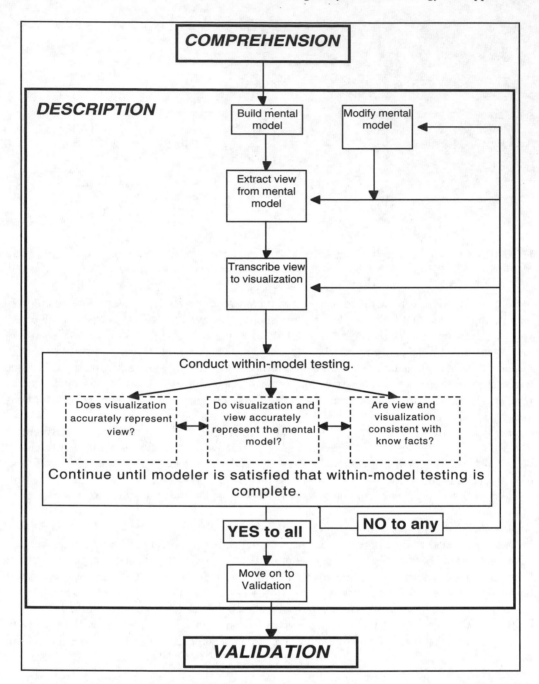

FIGURE I.3.5   The process of model formulation with visualization, which takes place entirely within Johnson-Laird's phase of *Description*.

the cognitive processes of building up a mathematical argument to the act of recognizing a human face. The act of recognizing a face is a pattern recognition task that involves matching one's internal image with an external representation. The visualization in modeling theory also involves the process of comparing internal images with external visualizations. Hadamard said of the mental imagery he used for each step of the proof of the existence of an unlimited sequence of prime numbers:

I need [this imagery] in order to have a simultaneous view of all elements of the argument, to hold them together, to make a whole of them—in short, to achieve ... synthesis (Hadamard, 1945, p. 77).

Physicists are well known for their use of visual imagery. Conceptualizing and carrying out complex experimental physics requires "that the experimenter starts out with a well-structured image of the actual connections between the events taking place" (Deutsch, 1981, p. 358). Deutsch also discusses "the striking degree to which an experimenter's preconceived image of the process which he is investigating determines the outcome of his observations" (p. 354). During the early days of quantum mechanics, "physicists experienced despair and helplessness because of their loss of visualization" (Miller, 1978, p. 73) as they ventured into a world where the electron became unvisualizable. The collective depression lifted when Shroedinger's wave function provided a new basis for visualization. More recently, as mentioned earlier, Richard Feynman's diagrammatic system revolutionized representations of particle interactions (Gleick, 1992).

Einstein described his dependence on visual imagery in his work in a letter to Hadamard (Hadamard, 1945, p. 142):

The words or the language, as they are written or spoken, do not seem to play any role in my mechanism of thought. The psychical entities which seem to serve as elements in thought are certain signs and more or less clear images which can be "voluntarily" reproduced and combined ... this combinatory play seems to be the essential feature in productive thought."

"Images which can be 'voluntarily' reproduced and combined" correspond to the imagistic views in the visualization in modeling theory, which can be extracted from the mental model from a variety of mental "angles." In his autobiography, Einstein frequently used the German words *Bild* (picture) and *Spiel* (play) (Holton, 1981). "Playing with pictures" is a fair description of the theory of visualization in modeling.

Nersessian's description of analogical reasoning as problem solving also supports the visualization in modeling theory (Nersessian, 1992, pp. 19–20):

... analogy is a primary means through which we transfer knowledge from one domain to another ... the creative heart of analogical reasoning is a modeling process in which relational structures [are abstracted] from existing models of representation and problem solutions are abstracted from a source domain and are fitted to the constraints of the new problem domain ... the analogical reasoning process often creates an abstraction or 'schema' common to both domains that can be used in further problem solving ... analogies themselves do the inferential work and generate the problem solution.

Abstracting relational structures from existing models of representation sounds very much like extracting a view from a mental model. Nersessian describes concept formation by analogical reasoning as "a process of abstraction from existing representations with increasing constraint satisfaction" and suggests that "this kind of imagistic reasoning [is] a species of analogical reasoning" (1992, p. 24). This description suggests the process of modifying the model based on internal images, or views, and external pictures, or visualizations, until the modeler is satisfied with the results.

3.4.3 Applying the Theoretical Framework to Darwin's Theory of Evolution

Darwin's Theory of Evolution is an example of a nonmathematical model. Darwin drew and redrew tree diagrams over the years as he developed the Theory of Evolution. Gruber (1978) calls this kind of persistent significant image of creative work an "image of wide scope." The development of an

FIGURE I.3.6   Darwin's first tree diagram.

FIGURE I.3.7   Darwin's second tree diagram.

image of wide scope appears to follow much of the process described by the visualization in modeling theory (Gruber and Barrett, 1974, p. 13):

> *[The images] are not simply chosen, but constructed, winnowed out, criticized, and reconstructed. They are the product of hard, imaginative, and reflective work, and in their turn they regulate the future course of that work. The scientist needs them in order to comprehend what is known and to guide the search for what is not yet known.*

The construction and reconstruction of images suggest repeated extractions of views from the mental model.

Three diagrams (Figures I.3.6, I.3.7, and I.3.8) found in Darwin's manuscripts illustrate the process of modeling with visualization in the development of the early model. It is possible to use these three diagrams as a detailed illustration of the visualization in modeling theory. The analysis of Darwin's thought processes as described in the next paragraph is based on interpretations of Gruber and Barrett (1974). Our own interpretations of Darwin's notes and sketches, which concur with those of Gruber and Barrett, follow.

In July 1837, Darwin conceived of a model of the development of species that included the concepts of monadism, adaptive equilibrium, and an "irregularly branching tree of nature" (Gruber and Barrett, 1974). Monadism refers to the spontaneous generation of life from inanimate matter, such as the

FIGURE I.3.8   Darwin's third tree diagram.

belief that rotting meat produces maggots. Adaptive equilibrium simply means that organisms adapt as needed to changing environments. Darwin extracted a view of a tree structure to visually express his model. The first tree diagram (Figure I.3.6) has three branches and depicts adaptive equilibrium by having separate branches and monadism by a point of origin for the tree. Upon examining this first diagram (comparing the visualization with the view, in the context of the mental model), Darwin realized that this picture could be made to account for observed discontinuities in nature (some species appear to have a traceable history; others do not). According to Gruber's analysis of Darwin's thought processes, it appears that Darwin was comparing the first tree diagram to his mental image of the tree diagram and also checking to see that the diagram accounted for various features of his model. Upon realizing that the discontinuities were unaccounted for in the diagram, Darwin drew the second diagram—a re-visualization of a view of the mental model—which shows that some species can be traced directly back to simpler organisms, while others cannot (Figure I.3.7). Looking at the second diagram, Darwin realized that if monadism and adaptive equilibrium coexist, then extinction must occur regularly, and so he drew the third diagram (Figure I.3.8), which incorporates the concept of extinction.

Prior to drawing the first tree diagram (Figure I.3.6), Darwin wrote in his notes (Gruber and Barrett, 1974, p. 442),

> *Would there not be a triple branching in the tree of life owing to three elements air, land & water, & the endeavour of each typical class to extend his domain into the other domains & subdivisions . . . ? The tree of life should perhaps be called the coral of life, based on branches dead, so that passages cannot be seen.*

With these words, Darwin appears to be translating his mental model into a view, which he then transcribes as the visualization in his first tree diagram, shown in Figure I.3.6. Immediately following the first diagram and immediately preceding the second (Figure I.3.7), Darwin states (Gruber and Barrett, 1974, p. 442), "Is it thus fish can be traced right down to simple organization—birds—not."

He appears to have compared either his view or his visualization with his mental model and found it lacking in this area. He then re-extracted the view and/or regenerated the visualization to accommodate this concept, resulting in the second tree diagram. Just before and surrounding the third tree diagram (Figure I.3.8) are the observations (Gruber and Barrett, 1974, p. 442):

> *If we grant similarity of animals in one country owing to springing from one branch, & the monucle has definite life, then all die at one period, which is not case ... Case must be that one generation then should have as many living as now. To do this & to have many species in same genus (as is) REQUIRES extinction.*

In other words, Darwin has extracted a propositional view from his mental model. This view contains a chain of reasoning leading to the conclusion that extinction of species must take place. Comparing his second tree diagram with this latest view, Darwin saw that the visualization could be modified to include the view containing the extinction reasoning and thus drew the third tree diagram. The process of drawing and modifying the tree diagrams in response to the changing mental models clearly was an integral part of Darwin's creative thought processes.

3.5   The Role of Visualization in Modeling

Having examined how visualization is believed to contribute to cognitive process and the characteristics of the modeling process itself, we propose a model of how visualization can support modeling. Figure I.3.9 represents an environment for the development and use of computational models for intelligent systems. We have made the assumption that to solve a given problem, the modeler must discover and understand significant relationships in a large amount of abstract data. By *abstract*, we mean that the significant relationships in the data have no physical analog. Therefore, there is no readily apparent real-world representation on which an intuitive visualization could be based. It is the role of the modeler to build a robust model incorporating discovered relationships to some end, i.e., predicting, explaining, and ensure that the model is correct, understandable, and usable.

Three primary uses of visualization are represented in Figure I.3.9. All are potentially interactive, as shown by the dashed motor control arrows in the diagram. Exploratory data visualization tools assist the modeler in discovering and understanding relationships within the data. This exploration leads to the creation or enhancement of a mental model that captures the implications of the data relative to the modeling objective. If the visualization and modeling tools are well integrated, this exploration can lead directly to the capture of relationships in the computational model. Otherwise, the modeler must construct the computational model manually by interacting with the model visualization software. In either case, the visualization of the computational model allows the user to compare the modeler's mental model, or a view of that mental model, with the computational model, and to interact with the model, the model output, and the data until a satisfactory model is obtained. Through the iterative process depicted in Figure I.3.9, the modeler focuses on the "missing but important" or the "present but unimportant," as evidenced by the data, and is assisted in building and refining effective models. Visualizations extracted from the data and the model facilitate cooperative work among modelers and communication of the model to others, including customers and end-users.

Table I.3.7 identifies several dimensions used to describe the modeling process and posits ways that visualization may be used. In the case of the modeling topics proposed by Willemain (1995), recent research provides insights into the use of diagrammatic and sentential representations by expert modelers according to modeling topic (Mosteller and Tukey, 1977). Interestingly, the use of diagrams varied significantly by modeler ($p = .0001$). By surveying users of the Modeler's Assistant software (1998), Waisel (1998) substantiated this individualized modeling style.

The use of diagrams also varied by modeling topic ($p = .0148$). In particular, model context had significantly fewer diagrams than did model realization. The use of diagrams was not significantly correlated to the particular problem being solved ($p = .2160$). Sentential representations were used sooner in the modeling process than were diagrams ($p = .0005$), and visualizations were used sooner for some problems (identified as harder problems) than for others ($p = .0011$). These data are consistent with the hypotheses that the expressiveness of sentential representations works well in the ambiguous early states of model development, and that the more difficult the problem, the more need modelers have of visualization.

Table I.3.8 suggests the types of visual representations expected to be most useful to modelers during each of Willemain's (1995) modeling topics. These suggestions are based on the results of analysis of expert modelers' activities and the general cognitive theory reviewed above (Willemain, 1995, Waisel, 1998). Both Tables I.3.6 and I.3.8 suggest areas of additional research to establish a clearer understanding of the ways modelers prefer to use visualizations, of when and how visualization can make the greatest contribution to model correctness, and of how the set of potentially effective modelers may be expanded. In particular, preattentive processing offers promise because of the unique way in which the capabilities of the human visual system are brought to bear on a problem without the modeler sensing any additional effort. How to effectively use preattention is not necessarily clear, however. In general, preattentive attributes do not combine because of the neurological structure at their root. More work like that of Healey et al. (1996) will be required to begin to understand the best use of preattention.

Norman (1993) suggests that cognitive tasks be classified as *experiential* or *reflective. Experiential cognition* involves data-driven processing. The information necessary (besides sensory input) must already be in our brains. It is like a reflex, a "knee-jerk," a real-time reaction. It is the cognition of a domain expert who sees the data and intuitively knows what it means.

Reflective reasoning, on the other hand, is slow and laborious. The use of external aids facilitates the reflective process by acting as external memory storage, allowing deeper chains of reasoning

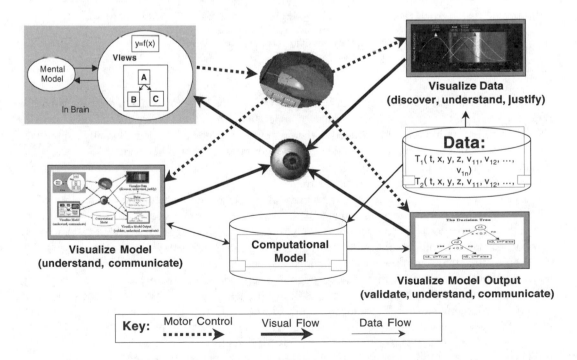

FIGURE I.3.9 Computer-based visualization in modeling.

TABLE I.3.7   Modeling Dimensions and Posited Interactions with Visualization

Modeling Dimension	Range	Comments
Subprocess[1]	Understanding, search	Search for significant relationships supported by data visualization, exploratory data analysis. Search for modeling approach supported by model visualization. Understanding supported by model visualization and output. Early understanding likely to be expressed sententially.
Context[2]	Discovery, Justification	Discovery similar to search above. Justification involves model output and comparison with data analysis.
Modeling topic[3]	Context, structure, realization, assessment, implementation	Visualization found to be significantly related to individual expert modeler and to modeling topic but not to specific problem.[5] (Waisel, 1998). See also Table I.3.8.
Objective[4]	Exploration, analysis, presentation	Exploration can be of data and of modeling options. Analysis of data and of model performance (model output). Presentation of relationships (data and model) and of model output.
Modeling skill	Continuum from novice to expert	Visualization of data, model, and model output, combined with meta-cognitive aids for managing complex problem decomposition, can combine with tutorial systems to help less experienced modelers.
Tool skill	Continuum from novice to expert	Visualization tools, like other software, should be easily learned and become "invisible," i.e., utilization should not require continued cognitive attention. Advanced functionality should be available when needed without being burdensome otherwise.

Sources: [1] Newell, A. and Simon, H.A., *Human Problem Solving*, Prentice-Hall, Englewood Cliffs, NJ, 1972.
[2] Morris, W.T., On the art of modeling, *Management Sci.*, 13:12, 1967, B707–B717.
[3] Willemain, T.R., Model formation: what experts think about and when, *Operations Res.*, 43(6), 916–932, 1995.
[4] Earnshaw, R. et al., *Visualization and Modeling*, Academic Press, London, 1997.
[5] Waisel, L.B., Modeler's Assistant V.I.0, Mac OS 8 and Win 95/98, Troy, NY, 1998.

over longer periods of time than would be possible without the aids. *Reflective reasoning* generally requires the aid not just of writing—books, computational tools, etc.—but also of other people. External representations should be tuned to the task at hand if they are to be maximally supportive. The environment should be quieting, devoid of material save what is relevant to the task. Rich, dynamic, continually present environments can interfere with reflection and are more compatible with experiential mode, event-driven processing.

Building on this concept, one can imagine that a modeling assistant should use preattentive visual stimuli to bring important information that will be immediately meaningful to the modeler's attention. When the software does not "know" what is important but is responding to the modeler's requests to display data or model output for user interpretation/exploration, the display should not be preattentive but should allow the user to reflect with minimal distraction.

Another promising area is that of meta-cognitive support. Whether novices fail to continually assess their own progress because they do not have sufficient cognitive resources or for some other reason, visualization can help. Visualization offers an effective way to help the modeler maintain a hierarchical set of submodels, to maintain the various submodels at different stages of development, and to move from one level to another without loss of cognitive effectiveness. Parker et al. (1998) refer to this as being able to change focus without losing context. The sketches of Willemain's

TABLE I.3.8   Expected Representational Uses by Modeling Topic

Topic	Representational Expectations	Comments
Understand model context	Sentential statements of objectives, visual investigation of data to discover or verify relationships; any model visualizations will likely be related to high-level requirements or architecture. Data analysis and exploration is most likely a dominant type of visualization. Modeling objective may also be represented visually, i.e., model inputs and outputs in a block diagram.	The ambiguity possible with sentential representations (effectiveness) can be an advantage as it avoids struggling to make the unknown explicit on the first pass.
Define model structure	Problem decomposition and specification of variables and relationships supported by model visualization; visual data analysis continues. Representation of the computational model will probably be the dominant type of visualization.	This activity will normally receive a major portion of total resources. Visual tools range from "balloons and strings" representations of "relatedness" to database schema, spreadsheets, and visual programming languages.
Model realization	Model is made more concrete as parameters are estimated and solution techniques are identified. Visualization of the model will continue to dominate.	Visualization of different types and at multiple levels supports hierarchical problem decomposition into "chunks."
Model assessment	Validation of model correctness (assumptions met?), feasibility, and acceptability. Visualization of model output and data will assist in determining adequacy.	This is important to Morris' (1967) justification context. The target audience can be the modeler, her colleagues, and/or the customer and future users.
Model implementation	Implement model; manage its transition into active use Debugging is aided by visualization of the model and its output. Visualization also helps in user understanding and acceptance.	Turnover in personnel means new users of the model must be brought up to speed and brought to see the benefits of using the model. Visualization will continue to assist "marketing" and training.

expert modelers demonstrate quite clearly the use of diagrams to visualize problem decomposition (Waisel, 1998).

3.6   Summary and Conclusions

Modeling is a cognitive process, limited by the working memory of the practitioner. External representation is the primary method of extending working memory. While information technology has expanded our ability to gather and analyze data into increasingly abstract domains, and the quantity of data available is exploding, we struggle to understand what the data mean and to develop robust models that make the newly acquired understanding useful and comprehensible by others.

Visualization can be an important part of exploring the data, building the model, and collaborating, but to successfully exploit visualization, we need to understand the modeling process and visualization itself. We have summarized foundational work in these areas and have proposed a general model of how visualization by expert modelers provides evidence relating the type of visualization to the modeling task, but many questions remain unanswered. We suggest additional interactions between visualization and the modeler and modeling process, and we hope that the proposed framework will help identify fruitful areas for future research.

Acknowledgments

This research was supported by the National Science Foundation grants SES-9012094, SBS-9730465, and SES-9730465.

This chapter is a revised and expanded version of a paper by the same authors entitled "Visualization and the Process of Modeling: A Cognitive Theoretic View," *Proceedings of KDD-2000: The Sixth ACM SIGKDD International Conference on Knowledge Discovery and Data Mining*, pp. 218–226, Boston, MA, August 20–23, 2000.

References

Ackoff, R.L. and Sasieni, M.W., *Fundamentals of Operations Research*, John Wiley & Sons, New York, 1968.

Alabastro, M.S., et al., The use of visual modeling in designing a manufacturing process for advanced composite structures, *IEEE Trans. Engineer. Management*, 42:3, 1995, 233–242.

Anderson, R.E. and Helstrup, T., Multiple perspectives on discovery and creativity in mind and on paper, *Imagery, Creativity, and Discovery: A Cognitive Perspective* (vol. 98), Roskos-Ewoldsen, B. et al., Eds., Elsevier Science Publishers B.V., Amsterdam, 223–254.

Bartee, E.M., A holistic view of problem solving, *Management Sci.*, 20(4), 1973, 439–448.

Brightman, H.J., *Problem Solving: A Logical and Creative Approach*, College of Business Administration, Georgia State University, Atlanta, 1980.

Card, S.K., Mackinlay, J.D., and Shneiderman, B., *Readings in Information Visualization: Using Vision to Think*, Morgan Kaufmann Publishers, San Francisco, 1999.

Churchman, C.W., Ackoff, R.L., and Arnoff, E.L., *Introduction to Operations Research*, John Wiley & Sons, New York, 1957.

Cleveland, W.S., *The Elements of Graphing Data*, Wadsworth Advanced Books and Software, Monterey, California, 1985.

Cleveland, W.S., Ed., *The Collected Works of John W. Tukey*, Chapman and Hall, New York, 1988.

Cleveland, W.S. and McGill, M.E., *Dynamic Graphics for Statistics*, Wadsworth, Inc., Belmont, CA, 1988.

Couger, J.D., *Creative Problem Solving and Opportunity Finding*, Boyd and Fraser, Danvers, MA, 1995.

Cowan, D.A., Developing a process model of problem recognition, *Acad. Management Rev.*, 11(4), 1986, 763–776.

Craik, K.J.W., *The Nature of Explanation*, University Press, Cambridge, U.K., 1952.

Deutsch, M., Imagery and inference in physical research, in *On Scientific Thinking*, Tweney, R.D., Doherty, M.E. and Mynatt, C.R., Eds., Columbia University Press, New York, 1981, 354–360.

Earnshaw, R., Vince, J., and Jones, H. (Eds.), *Visualization and Modeling*, Academic Press, London, 1997.

Eilon, S., Structuring unstructured decisions, *Omega, Int. J. Management Sci.*, 13(5), 1985, 369–377.

Evans, J.R., *Creative Thinking in the Decision and Management Sciences*, South-Western Publ. Co., Cincinnati, OH, 1991.

Finke, R., *Creative Imagery: Discoveries and Inventions in Visualization*, Lawrence Erlbaum Associates, Hillsdale, NJ, 1990.

Finke, R.A. and Slayton, K., Explorations of creative visual synthesis in mental imagery, *Memory and Cognition*, 16, 1988, 252–257.

Gardner, H., *The Mind's New Science*, Basic Books, Inc., New York, 1985.

Gathercole, S.E. and Baddeley, A.D., *Working Memory and Language*, Lawrence Erlbaum Associates, Hove, U.K., 1993.

Gleick, L., *Genius: The Life and Science of Richard Feynman*, Pantheon Books, New York, 1992.

Gruber, H.E., Darwin's "tree of nature" and other images of wide scope, in *On Aesthetics in Science*, Wechsler, J., Ed., MIT Press, Cambridge, MA, 1978, 121–140.

Gruber, H.E. and Barrett, P.H., *Darwin on Man*, E.P. Dutton & Co., New York, 1974.

Guilford, J.P., A revised structure of intellect, *Report of Psychology*, 19, 1957, 1–63.

Gurr, C.A., On the isomorphism, or lack of it, of representations, in *Theory of Visual Languages*, Marriot, K. and Meyer, B., Eds., Springer-Verlag, 1997, 288–301.

Gurr, C., Lee, J., and Stenning, K., Theories of Diagrammatic reasoning: Distinguishing component problems, *Minds and Machines*, 8, 1998, 533–557.

Hadamard, J., *The Psychology of Invention in the Mathematical Field*, Princeton University Press, Princeton, NJ, 1945.

Healey, C.G., Booth, S., and Enns, J.T., High-speed visual estimation using preattentive processing, *ACM Transactions on Computer-Human Interaction*, 3:2, 1996, 107–135.

Hillier, F.S. and Lieberman, G.J., *Operations Research*, 2nd ed., Holden-Day, San Francisco, CA, 1974.

Hoaglin, D.C., Mosteller, F., and Tukey, J.W., *Exploring Data Tables, Trends, and Shapes*, John Wiley & Sons, New York, 1985.

Hoaglin, D.C., Mosteller, F., and Tukey, J.W., *Understanding Robust and Exploratory Data Analysis*, John Wiley & Sons, New York, 1983.

Holton, G., What, precisely, is thinking? in *On Scientific Thinking*, Tweney, R.D., Doherty, M.E., and Mynatt, C.R. (Eds.) Columbia University Press, New York, 1981.

Hong, E.-S. and O'Neil, Jr., H.F., Instructional strategies to help learners build relevant mental models in inferential statistics, *J. Ed. Psychol.*, 84:2, 1992, 150–159.

Johnson-Laird, P.N., *The Computer and the Mind*, Harvard University Press, Cambridge, MA, 1988.

Johnson-Laird, P.N., *Mental Models: Towards a Cognitive Science of Language, Inference, and Consciousness*, Harvard University Press, Cambridge, MA, 1983.

Johnson-Laird, P.N. and Byrne, R.M.J., *Deduction*, Lawrence Erlbaum Associates, Hove, U.K., 1991.

Jones, C.V., *Visualization and Optimization*, Kluwer Academic Publishers, Boston, MA, 1996.

Kieras, D.E. and Myer, D.E., *EPIC: A Cognitive Architecture for Computational Modeling of Human Performance*, available online at http://ai.eecs.umich.edu/people/kieras/epic.html.

Kosslyn, S.M., Sukel, K.E., and Bly, B.M., 1999. Squinting with the mind's eye: Effects of stimulus resolution on imaginal and perceptual comparisons, *Memory and Cognition*, 27: 2, 1999, 276–287.

Larkin, J.H. and Simon, H.A., Why a diagram is (sometimes) worth ten thousand words, *Cognitive Science*, 11, 1987, 65–99.

Mackinlay, J.D., Automating the design of graphical presentations of relational information, *ACM Trans. Graphics*, 5: 2, 1986, 111–141.

Marr, D., *Vision: A Computational Investigation into the Human Representation and Processing of Visual Information*, W.H. Freeman and Company, San Francisco, 1982.

Massey, A.P. and Wallace, W.A., Understanding and facilitating group problem structuring and formulation: Mental representations, interaction, and representation aids, *Decision Support Systems*, 17, 253–274, 1996.

McPherson, J.H., The people, the problems, and the problem-solving methods, *J. Creative Behavior*, 2(2), 1968, 103–110.

Miller, A.I., Visualization lost and regained: The genesis of the quantum theory in the period 1913–27, in *On Aesthetics in Science*, J. Wechsler, Ed., MIT Press, Cambridge, MA, 1978, 72–102.

Morris, W.T., On the art of modeling, *Management Sci.*, 13: 12, 1967, B-707–B-717.

Mosteller, F. and Tukey, J.W., *Data Analysis and Regression: A Second Course in Statistics*, Addison-Wesley, Reading, MA, 1977.

Nersessian, N.J., How do scientists think? Capturing the dynamics of conceptual change in science, in *Cognitive Models of Science, Minnesota Studies in the Philosophy of Science*, Giere, R.N., Ed., University of Minnesota Press, Minneapolis, MN, 15, 1992, 3–44.

Newell, A., *Unified Theories of Cognition*, Harvard University Press, Cambridge, MA, 1990.

Newell, A., Shaw, J.C., and Simon, H.A., The processes of creative thinking, in Gruber, H.E., Terrell, G., and Wertheimer, M., Eds., *Contemporary Approaches to Creative Thinking: A Symposium Held at the University of Colorado*, Atherton Press, New York, 1962, 63–119.

Newell, A. and Simon, H.A., *Human Problem Solving*, Prentice-Hall, Englewood Cliffs, NJ, 1972.

Norman, D.A., *Things That Make Us Smart: Defending Human Attributes in the Age of the Machine*, Addison-Wesley Publishing Company, Reading, MA, 1993.

Parker, G., Franck, G., and Ware, C., Visualization of large nested graphs in 3D: Navigation and interaction, *J. Visual Languages and Computing*, 9, 1998, 299–317.

Parnes, S.J., Noller, R.B., and Biondi, A.M., *Guide to Creative Action*, Charles Scribner's Sons, New York, 1977.

Pinker, S., *How The Mind Works*, W.W. Norton and Company, New York, 1997.

Pinker, S., Ed., *Visual Cognition*, Elsevier Science Publishers B. V., Amsterdam, 1984.

Powell, S.G., The teachers' forum: Teaching the art of modeling to MBA students, *Interfaces*, 25: 3, 1995, 88–94.

Reisberg, D. and Logie, R., The ins and outs of working memory, in *Imagery, Creativity, and Discovery: A Cognitive Perspective*, Roskos-Ewoldsen, B., Intons-Peterson, M.J., and Anderson, R.E., Eds., Elsevier Science Publishers B.V., Amsterdam, 1993, 39–76.

Rivett, P., *Principles of Model Building*, John Wiley & Sons, New York, 1972.

Roskos-Ewoldsen, B., Intons-Peterson, M.J., and Anderson, R.E., *Imagery, Creativity, and Discovery: A Cognitive Perspective*, North-Holland Elsevier Science Publishers B.V., Amsterdam, 1993.

Savage, S.L., Innovative use of spreadsheets in teaching, *OR/MS Today*, 23(5), 1996, 41.

Savage, S.L., Weighing the pros and cons of decision technology in spreadsheets, *OR/MS Today*, 24:1, 1997.

Schoenfeld, A.H., Ed., *Cognitive Science and Mathematics Education*, Lawrence Erlbaum Associates, Hillsdale, NJ, 1987.

Schwenk, C. and Thomas, H., Formulating the mess: The role of decision aids in problem formulation, *Omega, Int. J. Management Sci.*, 11(3), 1983, 239–252.

Simon, H.A., *The New Science of Management Decision*, Harper & Brothers, New York, 1960.

Smith, G.F., Defining managerial problems: A framework for prescriptive theorizing, *Management Sci.*, 35(8), 1989, 963–981.

Smith, G.F., Towards a heuristic theory of problem structuring, *Management Sci.*, 34(12), 1988, 1489–1506.

Stenning, K. and Gurr, C., Human-formalism interaction: Studies in communication through formalism, *Interacting with Computers*, 9, 1997, 111–128.

Tegarden, D.P., Business information visualization, *Comm. Assoc. Information Syst.*, 1: 4, 1999.

Tersine, R. and Grasso, E.T., Models: A structure for managerial decision making, *Ind. Management*, 21, 1979, 6–11.

Todd, P. and Benbasat, I., Evaluating the impact of DSS, cognitive effort, and incentives on strategy selection, *Information Syst. Res.*, 10: 4, 1999, 356–374.

Tufte, E.R., Envisioning information, Graphics Press, Cheshire, CT, 1990.

Tufte, E.R., *The Visual Display of Quantitative Information*, Graphics Press, Cheshire CT, 1983.

Tufte, E.R., *Visual Explanations*, Graphics Press, Cheshire, CT, 1997.

Tukey, J.W., *Exploratory Data Analysis*, Addison-Wesley, Reading, MA, 1977.

Urban, G.L., Building models for decision makers, *Interfaces*, 4(3), 1974, 1–11.

VanGundy, A.B., *Techniques of Structured Problem Solving*, Van Nostrand Reinhold Company, New York, 1988.

VanLehn, K., Problem solving and cognitive skill acquisition, *Tech. Rep. AIP-32*, submitted to the Computer Sciences Division, Office of Naval Research, Arlington, VA, by the Department of Psychology and Computer Science, Carnegie Mellon University, Pittsburgh, PA.

Vessey, I., Cognitive fit: A theory-based analysis of the graphs vs. the tables literature, *Decision Sciences*, 22: 2, 1991, 210–241.

Wade, N.J. and Swanston, M., *Visual Perception*, Routledge, London, 1991.

Waisel, L.B., *The Cognitive Role of Visualization in Modeling*, Ph.D. Thesis, Rensselaer Polytechnic Institute, Troy, NY, 1998.

Waisel, L.B., Modeler's Assistant V.1.0, Mac OS 8 and Win 95/98, Troy, NY, 1998.

Waisel, L.B., Wallace, W.A., and Willemain, T.R., *Using diagrammatic representations in mathematical modeling: The sketches of expert modelers*, Proc.: AAAI 1997 Fall Symp. on Reasoning with diagrammatic representations (Tech. Rep. No. 97–03), MIT, AAAI Press, Menlo Park, CA, 1997.

Waisel, L.B., Wallace, W.A., and Willemain, T.R., Visualizing modeling heuristics: An exploratory study, *Proc. 1999 Int. Conf. Information Systems* (ICIS), Charlotte, NC, 1999.

Wallas, G., *The Art of Thought*, Harcourt, Brace and Company, New York, 1926.

Ware, C., *Information Visualization: Perception for Design*, Academic Press, San Diego, CA, 2000.

Willemain, T.R., Insights on modeling from a dozen experts, *Operations Res.*, 42: 2, 1994, 213–222.

Willemain, T.R., Model formulation: what experts think about and when, *Operations Res.*, 43(6), 916–932, 1995.

4

Online Adaptation of Intelligent Decision-Making Systems

Hisao Ishibuchi
Osaka Prefecture University

Ryoji Sakamoto
Osaka Prefecture University

Tomoharu Nakashima
Osaka Prefecture University

4.1 Introduction

This chapter discusses the online adaptation of decision-making systems that choose an action from alternative ones for handling the situation at hand in a dynamically changing environment. When we have previous cases handled by human experts, they are used as training data for designing decision-making systems. Typically each case consists of a set of input values (i.e., input vector) and a selected action. The input vector corresponds to the available information for decision making. The choice of an action can be viewed as a kind of classification task where each input vector is to be classified into one of given classes. Each class in classification problems corresponds to a single action in decision-making problems. We can use various techniques for classification problems such as statistical methods,[1] machine learning techniques,[2-4] neural networks,[5,6] and fuzzy systems.[7,8] These techniques usually handle static tasks where training data are given in advance. Our task in this chapter is dynamic because training data are successively obtained from a series of decision making systems. That is, the amount of available information for designing decision-making systems continues to increase. Decision-making systems are continually adjusted whenever a piece of new information is obtained. The current adaptation affects the generation of available information in the future.

0-8493-1121-7/03/$0.00+$1.50
© 2003 by CRC Press LLC

As adaptable decision-making systems, we use neural networks and fuzzy systems. Their online adaptation is explained through computer simulations on a multi-agent system. Each agent is an individual decision maker which tries to maximize its own payoff. Each agent can be viewed as a producer of a product that is to be sold in a market. The decision of each agent is to choose a single market from several ones where its product is sold. The market price at each market is determined by the number of agents that choose that market. If many agents choose a particular market, the market price at that market becomes low due to the concentration of products. On the other hand, the market price becomes high if the market is chosen by only a small number of agents. The point of the decision making is to choose a market that is not chosen by many agents. Our multi-agent system was first formulated in Ishibuchi et al.[9] as a noncooperative repeated game with 100 agents and five markets where each agent is supposed to simultaneously choose a single market in every round. Several strategies for this market selection game were examined in Ishibuchi et al.[10,11] Our market selection game is much more complicated than the well-known iterated Prisoner's Dilemma game.[12,13] This is because a large number of inhomogeneous agents are involved in our market selection game. Moreover, the payoff of each agent cannot be represented in a simple tabular form.

The main characteristic feature of the payoff mechanism in our multi-agent system is that a high payoff cannot be obtained from an action that is chosen by many agents. This means that the majority of agents receive a low payoff with respect to their actions. A high payoff can be obtained from an action that is chosen by only a small number of agents. Such deterioration in payoff due to the concentration of agents (i.e., due to the same decision by many agents) can be observed in many everyday situations. For example, the chance to pass the entrance examination of a particular department of a university may decrease as the number of applicants to that department increases. Various choices related to plans for summer vacation also have similar payoff mechanisms. For example, the choice of the same highway to a summer resort by many drivers may decrease their payoff due to heavy traffic jams. The choice of the same resort by many families may also decrease their payoff due to several negative effects such as the hike in travel expenses and difficulty in booking rooms.

Another important characteristic feature shared by our multi-agent system rooms and these everyday situations is the dependence of future decision making on previous results. For example, if the competition to pass the entrance examination of a particular department is unusually tough this year, the number of applicants to that department may decrease next year. As a result, the competition for that department becomes easier. If many drivers have a hard time on a highway due to a heavy traffic jam, they will try to avoid the traffic jam on that highway next time. This may somewhat ease the traffic jam on that highway. As shown by these discussions, our multi-agent system shares some interesting features with many everyday decision-making problems.

4.2 Problem Description

Our multi-agent system involves n agents and m markets. All agents and markets are located in a two-dimensional world. We use a market selection game in Figure I.4.1 with 121 agents ($n = 121$) and six markets ($m = 6$) as an example of our multi-agent system. In Figure I.4.1, the location of each agent is a grid point in the homogeneously partitioned 11×11 grid of the two-dimensional square $[0, 100] \times [0, 100]$. The location of each market is as follows. Market 1: (15, 85), Market 2: (35, 15), Market 3: (55, 15), Market 4: (55, 55), Market 5: (75, 75), and Market 6: (85, 35). Every agent i ($i = 1, 2, \ldots, n$) is supposed to simultaneously choose one of the m markets in every round of our market selection game. An example of the market selection by the 121 agents is shown in Figure I.4.2 where each agent chooses its nearest market. Let s_i be the action of the i-th agent. The action s_i is to choose one of the given m markets: $s_i \in \{1, 2, \ldots, m\}$. The simultaneous market selection by the n agents is iterated for a pre-specified

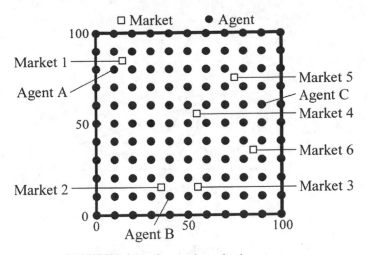

FIGURE I.4.1 Our market selection game.

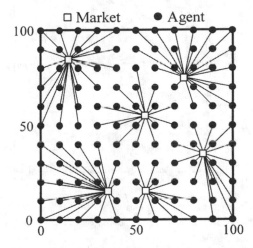

FIGURE I.4.2 An example of the market selection by 121 agents where each agent chooses its nearest market.

number of rounds because our market selection game is a repeated game as is the iterated Prisoner's Dilemma game.

Every agent has a single product to be sold in a market in every round of our market selection game. A fixed transportation cost c_{ij} is required for the transportation of the product from the i-th agent's location to the j-th market. We simply define c_{ij} by the Euclidean distance from the i-th agent's location to the j-th market. For example, the transportation cost c_{A1} of the product of Agent A in Figure I.4.1 to Market 1 is calculated as $C_{A1} = \sqrt{(15-10)^2 + (85-80)^2}$. When the j-th market is selected by the i-th agent, the agent's payoff r_{ij} is defined by the market price p_j at the selected market and the transportation cost c_{ij} to that market as $r_{ij} = p_j - c_{ij}$. It is assumed that the market price at each market is determined by a linear decreasing function of the number of agents who choose that market. Let X_j be the number of agents who choose the j-th market. It should be noted that the equality $X_1 + X_2 + \cdots + X_m = n$ holds from the definition of our market selection game. In Figure I.4.1, $X_1 + X_2 + \cdots + X_6 = 121$. The market price p_j at the j-th market is defined as:

$$p_j = a_j - b_j \cdot X_j, \tag{I.4.1}$$

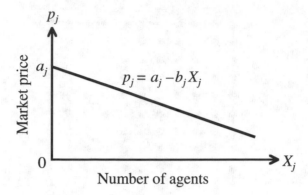

FIGURE I.4.3 Market price at each market.

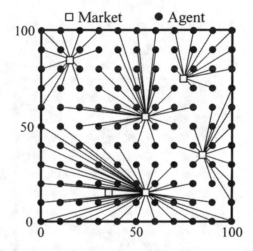

FIGURE I.4.4 Best choice by each agent for the actions of the other agents in Figure I.4.2. While the choice of each agent is optimal for the actions of the other agents in Figure I.4.2, it is not optimal for the current actions of the other agents in this figure.

where a_j and b_j are positive constants that specify the market price mechanism in the j-th market (Figure I.4.3). In our market selection game in Figure I.4.1, we use the following linear decreasing function for all the six markets:

$$p_j = 200 - 3X_j \text{ for } j = 1, 2, 3, 4, 5, 6. \tag{I.4.2}$$

The payoff of the i-th agent who chooses the market s_i (i.e., the i-th agent with the action s_i) is defined as:

$$r_{i(s_i)} = p_{(s_i)} - c_{i(s_i)} = a_{(s_i)} - b_{(s_i)} \cdot X_{(s_i)} - c_{i(s_i)}. \tag{I.4.3}$$

In this formulation, $X_{(s_i)}$ is the number of agents who choose the market s_i. Thus the payoff of the i-th agent depends on the actions of all agents. From the payoff mechanism in Equation (I.4.3), we can see that a high payoff cannot be obtained from an action that is also chosen by many agents. This means that a majority of agents receive a low payoff with respect to their actions. A high payoff can be obtained from an action that is chosen by only a small number of agents. Thus the point of the market selection is to choose a market that is not chosen by many agents. Of course, the transportation cost to each market should be taken into account. For example, the best choice for

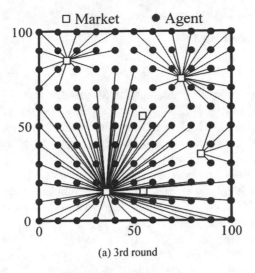

(a) 3rd round

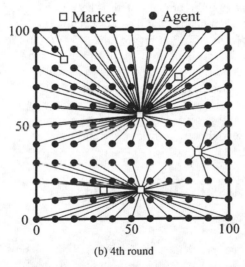

(b) 4th round

FIGURE I.4.5   Market selection using the optimal strategy for the previous actions.

Agent A in Figure I.4.1 almost always may be Market 1 because the other markets are far from this agent. On the other hand, the choice between Market 2 and Market 3 by Agent B totally depends on the actions of the other agents because the transportation costs to these markets are not large for Agent B. Agent C may have more flexibility at the market selection because there are three markets at almost the same distance from this agent.

In Figure I.4.2, each agent chooses its nearest market. The number of agents and the market price at each market are as follows in Figure I.4.2:

Market 1: Market price is 125 (25 agents)
Market 2: Market price is 128 (24 agents)
Market 3: Market price is 161 (13 agents)
Market 4: Market price is 149 (17 agents)
Market 5: Market price is 128 (24 agents)
Market 6: Market price is 146 (18 agents)

The payoff of Agents A, B, and C is calculated as follows:

Agent A: 117.93 from Market 1
Agent B: 120.93 from Market 2
Agent C: 106.79 from Market 5

If Agent B changes the choice from Market 2 to Market 3 with a higher market price, the payoff increases from 120.93 to 142.19 ($142.19 = 200 - 3(13 + 1) - 15.81$) if the other agents do not change their actions in Figure I.4.2. Thus, Market 3 is a better choice for Agent B than Market 2. By examining the other markets, we can see that Market 3 is the best choice for Agent B if the actions of the other agents are given as in Figure I.4.2. Of course, some of the other agents may change their actions. Thus, no agent knows its best action in advance. Every agent knows its best action only when the current round of the market selection game is completed. While such information can be used in the market selection in the next round, no agent knows whether the best action in the current round is still a good choice in the next action.

In Figure I.4.4, we show the market selection by the 121 agents where every agent chooses the best market for the actions in Figure I.4.2. As we have already explained, the payoff of Agent B would increase from 120.93 to 142.19 by changing the choice from Market 2 to Market 3 if the other agents did not change their actions in Figure I.4.2. The payoff of Agent B, however, decreases from 120.93 in Figure I.4.2 to 49.19 in Figure I.4.4. This is because many agents also choose Market 3 in Figure I.4.4. If Agent B did not change the market selection (i.e., chose Market 2 in Figure I.4.4), the payoff would increase from 120.93 in Figure I.4.2 to 189.93 in Figure I.4.4. As shown by these discussions about the payoff for Agent B, the payoff for each agent totally depends on the actions of the other agents. No agent, however, knows the current actions of the other agents because each agent is supposed to simultaneously perform the market selection. Moreover, the current action of each agent may affect the choice of future actions by the other agents. Thus, our market selection game is a complicated dynamic system where theoretical analysis is very difficult. In this chapter, we explain the online adaptation of decision-making systems using computer simulations on the market selection game in Figure I.4.1.

4.3   Handling as Classification Problems

4.3.1   Data Acquisition

As we have already mentioned, each agent can know the best market when the current round of the market selection game is completed. For identifying the best market for the completed t-th round, we calculate the potential payoff from each market that was not actually selected in the t-th round (i.e., market j such that $s_i \neq j$ in the t-th round). Let r_{ijt} be the actual or potential payoff from the j-th market in the t-th round for the i-th agent. When the j-th market was actually selected in the t-th round by the i-th agent, r_{ijt} is the same as the actual payoff in Equation (I.4.3). For the other markets that were not actually selected, r_{ijt} is calculated as follows:

$$r_{ijt} = a_j - b_j \cdot (X_j + 1) - c_{ij}, \tag{I.4.4}$$

where X_j is the number of agents who actually chose the j-th market in the t-th round. The i-th agent is added to X_j as $(X_j + 1)$ in Equation (I.4.4) for calculating the potential payoff, because it did not select the j-th market in the t-th round (i.e., because X_j does not include the i-th agent). The best market in the t-th round for the i-th agent is the market with the maximum value of r_{ijt} among the m markets.

Note that the best market for the t-th round can be identified only when that round is completed. Thus, the information about the best market for the t-th round cannot be utilized in the market selection for that round. A simple way to utilize this information for market selection is that the best market for the t-th round is used as the market to be selected in the $(t + 1)$-th round. Of course, this strategy called "the optimal strategy for the previous actions"[9–11] is not always optimal for the current round. When all the other agents do not change their actions, this strategy is actually optimal for the current round as well as for the previous round. On the other hand, very low payoff is obtained when many agents use this strategy. For example, when all agents use this strategy, undesired concentration of agents to a few markets is evolved as shown in Figure I.4.4. This figure shows the market selection when every agent chooses the nearest market in the first round (Figure I.4.2) and uses the optimal strategy for the previous actions in the second round. Figure I.4.5 shows the market selection when this strategy is also used by all agents in the third and fourth rounds after the second round in Figure I.4.4. In Figure I.4.5, we can observe the synchronized oscillation of the market selection that leads to very poor average payoff over all agents.

From Figures I.4.4 and I.4.5, we can see that the information about the best market in the previous round cannot be directly used for the market selection in the current round. The optimal strategy for the previous actions works very well only when the number of agents with this strategy is small. When many agents use this strategy, they cannot obtain high payoff.

We use the information about the best market in the previous round as training data for the online learning of decision-making systems. The best market in the t-th round is used in the learning as the target corresponding to the market price vector in the $(t - 1)$-th round. This means that our decision-making systems choose a market for the current round using the market price of each market in the previous round. That is, the input vector to our decision-making systems consists of the market prices at the m markets in the previous round. The output is a market to be selected in the current round.

We denote the market price vector in the t-th round as $\boldsymbol{p}_t = (p_{t1}, p_{t2}, \ldots, p_{tm})$. Let $c_{i(t+1)}$ be the best market for the i-th agent in the $(t + 1)$-th round. When the $(t + 1)$-th round is completed, we can identify the best market $c_{i(t+1)}$. Thus we generate a labeled pattern $(\boldsymbol{p}_t, c_{i(t+1)})$. This newly obtained information is used for the adaptation of our decision-making systems before the market selection for the $(t + 2)$-th round. The first labeled pattern $(\boldsymbol{p}_1, c_{i2})$ is obtained when the second round is completed. This pattern is utilized for the adaptation of our decision-making systems before the market selection in the third round. This means that we have no training patterns for the market selection in the first two rounds. When the third round is completed, the second labeled pattern $(\boldsymbol{p}_2, c_{i3})$ is obtained. In this manner, we have $(t - 1)$ labeled patterns when the t-th round of our market selection game is completed. Those labeled patterns $\{(\boldsymbol{p}_1, c_{i2}), (\boldsymbol{p}_2, c_{i3}), \ldots, (\boldsymbol{p}_{t-1}, c_{it})\}$ can be used for the adaptation of our decision-making systems for the market selection by the i-th agent in the $(t + 1)$-th round.

4.3.2 Neural Network Learning

Multi-layer feedforward neural networks are one of the most well-known adaptive systems. We use a three-layer feedforward neural network with sigmoid activation functions in the hidden and output layers[5] as a decision-making system for our market selection game. Since we use the market prices as inputs to our neural network, the number of input units is the same as the number of markets (i.e., m). The number of hidden units can be arbitrarily specified. In our computer simulations, we use five hidden units. The number of output units is also m because each of them corresponds to a single market. For the market selection in the current round, the previous market prices are presented to our neural network. Then the output unit with the maximum output value is identified. The corresponding market is chosen for the market selection in the current round.

When the labeled pattern $(\boldsymbol{p}_{t-1}, c_{it})$ is obtained (i.e., when the t-th round is completed), this pattern is used for adjusting our neural network. In the learning, the market price vector $\boldsymbol{p}_{t-1}$ is presented to our neural network for calculating the output vector $\boldsymbol{o}_t = (o_{t1}, o_{t2}, \ldots, o_{tm})$. The corresponding target vector $\boldsymbol{u}_t = (u_{t1}, u_{t2}, \ldots, u_{tm})$ is defined by the best market c_{it} in the labeled pattern $(\boldsymbol{p}_{t-1}, c_{it})$ as:

$$u_{tj} = \begin{cases} 1, & \text{if } j = c_{it}, \\ 0, & \text{if } j \neq c_{it}. \end{cases} \qquad (I.4.5)$$

The cost function to be minimized is defined as:

$$e_t = \sum_{j=1}^{m} (u_{tj} - o_{tj})^2 / 2. \qquad (I.4.6)$$

The connection weights and biases of our neural network are adjusted by the well-known back-propagation algorithm with the momentum term[5] for minimizing the above cost function. In our computer simulations, first the connection weights and biases are initialized using random real numbers in the closed interval $[-0.5, 0.5]$. The market selection in the first and second rounds is randomly performed because we have no training patterns before the second round is completed. When the first training pattern $(\boldsymbol{p}_1, c_{i2})$ is obtained from the second round, the neural network is adjusted using this training pattern. The back-propagation algorithm is iterated just once. Then the adjusted neural network is used for the market selection in the third round. After the third round is completed, the neural network is further adjusted by a single iteration of the back-propagation algorithm using the second training pattern $(\boldsymbol{p}_2, c_{i3})$. When the t-th round is completed, our neural network is adjusted by the $(t-1)$-th training pattern $(\boldsymbol{p}_{t-1}, c_{it})$. In such an incremental manner, our neural network is continually adjusted during the iterative execution of our market selection game.

In Figure I.4.6, we show the market selection by the 121 agents, each of which uses its own neural network for the market selection. The market selection in the first two rounds is random as mentioned above. The market selection in the third round is performed by the adjusted neural networks. From Figure I.4.6, we can see that the market selection by the 121 agents is gradually coordinated by the learning of the neural networks. In this computer simulation, we specify the learning rate and the momentum constant in the back-propagation algorithm as 0.2 and 0.9, respectively. The market selection game was iterated until the 1000th round. This computer simulation was performed 100 times. The average payoff at each round over the 100 trials is summarized in Figure I.4.7. This figure shows how neural networks were adjusted during the execution of our market selection game.

4.3.3 Fuzzy System Learning

Fuzzy rule-based classification systems[14–18] can be used for our market selection game in the same manner as neural network-based classification systems. Fuzzy rule-based systems choose a single market for the current round using the market prices at the m markets in the previous round. We use the following fuzzy if-then rules for the market selection:

Rule R_k : If p_1 is A_{k1} and $\ldots$ and p_m is A_{km} then c_k with CF_k, $k = 1, 2, \ldots, K$, $\qquad$ (I.4.7)

where A_{kj} is an antecedent fuzzy set with a linguistic label, c_k is a consequent market, CF_k is a certainty grade, and K is the number of fuzzy if-then rules. The fuzzy if-then rule R_k in Equation (I.4.7) is interpreted as "*If the market prices in the previous round are* $(A_{k1}, \ldots, A_{km})$ *then choose the market* c_k *in the current round.*" The certainty grade CF_k is used for representing the weight of the fuzzy if-then rule R_k (see Ishibuchi and Nakashima[16] for the effect of rule weights on the performance of fuzzy rule-based classification systems).

In our computer simulations, we use two antecedent fuzzy sets, "*low*" and "*high*," for all the six markets. Thus, $2^6 = 64$ fuzzy if-then rules are generated by combining these two antecedent fuzzy

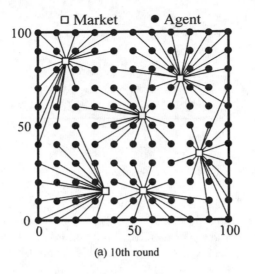

(a) 10th round

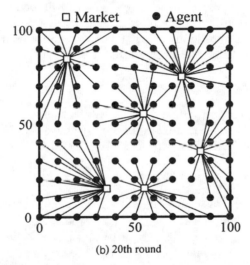

(b) 20th round

FIGURE I.4.6 Market selection by neural network-based classification systems. Each of the 121 agents uses its own neural network for the market selection.

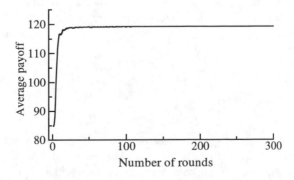

FIGURE I.4.7 Average payoff at each round when neural network-based classification systems were used by all the 121 agents.

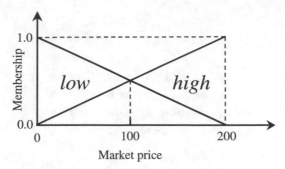

FIGURE I.4.8   Two antecedent fuzzy sets: *"low"* and *"high."*

sets for the six markets in the antecedent part. Those rules are used for the market selection of each agent. That is, the fuzzy rule-based classification system for each agent consists of 64 fuzzy if-then rules in our computer simulations. The membership functions of the two antecedent fuzzy sets are specified by simply dividing the possible range $[0, 200]$ of the market price into two triangular fuzzy sets as shown in Figure I.4.8.

The market price vector $\boldsymbol{p}_t = (p_{t1}, p_{t2}, \ldots, p_{tm})$ in the t-th round is presented to our decision-making system with the K fuzzy if-then rules in Equation (I.4.7) for the market selection in the $(t + 1)$-th round. Then the compatibility grade of the market price vector $\boldsymbol{p}_t$ with each fuzzy if-then rule R_k is calculated using the product operation as:

$$\mu_k(\boldsymbol{p}_t) = \mu_{k1}(p_{t1}) \cdot \mu_{k2}(p_{t2}) \cdot \cdots \cdot \mu_{km}(p_{tm}), \tag{I.4.8}$$

where $\mu_{kj}(\cdot)$ is the membership function of the antecedent fuzzy set A_{kj}. The winner rule R_{k*} is identified by calculating the product of the compatibility grade $\mu_k(\boldsymbol{p}_t)$ and the certainty grade CF_k for each fuzzy if-then rule. That is, the winner rule R_{k*} has the maximum product:

$$\mu_{k*}(\boldsymbol{p}_t) \cdot CF_{k*} = \max\{\mu_k(\boldsymbol{p}_t) \cdot CF_k : k = 1, 2, \ldots, K\}. \tag{I.4.9}$$

The consequent market c_{k*} of the winner rule R_{k*} is chosen for the market selection in the $(t + 1)$-th round. When multiple rules have the same maximum product in Equation (I.4.9), the winner rule cannot be uniquely specified. In this case, a single rule is randomly chosen from those rules with the maximum product as the winner rule R_{k*} for the market selection in the $(t + 1)$-th round. When no rule has a positive product (i.e., the maximum product in Equation (I.4.9) is zero), a single market is randomly selected from the m markets because no rule can be used for the market selection.

An incremental learning algorithm of our fuzzy rule-based classification system[11] is based on a heuristic fuzzy rule generation procedure[14–16] for pattern classification problems. Let β_{kj} be the discounted sum of the compatibility grades between the fuzzy if-then rule R_k and training patterns labeled as the j-th market (i.e., $c_{it} = j$). When the t-th round is completed (i.e., when a training pattern $(\boldsymbol{p}_{t-1}, c_{it})$ is obtained), β_{kj} is updated as:

$$\beta_{kj}^{\text{New}} = \begin{cases} \gamma \cdot \beta_{kj}^{\text{Old}} + \mu_k(\boldsymbol{p}_{t-1}), & \text{if } j = c_{it}, \\ \gamma \cdot \beta_{kj}^{\text{Old}}, & \text{if } j \neq c_{it}, \end{cases} \tag{I.4.10}$$

where γ is a kind of discount rate $(0 \leq \gamma \leq 1)$ introduced for discounting the effect of the previously obtained training patterns on the decision making in the current round. When $\gamma = 1$, the compatibility grades of the previously obtained training patterns are not discounted in Equation (I.4.10). This means that the effect of old training patterns is the same as that of new training patterns. Usually γ is a positive real number such that $0 < \gamma < 1$. In this case, the older a training pattern is, the smaller its effect is. When $\gamma = 0$, β_{kj} is calculated only from the latest single training pattern $(\boldsymbol{p}_{t-1}, c_{it})$. In this extreme case, only a single market has a positive value of β_{kj} among the m markets (i.e., among β_{kj}'s for $j = 1, 2, \ldots, m$).

The consequent market c_k of the fuzzy if-then rule R_k is determined in the same manner as a heuristic fuzzy rule generation procedure:[14–16]

$$\beta_{k(c_k)} = \max\{\beta_{kj} : j = 1, 2, \ldots, m\}. \tag{I.4.11}$$

That is, the consequent market c_k has the maximum discounted sum of the compatibility grades among the m markets. When the consequent market c_k cannot be uniquely determined (i.e., multiple markets have the same maximum value in Equation (I.4.11)), we specify c_k as $c_k = \phi$ for indicating that the fuzzy if-then rule R_k is a dummy rule with no effect on the market selection. For the dummy rule R_k, we specify the certainty grade CF_k as $CF_k = 0$. From the definition of the winner rule in Equation (I.4.9), we can see that any dummy rule with $CF_k = 0$ is never selected as the winner rule.

When R_k is not a dummy rule, its certainty grade CF_k is defined from β_{kj} and c_k as follows:[14–16]

$$CF_k = \frac{\beta_{k(c_k)} - \overline{\beta}}{\sum_{j=1}^{m} \beta_{kj}}, \tag{I.4.12}$$

where

$$\overline{\beta} = \sum_{\substack{j=1 \\ j \neq c_k}}^{m} \frac{\beta_{kj}}{(m-1)}. \tag{I.4.13}$$

Alternatively, the certainty grade CF_k can be defined as:

$$CF_k = \frac{\beta_{k(c_k)}}{\sum_{j=1}^{m} \beta_{kj}}. \tag{I.4.14}$$

This definition has also been used in fuzzy rule-based classification systems.[17] It is shown in Ishibuchi, Yamamoto, and Nakashima[18] that the definition in Equations (I.4.12) and (I.4.13) leads to better classification performance than Equation (I.4.14). Thus, we use the definition in Equations (I.4.12) and (I.4.13) in our fuzzy rule-based classification system for the market selection game. While the definition of the certainty grade in Equations (I.4.12) and (I.4.13) seems to be somewhat complicated, it can be easily understood if we consider a two-class pattern classification problem (i.e., market selection with two markets). For example, when $\beta_{k1} > \beta_{k2}$, the consequent market c_k and the certainty grade CF_k are determined as $c_k = 1$ and $CF_k = (\beta_{k1} - \beta_{k2})/(\beta_{k1} + \beta_{k2})$, respectively.

Our fuzzy rule-based classification system is adjusted when every round of our market selection game is completed (i.e., whenever a piece of new information is obtained). This is done by updating β_{kj} using Equation (I.4.10). Each fuzzy if-then rule is modified by the updated β_{kj} using Equations (I.4.11)–(I.4.13). That is, the consequent market and the certainty grade of each fuzzy if-then rule are redefined after every round. As in the previous subsection with neural networks, the market selection in the first two rounds is performed randomly because the first training pattern is obtained after the second rounds. When the second round is completed, each fuzzy if-then rule is adjusted. Then the market selection in the third round is performed using the adjusted fuzzy rule-based classification system. Performance of fuzzy rule-based classification systems for our market selection game strongly depends on the specification of the initial value of β_{kj}, especially when all agents use fuzzy rule-based classification systems. In Ishibuchi et al.,[11] we specified the initial value of β_{kj} as $\beta_{kj} = 0$. This specification led to simulation results similar to the optimal strategy for the

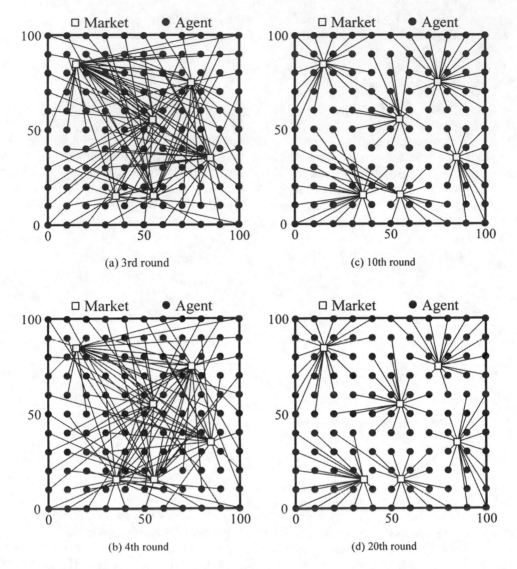

FIGURE I.4.9   Market selection by fuzzy rule-based classification systems. Each of the 121 agents uses its own fuzzy rule-based system for the market selection.

previous actions. That is, poor average payoff was obtained due to synchronized oscillation of the market selection. Thus, we specify the initial value of β_{kj} by a random real number in the closed interval [0, 1]. Poor results were also obtained due to the same reason when the discount rate γ was small. In this case, simulation results by fuzzy rule-based classification systems were also similar to the optimal strategy for the previous actions. Thus, we specify the discount rate γ as $\gamma = 0.9$.

In Figure I.4.9, we show the market selection by the 121 agents, each of which uses its own fuzzy rule-based classification system for the market selection. The market selection in the first two rounds is random as mentioned above. The market selection in the third round is performed by each agent using the adjusted fuzzy rule-based classification system. From Figure I.4.9, we can see that the market selection by the 121 agents is gradually coordinated by the learning of the fuzzy rule-based classification system of each agent. The market selection game was iterated until the 1000th round.

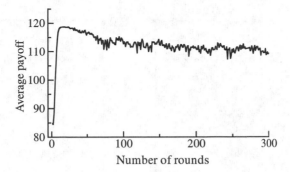

FIGURE I.4.10   Average payoff at each round when fuzzy rule-based systems were used by all the 121 agents.

This computer simulation was performed 100 times using different market selection in the first two rounds. The average payoff over the 100 trials at each round is summarized in Figure I.4.10. This figure shows how fuzzy rule-based classification systems were adjusted during the execution of our market selection game. The market selection by fuzzy rule-based classification systems is not stable as shown in Figure I.4.10 if compared with the case of neural network-based classification systems in Figure I.4.7.

4.4   Handling as Approximation Problems

4.4.1   Data Acquisition

In the previous section, the market selection game was handled as a pattern classification problem. In this section, we explain how the market selection game can be handled as a function approximation problem where the expected payoff from each market is estimated using available information obtained from the previous rounds. The market with the maximum expected payoff is selected by each agent.

Let v_{ij} be the expected payoff from the j-th market for the i-th agent. A simple way for estimating v_{ij} from the previous results is to use the following update rule when the t-th round of the market selection game is completed:

$$v_{ij}^{New} = (1 - \alpha) \cdot v_{ij}^{Old} + \alpha \cdot r_{ijt} \text{ for } j = 1, 2, \dots, m, \tag{I.4.15}$$

where α is a learning rate $(0 < \alpha \leq 1)$, and r_{ijt} is the actual or potential payoff of the i-th agent from the j-th market in the t-th round. As we have already explained in the previous section, r_{ijt} is calculated by Equation (I.4.3) when the j-th market was actually selected by the i-th agent in the t-th round. Otherwise it is calculated by Equation (I.4.4) as the potential payoff. The market with the largest expected payoff v_{ij} is selected among the m markets. When multiple markets have the same largest expected payoff, one market is randomly selected from those markets. This strategy was called "the maximum expected payoff strategy" in Ishibuchi et al.[11] In this strategy, the above-mentioned greedy method is used for the market selection because the expected payoff of all markets is updated by Equation (I.4.15). If the expected payoff of only the actually selected market is updated, some exploration mechanism should be included in this strategy, because the greedy method does not work well in the framework of reinforcement learning.

The market selection in the first round is performed randomly because the expected payoff from each market cannot be calculated before the first round is completed. The initial value of the expected payoff v_{ij} from each market is specified as a random real number in the closed interval [0, 200]. The

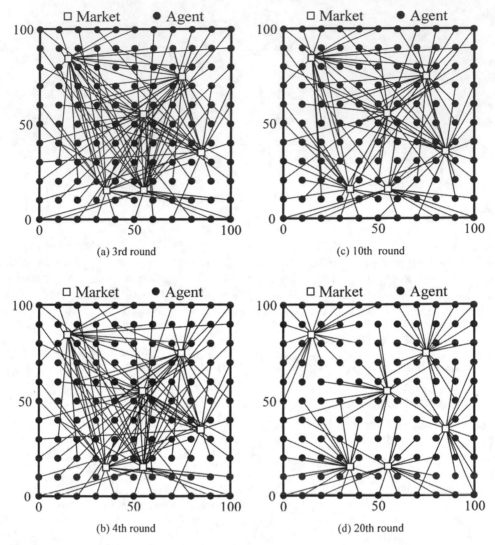

FIGURE I.4.11   Market selection by the 121 agents with the maximum expected payoff strategy.

specification of the initial value of each action is very important for facilitating the exploration of actions in reinforcement learning. The effect of the initial value of each action, however, is limited to early rounds of our market selection games in our maximum expected payoff strategy, because we do not use the framework of reinforcement learning. When the first round is completed, the expected payoff from each market is updated by Equation (I.4.15).

The learning rate α in the update rule of the expected payoff in Equation (I.4.15) is a user-definable parameter. The value of α shows the importance of the latest information obtained from the current round (i.e., r_{ijt} obtained from the t-th round). The larger the value of α is, the larger the effect of the latest information on the market selection is. In the extreme case with $\alpha = 1$, the market selection by the maximum expected payoff strategy is the same as the optimal strategy for the previous actions. This is because the expected payoff is calculated only from the previous single round. That is, the expected payoff is the same as the actual or potential payoff in the previous round. In this case, poor simulation results are obtained due to the synchronized oscillation of the market selection. On the other hand, when the value of α is very small, the maximum expected payoff strategy cannot be

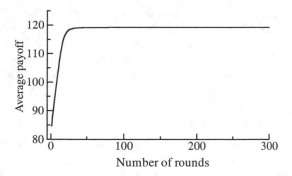

FIGURE I.4.12 Average payoff at each round when the maximum expected payoff strategy was selected by all the 121 agents.

rapidly adjusted to sudden changes of environment. In our computer simulations, we specify the value of α as $\alpha = 0.1$. Figure I.4.11 shows the market selection by the maximum expected payoff strategy. In this figure, all the 121 agents use this strategy.

As mentioned above, the market selection is random in the first round. The market selection in the second round is based on the expected payoff adjusted by the results in the first round. When the second round is completed, the expected payoff is further adjusted. The adjusted expected payoff is used for the market selection in the third round. In this manner, the adjustment of the expected payoff and the market selection based on the adjusted payoff are iterated. From Figure I.4.11, we can see that the market selection is gradually coordinated. The market selection game was iterated until the 1000th round. This computer simulation was performed 100 times. The average payoff over the 100 trials at each round is summarized in Figure I.4.12. In our computer simulations, the initial value of the expected payoff v_{ij} from each market was specified as a random real number in the closed interval [0, 200]. If we use a random initial value in a much narrower interval, the average payoff in Figure I.4.12 may increase more rapidly. About 20 rounds were required in Figure I.4.12 for tuning random initial values in [0, 200].

In some cases, the expected payoff from each market can be estimated as a function of the market prices in the previous round. That is, the expected payoff from each market for the market selection in the $(t + 1)$-th round is represented as a function of the market price vector $\boldsymbol{p}_t = (p_{t1}, \ldots, p_{tm})$ in the t-th round. Neural networks and fuzzy rule-based systems are used as function approximation systems for approximately representing the mapping from the market price vector to the expected payoff. The input vector to such a function approximation system is the market price vector $\boldsymbol{p}_{t-1}$ in the previous round. The output value corresponding to $\boldsymbol{p}_{t-1}$ is the expected payoff v_{ij} in the current round. The target for this output value is the actual or potential payoff r_{ijt}, which is obtained when the current round is completed. The first input-output pair $(\boldsymbol{p}_1, r_{ij2})$ is obtained for the approximation of the expected payoff of the i-th agent from the j-th market when the second round of the market selection game is completed. Such an input-output pair is obtained for each market. That is, a set of m input-output pairs is obtained for the learning of m function approximation systems of each agent. The m input-output pairs can be viewed as a single input-output pair of the market price vector and the payoff vector. In the same manner, m function approximation systems can be viewed as a single approximation system of the mapping from the market price vector to the payoff vector (i.e., a single m-input and m-output system).

In this section, we explain the market selection by each agent as an approximation problem involving m approximation systems with m inputs and a single output (i.e., m-input and single-output systems). The market selection in the third round is performed using the adjusted m approximation systems. When the third round is completed, the second input-output pair $(\boldsymbol{p}_2, r_{ij3})$ is obtained from the j-th market for the market selection in the fourth round. In this manner, $(t-1)$ input-output pairs $\{(\boldsymbol{p}_1, r_{ij2}), (\boldsymbol{p}_2, r_{ij3}), \ldots, (\boldsymbol{p}_{t-1}, r_{ijt})\}$ are available for the market selection in the $(t+1)$-th round. Those input-output pairs are used as training data for the approximation of the expected payoff of the i-th agent from the j-th market. That is, they are used for the learning of neural networks and fuzzy rule-based systems before the market selection in the $(t+1)$-th round.

4.4.2   Neural Network Learning

We use a three-layer feedforward neural network with sigmoid activation functions in the hidden and output layers as an m-input and single-output function approximation system. Thus, our neural network has m input units and a single output unit. The number of hidden units can be arbitrarily specified. In our computer simulation, the number of hidden units is specified as five. Such a neural network is used for the approximation of the expected payoff from each market. Thus, each agent has m neural networks. The j-th neural network is used for approximating the expected payoff from the j-th market ($j = 1, 2, \ldots, m$).

The market selection in the first two rounds is randomly performed because there is no training data. The first input-output pair $(\boldsymbol{p}_1, r_{ij2})$ is obtained when the second round is completed. The j-th neural network of the i-th agent is adjusted by a single iteration of the back-propagation algorithm using this input-output pair before the market selection in the third round. Each of the other $(m-1)$ neural networks of the i-th agent is also adjusted in the same manner using the corresponding input-output pair. The market selection in the third round is performed using the adjusted m neural networks. That is, the market with the maximum expected payoff is chosen for the third round. This selection is performed by identifying the neural network with the largest output value when the market price vector $\boldsymbol{p}_2$ in the second round is presented to each neural network. The second input-output pair $(\boldsymbol{p}_2, r_{ij3})$ is obtained for the learning of the j-th neural network after the third round is completed. The j-th neural network is adjusted by a single iteration of the back-propagation algorithm using this input-output pair before the market selection in the fourth round. Each of the other $(m-1)$ neural networks of the i-th agent is also adjusted in the same manner using the corresponding input-output pair obtained from the third round. The market selection in the fourth round is performed using the adjusted m neural networks. In this manner, we can iterate the market selection by each agent and the learning of its m neural networks.

In Figure I.4.13, we show the market selection by the 121 agents, each of which uses its own six neural networks for approximating the expected payoff from the six markets. Those neural networks are adjusted by the back-propagation algorithm with the learning rate 0.2 and the momentum constant 0.9. The market selection in the first two rounds is random, as mentioned above. The market selection in the third round is performed by the adjusted neural networks. From Figure I.4.13, we can see that the market selection by the 121 agents is gradually coordinated by the learning of the neural networks. The market selection game was iterated until the 1000th round. This computer simulation was performed 100 times using different market selection in the first two rounds and different initial values of connection weights and biases. Those initial values were specified as random real numbers in the closed interval $[-0.5, 0.5]$. The average payoff at each round over the 100 trials is summarized in Figure I.4.14. From Figure I.4.14, we can see that the average payoff rapidly improved as in the case of neural network-based classification systems in Figure I.4.7.

4.4.3   Fuzzy System Learning

Fuzzy rule-based systems can also be used for approximating the expected payoff v_{ij} of the i-th agent from the j-th market. As in the previous subsection with neural networks, fuzzy rule-based systems are used as m-input and single-output approximation systems. Inputs to each fuzzy rule-based system are m market prices in the previous round. The corresponding output is the expected payoff from

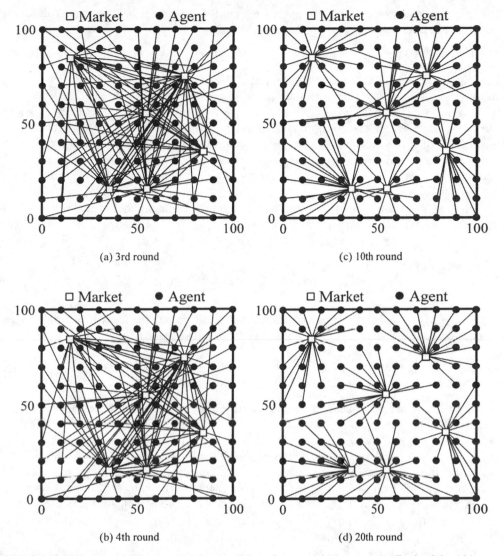

(a) 3rd round

(b) 4th round

(c) 10th round

(d) 20th round

FIGURE I.4.13   Market selection by neural network-based approximation systems. Each of the 121 agents uses its own six neural networks for the market selection.

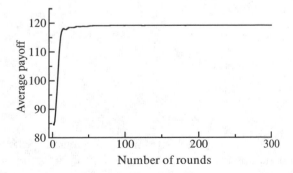

FIGURE I.4.14   Average payoff at each round when neural network-based approximation systems were used by all the 121 agents.

each market in the current round. For approximating the expected payoff v_{ij} of the i-th agent from the j-th market, we use the following fuzzy if-then rules:

$$\text{Rule } R_k : \text{If } p_1 \text{ is } A_{k1} \text{ and } \ldots \text{ and } p_m \text{ is } A_{km} \text{ then } v_{ij} \text{ is } v_{ijk}, \ k = 1, 2, \ldots, K, \tag{I.4.16}$$

where v_{ijk} is a consequent real number. These fuzzy if-then rules have the same antecedent part as those for pattern classification in the previous section. The consequent part is a real number for function approximation while it is a class label for pattern classification. Fuzzy if-then rules of the form in Equation (I.4.16) have been commonly used in many applications of fuzzy rule-based systems to control, modeling, and estimation. The fuzzy if-then rule R_k in Equation (I.4.16) is interpreted as *"If the market prices in the previous round are $(A_{k1}, \ldots, A_{km})$, then the expected payoff from the j-th market in the current round is v_{ijk}."*

We use the following weighted averaging scheme for calculating the output $\hat{v}_{ij}$ from the fuzzy rule-based approximation system with the K fuzzy if-then rules in Equation (I.4.16) when the input vector p_t (i.e., the market price vector in the t-th round) is presented:

$$\hat{v}_{ij} = \frac{\sum_{k=1}^{K} \mu_k(p_t) \cdot v_{ijk}}{\sum_{k=1}^{K} \mu_k(p_t)} = \sum_{k=1}^{K} \mu_k^*(p_t) \cdot v_{ijk}, \tag{I.4.17}$$

where $\mu_k(p_t)$ is the compatibility grade of the input vector p_t with the fuzzy if-then rule R_k defined by Equation (I.4.8), and $\mu_k^*(p_t)$ is the normalized compatibility grade:

$$\mu_k^*(p_t) = \frac{\mu_k(p_t)}{\sum_{k=1}^{K} \mu_k(p_t)}. \tag{I.4.18}$$

Each agent has m fuzzy rule-based approximation systems with K fuzzy if-then rules of the form in Equation (I.4.16). The j-th fuzzy rule-based system corresponds to the j-th market. For the market selection in the $(t + 1)$-th round, the output $\hat{v}_{ij}$ from each fuzzy rule-based system is calculated by Equation (I.4.17). Then the market with the largest output is chosen from the m markets. That is, the market with the maximum expected payoff is chosen. When multiple markets have the same largest output, one market is randomly selected from those markets. The same greedy method was used in the maximum expected payoff strategy and the neural network-based approximation systems.

When the t-th round is completed, the input-output pair (p_{t-1}, r_{ijt}) is obtained for the learning of the j-th fuzzy rule-based system of the i-th agent. The learning of the fuzzy rule-based system is performed by updating the consequent real number v_{ijk} of each fuzzy if-then rule using the actual or potential payoff r_{ijt}. The difference between the consequent real number v_{ijk} and the actual or potential payoff r_{ijt} is decreased by the learning. We use the following update rule:

$$\begin{aligned}
v_{ijk}^{\text{New}} &= (1 - \alpha \cdot \mu_k^*(p_{t-1})) \cdot v_{ijk}^{\text{Old}} + \alpha \cdot \mu_k^*(p_{t-1}) \cdot r_{ijt} \\
&= (1 - \alpha^*) \cdot v_{ijk}^{\text{Old}} + \alpha^* \cdot r_{ijt}, \tag{I.4.19}
\end{aligned}$$

where

$$\alpha^* = \alpha \cdot \mu_k^*(p_{t-1}). \tag{I.4.20}$$

This update rule is the same as Equation (I.4.15) in the maximum expected payoff strategy except that the learning rate α is multiplied by the normalized compatibility grade $\mu_k^*(p_{t-1})$. The amount of the adjustment is proportional to the normalized compatibility grade in Equation (I.4.19). Fuzzy if-then rules with small compatibility grades are slightly adjusted, while those with large compatibility grades are significantly adjusted.

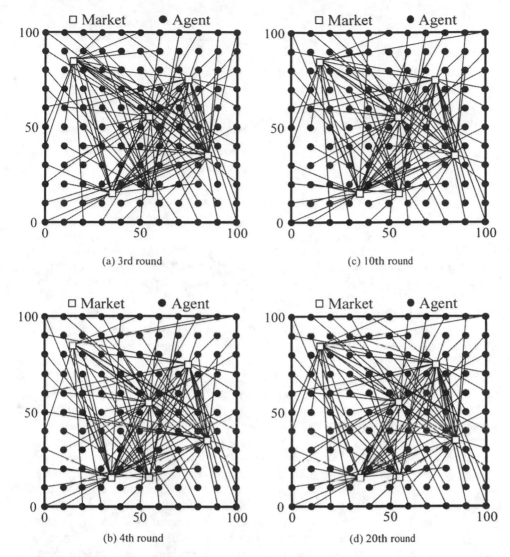

FIGURE I.4.15   Market selection by fuzzy rule-based approximation systems. Each of the 121 agents uses its own six fuzzy rule-based systems for the market selection.

The update rule in Equation (I.4.20) can be viewed as a local learning algorithm because the error between the estimated output $\hat{v}_{ij}$ and the target output r_{ijt} is not used in the learning of the fuzzy rule-based system. It is possible to derive a global learning algorithm from the following error measure:

$$E = (r_{ijt} - \hat{v}_{ij})^2/2. \tag{I.4.21}$$

In our computer simulations, we use the local learning rule in Equation (I.4.19). Different rule generation schemes based on local learning have been proposed for function approximation problems.[19,20] In general, local learning improves the interpretability of fuzzy if-then rules while global learning improve the accuracy of fuzzy rule-based systems (Yen et al.[21]).

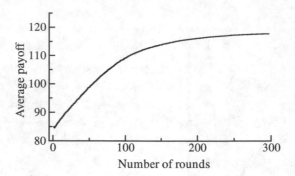

FIGURE I.4.16 Average payoff at each round when fuzzy rule-based approximation systems were used by all the 121 agents.

In Figure I.4.15, we show the market selection by the 121 agents, each of which uses its own six fuzzy rule-based systems for approximating the expected payoff from the six markets. Each fuzzy rule-based system consists of 64 fuzzy if-then rules generated from the two antecedent fuzzy sets "*low*" and "*high*" in Figure I.4.8. Those fuzzy if-then rules are adjusted by the update rule in Equation (I.4.19) with the learning rate $\alpha = 0.1$. The initial values of all the consequent real numbers are specified as random real numbers in the closed interval [0, 200]. In the next section, we also examine the constant initialization of v_{ijk} as $v_{ijk} = 200$.

The market selection in the first two rounds is random as mentioned above. The market selection in the third round is performed by the adjusted fuzzy rule-based systems. From Figure I.4.15, we can see that good coordination of the market selection by the 121 agents is not obtained in the first 20 rounds. The market selection game was iterated until the 1000th round. This computer simulation was performed 100 times using different market selection in the first two rounds. The average payoff at each round over the 100 trials is summarized in Figure I.4.16. We can see that the increase in the average payoff in Figure I.4.16 is much slower than Figure I.4.12 by the maximum expected payoff strategy. This is because the learning rate is multiplied by the normalized compatibility grade in the learning of fuzzy rules in Equations I.4.19 and I.4.20. The adjustment of random initial values of consequent real numbers r_{ijk}'s in the closed interval [0, 200] required about 300 rounds in Figure I.4.16, which is much longer than the case of the maximum expected payoff strategy in Figure I.4.12. As we will show in the next section, the performance of the fuzzy rule-based approximation strategy can be improved by using the same constant initial value for all consequent real numbers r_{ijk}'s.

4.5 Computer Simulations

4.5.1 Performance of Each Strategy

We have already shown some simulation results where a single strategy was assigned to all the 121 agents. In those computer simulations, the market selection game was iterated for 1000 rounds. Such a trial with 1000 rounds was performed 100 times using the same strategy. We calculated the average payoff from each strategy through those computer simulations. As the fuzzy rule-based classification strategy, we examined two versions. One version uses the same constant initial value for all β_{kj}'s (i.e., $\beta_{kj} = 0$ for $\forall(j, k)$), and the other version uses a random real number in the closed interval [0, 1] as the initial value of each β_{kj}. We also examined two versions of the fuzzy rule-based approximation strategy. One version uses the same constant initial value for all v_{ijk}'s (i.e., $v_{ijk} = 200$ for $\forall(i, j, k)$),

TABLE I.4.1 Average Payoff from Each Strategy when All the Agents Used the Same Strategy

Strategy	Average Payoff
Random selection	84.8
Minimum transportation cost	116.5
Optimal for previous actions	28.7
Neural network classification	119.0
Fuzzy classification (constant)	90.9
Fuzzy classification (random)	110.8
Maximum expected payoff	119.1
Neural network approximation	118.9
Fuzzy approximation (constant)	118.8
Fuzzy approximation (random)	116.1

and the other version uses a random real number in the closed interval $[0, 200]$ as the initial value of each v_{ijk}. Simulation results are summarized in Table I.4.1.

In this table, good results were obtained from the strategies based on the expected payoff (i.e., the maximum expected payoff strategy, two versions of the fuzzy rule-based approximation strategy, and the neural network-based approximation strategy). Some strategies based on the best market in the previous round did not work well. For example, the performance of the optimal strategy for the previous actions was very poor. The performance of the fuzzy rule-based classification strategy with the same constant initial value of β_{kj} was also very poor. This is because the synchronized oscillation of the market selection (i.e., periodical undesired concentration of agents) was evolved by these two strategies. The synchronized oscillation of the market selection was avoided in many trials by using a random initial value of β_{kj} in the fuzzy rule-based classification strategy. In the following computer simulations, we use random initial values in the two fuzzy rule-based strategies.

4.5.2   Competition between Two Strategies

The performance of each strategy strongly depends on strategies adopted by the other agents. We examined the performance of each strategy against another strategy by computer simulations where a single agent used one strategy and the other 120 agents used another strategy. That is, an agent with one strategy played against the other 120 agents with another strategy. This performance examination was executed for all combinations of the following eight strategies:

1. Rand: Random selection strategy.
2. Cost: Minimum transportation cost strategy.
3. Opt: Optimal strategy for the previous actions.
4. N-C: Neural network-based classification strategy.
5. F-C: Fuzzy rule-based classification strategy.
6. Exp: Maximum expected payoff strategy.
7. N-A: Neural network-based approximation strategy.
8. F-A: Fuzzy rule-based approximation strategy.

For each combination, our computer simulation with 1000 rounds was performed 121 times, so that all agents were selected as a minority agent with respect to its strategy. That is, when the performance of Strategy A was examined against Strategy B, first Agent 1 with Strategy A played against the other 120 agents with Strategy B for 1000 rounds. Next, Agent 2 with Strategy A played

TABLE I.4.2 Average Payoff of a Single Agent with a Minority Strategy when It Played against the Other 120 Agents with a Majority Strategy

Minority Strategy	Majority Strategy of the Other 120 Agents							
	Rand	Cost	Opt	N-C	F-C	Exp	N-A	F-A
Rand	(84.8)	84.7	85.1	84.8	84.8	84.8	84.8	84.8
Cost	117.5	(116.5)	125.3	118.9	119.6	118.9	118.8	119.1
Opt	113.9	122.1	(28.7)	119.2	115.1	119.3	119.2	119.4
N-C	117.1	122.1	165.8	(119.0)	115.6	119.1	119.0	119.2
F-C	117.0	122.1	167.0	119.0	(110.8)	119.1	119.0	119.3
Exp	117.0	122.0	131.2	118.9	119.6	(118.9)	118.9	119.1
N-A	116.8	122.0	165.0	119.0	120.3	119.1	(118.9)	119.2
F-A	114.2	119.3	164.8	115.8	117.2	115.8	115.8	(116.1)

against the other 120 agents with Strategy B. In this manner, the performance of Strategy A against Strategy B was evaluated by calculating the average payoff obtained by Strategy A over 121 trials with 1000 rounds. Such evaluation was performed for all combinations of the eight strategies.

Simulation results are summarized in Table I.4.2. From this table, we can see that very good results were obtained by the two fuzzy rule-based strategies and the two neural network-based strategies. The average payoff by these strategies was very high, especially when they played against 120 agents with the optimal strategy for the previous actions. This is because fuzzy rule-based systems and neural networks could learn the synchronized oscillation of the market selection by the optimal strategy for the previous actions (Figure I.4.5). That is, they learned to expect the actions of the other agents with the optimal strategy for the previous actions. Among the eight strategies in Table I.4.2, only the four strategies based on fuzzy rule-based systems and neural networks could learn to expect the actions of the other agents.

We also examined the performance of each strategy under a different situation where almost the same number of agents (i.e., about 60 agents) had one of two strategies. First we randomly assigned one of the two strategies to each agent. Then we iterated the market selection game for 1000 rounds. Such a trial was performed 100 times using different random assignment of the two strategies to each agent. During each trial, each agent continued to use the randomly assigned strategy through 1000 rounds. We examined all combinations of the eight strategies. Simulation results are summarized in Table I.4.3. From the comparison between Table I.4.2 and Table I.4.3, we can see that the average payoff obtained by each strategy in Table I.4.3 is smaller than Table I.4.2 in many cases. This observation suggests that the performance of each strategy decreases as the number of agents with that strategy increases. This is because the payoff of each agent is small when many agents choose the same market. The same observation can be obtained from the comparison between Table I.4.1 and I.4.3. All the 121 agents used the same strategy in Table I.4.1 while half of the 121 agents used the same strategy in Table I.4.3. The average payoff in Table I.4.3 is much larger than Table I.4.1 for some strategies.

4.5.3 Competition among Several Strategies

We also examined the competition among all the eight strategies. First, one of the eight strategies was randomly assigned to each agent. Then the market selection game was iterated for 1000 rounds. Such a trial was performed 100 times using different assignment of the eight strategies to each agent. During each trial, each agent continued to use the randomly assigned strategy through 1000 rounds. Average payoff obtained by each strategy over 100 trials is summarized in Table I.4.4.

TABLE I.4.3   Average Payoff Obtained from One Strategy against Another Strategy

Examined Strategy	Another Strategy that Played against the Examined Strategy							
	Rand	Cost	Opt	N-C	F-C	Exp	N-A	F-A
Rand	(84.8)	84.9	84.9	84.9	84.9	84.9	84.9	84.9
Cost	116.7	(116.5)	120.1	118.5	118.5	118.5	118.5	118.6
Opt	77.7	74.8	(28.7)	97.3	84.7	33.4	76.7	81.5
N-C	118.4	118.8	120.5	(119.0)	119.2	119.2	119.1	119.1
F-C	118.4	117.8	116.1	119.0	(110.8)	117.3	118.5	114.3
Exp	118.4	118.7	123.2	118.8	118.8	(119.1)	119.0	119.0
N-A	118.2	118.7	120.6	118.9	118.9	118.9	(118.9)	119.1
F-A	115.6	115.9	121.5	115.9	116.2	115.9	115.9	(116.1)

One of the two strategies was randomly assigned to each agent.

TABLE I.4.4   Average Payoff from Each Strategy when One of the Eight Strategies was Randomly Assigned to Each Agent

Strategy	Average Payoff
Random selection	84.7
Minimum transportation cost	118.1
Optimal for previous actions	117.1
Neural network classification	119.1
Fuzzy classification	119.1
Maximum expected payoff	119.0
Neural network approximation	118.8
Fuzzy approximation	116.2

From this table, we can see that good results were obtained from the seven strategies except for the random selection strategy. This is because the number of agents with the same strategy is small (i.e., about 15 agents on the average). We can also see that almost the same average payoff was obtained from all the seven strategies except for the random selection strategy. For example, the average payoff by the minimum transportation cost is 118.1 in Table I.4.4, while it was 116.5 in Table I.4.1. Note that each agent with this strategy always chooses its nearest market. Thus, the difference in the average payoff between Tables I.4.1 and I.4.4 was caused by the actions of the other agents.

In Table I.4.1, all the 121 agents used the same strategy. This means that the minimum transportation cost strategy did not try to avoid undesired concentration of agents as shown in Figure I.4.2. On the other hand, each of the eight strategies was used by almost the same number of agents in Table I.4.4. While the minimum transportation cost strategy did not try to avoid undesired concentration of agents, some strategies could do so. Thus, each market was selected by almost the same number of agents in computer simulations for Table I.4.4. As a result, many agents involved in the market selection game could enjoy high payoff except for agents with the random selection strategy.

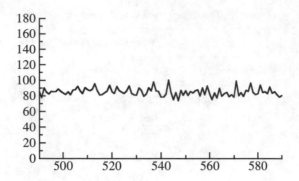

FIGURE I.4.17   Adaptation of the random selection strategy.

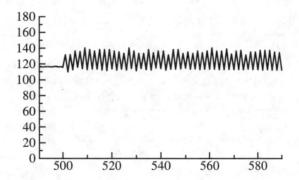

FIGURE I.4.18   Adaptation of the minimum transportation cost strategy.

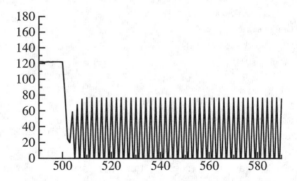

FIGURE I.4.19   Adaptation of the optimal strategy for the previous actions.

4.5.4 Adaptation to Changes of Environment

In this subsection, we examine the adaptability of each strategy to a sudden change of the environment. In our computer simulations, one of the eight strategies was assigned to a single agent and another agent was assigned to the other 120 agents. We suddenly changed the strategy of the other 120 agents after the 500th round from the minimum transportation cost strategy to the optimal strategy for the previous actions. That is, a single minority agent played against the other 120 agents with the minimum transportation cost strategy in the first 500 rounds, and it played against those with the optimal strategy for the previous actions in the last 500 rounds. The adaptability of each minority strategy used by a single agent was examined through 121 trials with 1000 rounds. In each trial, the examined strategy was assigned to a different agent. Over those 121 trials for examining a particular strategy, we

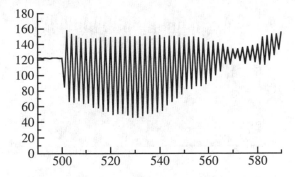

FIGURE I.14.20 Adaptation of the neural network-based classification strategy.

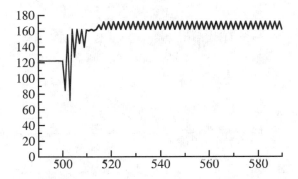

FIGURE I.4.21 Adaptation of the fuzzy rule-based classification strategy.

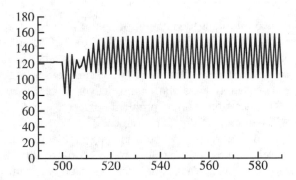

FIGURE I.4.22 Adaptation of the maximum expected payoff strategy.

calculated the average payoff obtained by the examined minority strategy in each round. Simulation results are summarized in Figures I.4.17–I.4.24. We can see from these figures that the two fuzzy rule-based strategies could quickly adapt to the change of the environment. While the two neural network-based strategies could also adapt to the change, the adaptation of neural networks was much slower than fuzzy systems.

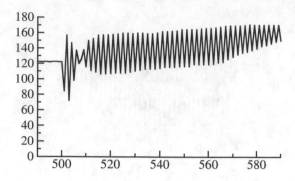

FIGURE I.4.23 Adaptation of the neural network-based approximation.

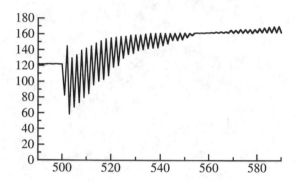

FIGURE I.4.24 Adaptation of the strategy fuzzy rule-based approximation strategy.

4.6 Conclusion

In this chapter, we examined the online adaptation of neural networks and fuzzy rule-based systems through computer simulations on a market selection game. Our market selection game is a multi-agent system formulated as a non-cooperative repeated game. Each agent is supposed to simultaneously choose a single market at every round of our market selection game for maximizing its own payoff. Fuzzy rule-based systems and neural networks were used as a decision-making system of each agent. The decision-making problem for each agent in our market selection game was formulated as a pattern classification problem and a function approximation problem. The main characteristic feature of such a decision-making problem is that the amount of available information successively increases. That is, a piece of information is obtained from every round of our market selection game. Thus, our decision-making system (i.e., neural networks and fuzzy rule-based systems) should be continually adjusted after every round. Through computer simulations, we demonstrated that neural networks and fuzzy rule-based systems have high adaptability.

References

1. Duda, R.O. and Hart, P.E., *Pattern Classification and Scene Analysis*, John Wiley & Sons, New York, 1973.
2. Weiss, S.M. and Kulikowski, C.A., *Computer Systems That Learn*, Morgan Kaufmann, San Mateo, 1991.

3. Quinlan, J.R., *C4.5: Programs for Machine Learning*, Morgan Kaufmann Publishers, San Mateo, CA, 1993.
4. Mitchell, T.M., *Machine Learning*, McGraw-Hill, Boston, 1997.
5. Rumelhart, D.E., McClelland, J.L., and the PDP Research Group, *Parallel Distributed Processing*, MIT Press, Cambridge, 1986.
6. Pandya, A.S. and Macy, R.B., *Pattern Recognition with Neural Networks in C++*, CRC Press, Boca Raton, 1996.
7. Bezdek, J.C. and Pal, S.K. (Eds.), *Fuzzy Models for Pattern Recognition: Methods That Search for Structures in Data*, IEEE Press, New York, 1992.
8. Kuncheva, L. I., *Fuzzy Classifier Design*, Physica-Verlag, Heidelberg, 2000.
9. Ishibuchi, H., et al., Fuzzy Q-learning for a multi-player non-cooperative repeated game, in *Proc. 6th IEEE Int. Conf. Fuzzy Systems*, Barcelona, 1997, pp. 1573–1579.
10. Ishibuchi, H., Sakamoto, R., and Nakashima, T., Evolution of unplanned coordination in a market selection game, *IEEE Trans. Evolutionary Computation*, vol. 5, pp. 524–534, 2001.
11. Ishibuchi, H., Sakamoto, R., and Nakashima, T., Learning fuzzy rules from iterative execution of games, *Fuzzy Sets and Systems* (to appear).
12. Axelrod, R., *The Evolution of Cooperation*, Basic Books, New York, 1984.
13. Fogel, D.B. (Ed.), Special Issue on the Prisoner's Dilemma, *BioSystems*, vol. 37, 1997.
14. Ishibuchi, H., Nozaki, K., and Tanaka, H., Distributed representation of fuzzy rules and its application to pattern classification, *Fuzzy Sets and Systems*, vol. 52, pp. 21–32, 1992.
15. Nozaki, K., Ishibuchi, H., and Tanaka, H., Adaptive fuzzy-rule-based classification systems, *IEEE Trans. Fuzzy Systems*, vol. 4, pp. 238–250, 1996.
16. Ishibuchi, H. and Nakashima, T., Effect of rule weights in fuzzy rule-based classification systems, *IEEE Trans. Fuzzy Systems*, vol. 9, no. 4, pp. 506 515, 2001.
17. Cordon, O., del Jesus, M.J., and Herrera, F., A proposal on reasoning methods in fuzzy rule-based classification systems, *Int. J. Approximate Reasoning*, vol. 20, pp. 21–45, 1999.
18. Ishibuchi, H., Yamamoto, T., and Nakashima, T., Fuzzy data mining: Effect of fuzzy discretization, in *Proc. 1st IEEE Int. Conf.* Data Mining, San Jose, pp. 241–248, 2001.
19. Wang, L.X. and Mendel, J.M., Generating fuzzy rules by learning from examples, *IEEE Trans. Systems, Man, and Cybernetics*, vol. 22, pp. 1414–1427, 1992.
20. Nozaki, K., Ishibuchi, H., and Tanaka, H., A simple but powerful heuristic method for generating fuzzy rules from numerical Data, *Fuzzy Sets and Systems*, vol. 86, pp. 251–270, 1997.
21. Yen, J., Wang, L., and Gillespie, C.W., Improving the interpretability of TSK fuzzy models by combining global learning and local learning, *IEEE Trans. Fuzzy Systems*, vol. 6, 530–537, 1998.

5

Knowledge-Intensive Collaborative Design Modeling and Decision Support Using Distributed Web-Based KBS Tools

Xuan F. Zha
Gintic Institute of Manufacturing Technology

He J. Du
Nanyang Technological University

Abstract.   This chapter presents a web-based knowledge-intensive collaborative environment for design modeling and decision support. The characteristics of WWW and CAD techniques supporting collaborative product design are first studied. Then, a unified design-with-objects and modules network scheme is proposed to model the distributed network-centric design processes. A web-based knowledge intensive distributed object modeling and evaluation (WebDOME) framework is proposed, which allows designers to build integrated models using both local and distributed resources and to collaborate by exchanging services. To facilitate the rapid construction of integrated models for design process, a web-based knowledge-intensive modeling and design support system, WebKIDSS, is implemented through a knowledge-server based WebDOME framework with concurrent integration of multiple cooperative knowledge sources and software using Java and CORBA technologies. The unified Java applet-based user interface of the WebKIDSS provides a cross-platform for distributed users to access to module servers throughout the network. By use of the developed prototype system, collaborative product design can be carried out simultaneously and intelligently in an integrated but open environment on the Internet/Intranet. Case applications are provided to demonstrate the functionality of the architecture and system, and illustrate how designers in different teams and organizations may participate and collaborate in design process.

Key words: World Wide Web, network-centric design, design-with-objects, knowledge-based systems, distributed object modeling and evaluation, integrated design and analysis, collaborative design, design support system.

5.1 Introduction

Design has been widely described as both an ill-structured activity, in that it lacks a well-defined objective, and also as one of the most demanding of all human activities (Gero, 1990). It requires the designer not only to be technically competent but also to have a well-developed sense of appreciation, including an awareness of spatial composition, form, line, and color and text issues. Given the "soft" nature in design process, especially in product design, AI strategies have been seen as particularly appropriate for the provision of computer support (Rodgers et al., 1999). Many of the issues that arise in design area are of a subjective nature and difficult to quantify, and the heuristic nature of AI and KBS technology offers a better means of representing such knowledge.

Design process in product development creates a design entity in accordance with the market-related, functional, technical, manufacturing, ergonomic, and aesthetic requirements. Many existing CAD software have incorporated some of the functional requirements necessary to support integrated design, e.g., direct interface to analytical, parametric feature-based design (such as Pro/Engineer and CADDS), and other design techniques. However, they still adopt the traditional product-process design method with various design activities carried out sequentially, and the liaison of the design activities have not been taken into account sufficiently. These tools are also used relatively late in the design process; in general, during the detailed design phase. This means that the designer has no CAD support during the early conceptual design phase; the downstream processes such as manufacturing or fabrication and assembly are considered very late in the design process, and the modifications imposed by these considerations are done when their economical consequences are worst. Concurrent engineering emphasizes the early consideration of downstream requirements in the early design stages. Critical to concurrent engineering is the maintenance of information consistency between participants and the rapid handling, exchange, and propagation of design information or events. It is imperative that an integrated design system addresses these issues. Further, in order to reduce training costs and support a broad spectrum of designers, the integrated system needs to provide a single system image that gives users a consistent and easy-to-use interface.

On the other hand, contemporary design problems embody significant levels of complexity, which make it unlikely that a single designer can work alone on a design problem. The continuing growth of knowledge and supporting information and ever-increasing complexity of design problems has led to increasing specialization. Computer tools for product design are generally stand-alone applications. However, design activities may involve many participants from different disciplines, and require a team of designers and engineers with different aspects of knowledge and experience to work together. Thus, there is a need to support and coordinate highly distributed and decentralized modeling activities in CAD systems. Wide-area networks and the internet-based WWW allow developers to provide remote web-based design servers, and CAD systems running on these servers can support a large-scale group of users who communicate over the network. As such, user interfaces based on the web protocols provide access to the remote web-based design servers, and users do not need special hardware or software to consult these services with appropriate web browsers.

The advantages of using WWW as an infrastructure to build a design system are threefold. First, web browsers can provide a nice human-interface for a design system, because they can display various media including hypertext, moving images, sounds, and three-dimensional (3D) graphics. They can also handle interactive operations of the media, such as manipulation with a mouse of a 3D object described in VRML. Second, hypertext transfer protocol (HTTP) can be a standard communication protocol of design system, since terminals connected to the internet can be accessed from any internet site via the protocol. Third, it becomes possible to use various hardware and software resources distributed over the internet together to accomplish a single mission. Therefore, it is natural to consider developing design support systems on the web.

This chapter aims to develop a web-based knowledge-intensive methodology and system for design modeling and decision support for collaborative design, which is intended to facilitate the rapid construction of integrated intelligent design models. The system provides a framework for capturing, storing, and retrieving design knowledge. It also provides systematic knowledge support for collaborative design. The proposed approach is based upon the assumption that the integrated design requires the skills of many designers and experts, and that each participant creates models and tools to provide information or simulation services to other participants given appropriate input information. A unified design-with-object scheme will be proposed and used for representing and modeling the distributed network-centric design processes. A web-based knowledge-intensive distributed object modeling and evaluation framework, called WebDOME, provides the basis for this concept, which is intended to link heterogeneous design models and tools, assist designer in evaluating the system performance of design alternatives, visualize trade-offs, find optimal solutions, and make decisions on the web. Building on the WebDOME framework, a client (browser)/knowledge server architecture that allows product design models/objects to be published and connected over the WWW to form integrated intelligent models is described. A prototype web-based knowledge intensive design support system, WebKIDSS, is implemented under the knowledge-server based WebDOME framework with concurrent integration of multiple cooperative knowledge sources and software.

The organization of this chapter is as follows. First, it begins with an overview of requirements for network-based design tools, and then reviews related work. Then, the description of the WebDOME framework is provided, and the system architecture for supporting different types of collaborative design activities in a distributed design environment is proposed and described. The remaining parts discuss how expert systems as knowledge servers technology can be used to assist in navigating design knowledge present on the WWW, drawing on an alternative view of knowledge-based design systems, and making decisions. The implementation of the WebKIDSS system exploits CORBA and Java technologies to achieve an interoperable and easily accessible environment. Case applications are also provided to demonstrate the functionality of the architecture and system, and to illustrate how designers in different teams and organizations may participate and collaborate in the design process.

5.2 Literature Review

Many researchers are working on enabling technologies or infrastructure to assist product designers in the computer network-centric design environment. Some are intended to help designers to collaborate or coordinate by sharing product information and manufacturing services through formal or informal interactions. Others propose frameworks that manage conflicts between design constraints, and assist designers in making decisions. In this section, we will review the work related to knowledge-intensive modeling and decision support for collaborative design.

5.2.1   Design Knowledge Modeling and Design Support

There is a large body of knowledge that designers call upon and use during the design process to match the ever-increasing complexity of design problems (Ullman, 1997). Typically, designers initially require broad, shallow knowledge to understand and analyze the problem, and to aid concept design development, materials selection, the specification of manufacturing processes, etc. Later in the design process, which is both nontrivial and nonlinear, the knowledge generated will be used for testing the performance of candidate solutions. The provision of timely and relevant knowledge to designers during the process is vital to the successful development of the product or system being designed, and to the future competitiveness of the company involved (Smith and Reinertson, 1991; Court et al., 1997). Moreover, design problems today require knowledge in many areas including ergonomics, packaging, management, manufacturing processes, etc.

Given that even the most routine of design tasks is dependent upon vast amounts of expert design knowledge, there is a need for some sort of support. The key focus is on the way in which designers capture, manage, access, and retrieve knowledge throughout the design process (Blessing, 1993; Bradley et al., 1994; Court et al., 1995; Wood III and Agogino, 1996; Dong and Agogino, 1996; Schott et al., 1997; Rodgers et al., 1999). During the design process, it has been estimated that designers spend between 30–40% of their time searching for and locating the "right" knowledge (Cave and Noble, 1986; Rodgers et al., 1999). Therefore, support, computer based or otherwise, is required to free designers from much of the drudgery involved in searching and locating relevant knowledge, so that they can concentrate on the more demanding and important activities involved in product design and development. Recent studies have been carried out for the knowledge requirements and design knowledge and information retrieval (Rodgers et al., 1999; Kuffner and Ullman, 1991). The findings from these studies conclude that designers rely heavily upon knowledge from their colleagues and from knowledge that is stored electronically (e.g., in databases, electronic reports, drawings, etc.) which are derived from more conventional paper-based materials. The role of the Internet in design practice is likely to be very significant in the future. This is likely to hold true for many design scenarios, as is the fact that large amounts of the knowledge that designers require access to is stored externally from their place of work. Therefore, a common framework is required that will enable designers to capture, store, and retrieve knowledge efficiently and effectively throughout the design process (Rodgers et al., 1999).

Generally, design knowledge can be classified into two categories: product (design object) knowledge and design process knowledge. Product knowledge has been fairly studied and a number of modeling techniques developed. Most of them are tailored to specific products or to specific aspects of the design activities. For example, the geometry modeling is mainly for supporting detailed design while the knowledge modeling is for supporting conceptual designs. Based on these techniques, the National Institute of Standards and Technology (NIST) in the U.S. is working on a design repository project (http://repos.mcs.drexel.edu/frameset.html). The project attempts to model three fundamental facets of an artifact representation: the physical layout of the artifact (form), an indication of the overall effect that the artifact creates (function), and a causal account of the operation of the artifact (behavior).

The importance of representation for design rationale has been recognized, but it is a more complex issue that extends beyond artifact function in the project. Design process knowledge can be described in two levels: design activities and design rationale. The Design Structure Matrix (DSM) has been used for modeling design process (activities), and some related researches have been conducted. For example, a web-based prototype system for modeling the product development process using a multi-tiered DSM was developed at Massachusetts Institute of Technology (MIT) in the U.S. (http://web.mit.edu/cipd/); Toshiba Corporation in Japan proposed a method for modeling and planning design process using the DSM. However, few researches have been found on design rationale (Pena-Mora et al., 1993, 1995).

In terms of representation scenarios, design knowledge can also be categorized into off-line and on-line knowledge representation. The former refers to existing knowledge representation, including design knowledge in handbook and design "know-how," etc.; the latter refers to the new design knowledge created in the course of design activities by designers themselves. For the off-line knowledge representation, there are two approaches to represent this knowledge. One is to highly abstract and categorize the existing knowledge, including experiences, into a series of design principles, rationales, and constraints. TRIZ is a good instance of this approach (http://www.triz-journal.com/). The other approach is to represent the collection of design knowledge into a certain case to describe the design knowledge. Case-based design is an example of this approach (Wood III and Agogino, 1996; Dong and Agogino, 1996).

The current research focus is on the computerization of the representation. For instance, Engineering Design Centre at Lancaster University (http://www.comp.lancs.ac.uk/edc/) established a unique knowledge representation methodology and knowledge base vocabulary based on the theory of domains, design principles, and computer modeling. They have developed a radical software tool for engineering knowledge management. The tool provides an engineering system designer with the capability to search a knowledge base of past solutions, and other known technologies to explore viable alternatives for product design. The on-line knowledge representation is to capture the dynamic design knowledge in a certain format for design reuse and archive. A few research efforts have been found in this area. Blessing (1993) proposes the PROcess-based SUpport System (PROSUS) based on a model of the design process rather than the product. It uses design matrix to represent the design process as a structured set of issues and activities. Together with the Common Product Data Model (CPDM), PROSUS supports the capture of all outputs of the design activity. The results show a promising approach.

In many design domains, particularly in product design, there are a number of elements involved that can be categorized as "subjective" (Roozenberg and Eekels, 1995). This means that it is difficult, if not impossible, to formulate rules that will be universally applicable, as there are elements inherent within design, such as color, texture, and shape, which are of a subjective nature (Glaze et al., 1996). Moreover, there exists high-level human interaction in design areas, particularly in the design of many domestic consumer products. It is difficult to formulate universal rules because of, e.g., national/regional differences in anthropometric data, differences in the backgrounds, education, and abilities of individuals, and also because of regional/local cultural expectations and requirements (Norman, 1989; Hubka and Eder, 1996). Research suggests, therefore, that there is a need to provide computer support that will supply clear and complete design knowledge, and also facilitate designer intervention and customization during the decision-making activities in the design process (Madni, 1988; Sanders and McCormick, 1992). Rodgers et al. (1999) describes a design support system WebCADET using distributed Web-based AI tools. The system can provide effective and efficient support for designers during their searches for design knowledge. WebCADET uses the "AI as text" approach, where KBSs can be seen as a medium to facilitate the communication of design knowledge between designers.

5.2.2 Network-Centric Design Framework

Most design processes are collaborative activities between designers from different disciplines who need to communicate and interact with one another. Design problems are assumed to be decomposed into sub-problems, and each team member focuses on a particular portion of the sub-problem. One of the key roles of an integrated modeling system is to mediate the information flow between the participants. In network-oriented computing, it is almost impossible to have a homogeneous environment. Each expert may have his/her own preferred computer system and specialized software tools. Therefore, it is crucial for a network-centric design system to support a heterogeneous computing environment. Further, design problems grow and participants change over time. Correspondingly, an open environment is needed so models from new participants can be added, allowing the design model and tools to evolve with the design problem (Pahng et al., 1998b; Zha, 1999). It is therefore important to provide designers with a means to compare design solution alternatives and support the decision-making process.

In a distributed design environment many designers may be working in collaboration, so there is a need to balance tradeoffs between competing viewpoints. Some design systems try to resolve the conflicts by notifying designers with conflicts so they can negotiate neutral solutions. A multiple attribute decision method was used to capture preferences and evaluate design alternatives from different viewpoints (Pahng et al., 1997, 1998a; Rodgers et al., 1999).

While many individuals and organizations may provide services so that an integrated product model can be constructed, it is not likely that each participant will disclose the full details or structure of their proprietary models and data. Providing a means for encapsulating expert knowledge or know-how is essential. An object-oriented approach provides a framework for such knowledge encapsulation (CORNAFION, 1985). Furthermore, an object-oriented architecture is also highly suited to a distributed computing environment (Toye et al., 1993). Distributed object technology, such as CORBA (Common Request Broker Architecture) and DCOM/ActiveX can be used to address the issue of distributed computing environment (Siegel, 1996; Chappel, 1996). A computer platform and language-independent interface definition allows software applications to communicate with each other, provided a neutral interface has been agreed upon. For example, the World Wide Web has gained its popularity and momentum through a platform-independent protocol (i.e., HTTP) and a language-independent scheme (i.e., HTML) for presenting information.

Distributed design systems have two distinct forms (Pahng et al., 1998a,b): distributed designers with access to centralized resources, or distributed designers with distributed resources (e.g., engineering models, databases, software applications, etc.). This work focuses on the latter architecture. Decentralized means that the coordination between design participants and models is not centrally modeled or controlled (analogous to the WWW). This is important because centrally controlling the interactions of all distributed resources may restrict system growth and flexibility. If there was a centralized control over the WWW for linking the hyper documents, it could not evolve so rapidly.

5.2.3 Collaborative Design Information Systems

When moving toward emerging distributed and virtual organizations, the role of social assistance in design knowledge access may be a problem (Rodgers et al., 1999). The communication technologies in general, and the Internet in particular, make these new working forms possible. The role of the expertise of colleagues in locating knowledge raises the issue of how such expertise can be made available and preserved in a distributed, ever-changing work team. Expert systems, as "embodiments of expertise," seem to offer a potential solution. Such agents can assist users in locating domain specific knowledge.

The SHARE project supports design teams by allowing them to gather, organize, re-access, and communicate design information over computer networks to establish a shared understanding of the design and development process (Toye et al., 1993). While SHARE is primarily directed toward interaction through integrated multimedia communication and groupware tools, the NEXT-LINK project incorporates agents to coordinate design decisions affected by specifications and constraints (Petrie et al., 1994). Another network-centric design system using interacting agents to integrate manufacturing services available over the network is also under development (Frost and Cutkosky, 1996).

The motivation and vision presented in this paper has similar themes, but emphasizes mathematical modeling, decision making, and search/optimization.

The Electronic Design Notebook is an interactive electronic document that maintains the look and feel of an engineering document to provide an integrated user interface for computer programs, design studies, planning documents, and databases (Lewis and Singh, 1995). Manufacturing tools and services are encapsulated in the hypertext documents and distributed through servers using HTTP (Sobolewski and Erkes, 1995).

A design information system proposed by Bliznakov et al. (1995) and Bliznakov (1996) incorporates a hybrid model for the representation of design information at several levels of formalization and granularity. It is intended to allow designers in a large virtual organization to indicate the status of tasks assigned to each designer or team so that other designers can follow their progress. A central database manages pointers and access methods for product and process information in the distributed environment.

Hardwick and Spooner (1995) proposes an information infrastructure architecture that enhances collaboration between design and manufacturing firms. This architecture uses the WWW for information sharing and the STEP standard (Owen, 1993) for product modeling. It utilizes the CORBA standard for interoperability between software applications in the virtual enterprise.

N-dim is a computer-based collaborative design environment for capturing, organizing, and sharing data (Westerberg et al., 1995). The system is the basis upon which applications can be added for the purpose of history maintenance, access control, and revision management. The primary focus of environment is on information modeling. The system provides a way for defining information types that capture the relations between data or models. There are also national-level efforts involving university and industry collaboration to make a variety of engineering services available over the Internet (MADEFast, 1999; NIIIP, 1999; RaDEO, 1998). The RaDEO program is concerned with comprehensive information modeling and design tools needed to support the rapid design of electro-mechanical systems. It supports engineers by improving their ability to explore, generate, track, store, and analyze design alternatives. The National Industrial Information Infrastructure Protocols (NIIIP) Consortium is attempting to develop open industry software protocols that will make it possible for manufacturers and their suppliers to effectively interoperate as if they were part of the same enterprise. The NIIIP goal is to help suppliers assemble virtual enterprises.

The Concurrent and Simultaneous Engineering System (CONSENS) is a European project that is exclusively devoted to concurrent engineering (Singh, 1995). It focuses on information infrastructure. Process, product, and project data management are supported by several engineering software applications integrated with the SIFRAME framework.

5.2.4   Design Rationale and Conflict Resolution

Case and Lu's (1996) Discourse Model provides software support for collaborative engineering design by treating interactions as a process of discourse. The model captures design commitments as opinions subject to review and revision by other designers. It also utilizes agents to identify conflicts between designers and to negotiate the resolution of conflicts. A computer-based design system developed by Sriram and Logcher (1993) provides a shared workspace where multiple designers work in separate engineering disciplines. In their Distributed and Integrated Environment for Computer-aided Engineering (DICE) program, an object-oriented database management system with a global control mechanism is utilized to resolve coordination and communication problems. Design rationale provided during the product design process is also used for resolving design conflicts (Pena-Mora et al., 1993, 1995).

5.3 Web-Based Knowledge-Intensive Collaborative Design Framework

Contemporary design process is knowledge intensive and collaborative. Knowledge-intensive support becomes critical in design process and recognized as a key solution toward future competitive advantages in product development. To improve design process and enhance design innovation, it is imperative to provide knowledge support in the design process and share design knowledge among distributed designers. The integrated design requires the skills of many designers and experts, whereby each participant creates models and tools to provide information or simulation services to other participants given appropriate input information. It is the goal that collectively, the network of participants exchanging services forms a concurrent model of the integrated design. In this section, a design knowledge modeling approach is first proposed. Then, a design-with-modules scheme is proposed for modeling network-centric design process, and a web-based distributed object modeling and evaluation framework is discussed based on the design with object scheme.

5.3.1 Design Knowledge Modeling

Design knowledge refers to the collection of information needed to support the design activities and decision making in design process. Successful capture of design knowledge, representing it efficiently, and making it easy for designers to access this knowledge is crucial to increase the design "science" contents compared to the "art" nature. In this chapter, both on-line and off-line design knowledge representations are dealt with, but more attention is paid to the design process knowledge. Since the design knowledge is very extensive, the focus is on some selected activities in the design processes. A systematic methodology and the relevant technologies are developed for knowledge modeling, i.e., capturing, representing, organizing, and managing knowledge in the design process.

Design knowledge, which is very domain specific, comes from a variety of sources such as corporate libraries, textbooks/handbooks, scientific literature, on-line resource, and, in particular, the experience of individual designers. It is stored in different forms including documents, drawings, electronic means, and memories of human beings. Therefore, design knowledge must be classified into different categories and represented in appropriate ways accordingly. Through analysis of design process, the design knowledge is abstracted and classified into four categories, off-line and on-line, product knowledge and process knowledge. Each category of knowledge is represented in different ways from multiple views. Product knowledge includes all information needed for the designing product throughout the design process, i.e., product specifications, concepts, structure, and geometry. In view of this, a new complete product information model is used, which consists of customer requirements, design specifications, function behavior, structure, and geometry.

Based on a combination of elements of semantic relationships with the object-oriented data model, a multi-level hybrid representation schema (meta-level, physical level, geometric level) is adopted to represent the design process knowledge in different design stages at different levels. To effectively manage and utilize the design knowledge, a generalized design knowledge matrix is proposed for its organization. Design knowledge is organized in terms of design process. All tasks in the design process are listed in the first column, while all information and design knowledge are categorized in the first row. The contents of design knowledge for each task are recorded in the corresponding cells of the design knowledge matrix with appropriate representations.

Both a knowledge-intensive framework and system are required to help designers carry out design modeling and decision support on the web, which cover the entire product design process from conceptual design to detailed design, and consider as early as possible in the process the different life-cycle phases. The framework has the following advantages: (i) providing a proper support of the entire design process, especially the early stage; and (ii) enabling real-time capture of design knowledge as part of the design process. Details about the framework and its implementation will be discussed below.

5.3.2 Network-Centric Design Process Modeling

The information processing in product design is inherently model based because the design object is structural in type. Therefore, object-oriented programming languages are desirable for declarative knowledge representation, object-oriented concepts, and fuzzy logic. As such, the object orientation scheme is employed, so that both calculating and reasoning work in design can be carried out. The hybrid design object model is, in fact, an attempt to set up a knowledge-intensive framework in such a way that it becomes possible to process various types of knowledge in a top-down design process (Gui, 1993; Zha, 1999; Zha et al., 2001a,b).

5.3.2.1 Design with Objects Scheme

The central design process inherent in design with objects scheme can be represented as the architecture as shown in Figure I.5.1 with five main types of objects involved: namely, design models (S), design objects (O), design algorithms (A), functions (requirements and constraints) (FRC), and the evaluation schema (E) (O'Grady and Liang, 1998). Object operators can express the relationship between these objects: inheritance, import, and message passing.

The architecture in Figure I.5.1 shows how the particular instance of a design model, S_1^k, is obtained from the design algorithm, evaluation schema, requirements, constraints, and the design model object. For pure formulation design or creative design, a new design model object S is defined that describes the form of the model. A specific instance, S_1^k, of this design model can then be created. For pure parametric design then, the design model object S has already been defined and the design process therefore only involves the determination of a specific instance, S_1^k, of the design model. Note that additional objects can be defined within the overall architecture.

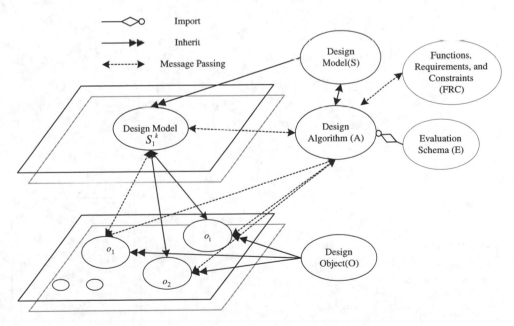

FIGURE I.5.1 Overall architecture of design with objects.

5.3.2.2 Distributed Design Modeling

The distributed object modeling and evaluation (DOME) framework asserts that multidisciplinary problems are decomposed into modular sub-problems (Pahng et al., 1997, 1998a,b). Modularity divides overall complexity and distributes knowledge and responsibility among designers. It also facilitates the reuse of modeling elements. Thus, DOME allows designers to define mathematical models or modules and integrate or interconnect them to form large system models. In this research, based on DOME framework, a web knowledge-server based DOME (WebDOME) is proposed for the module-based distributed design modeling.

Modules, as shown in Figure I.5.2(a), represent knowledge related to different aspects of the design in the form of variables and relations. The variables contained in the module are represented as interconnected circles. The directed arcs imply dependency. The outputs and inputs to the module constitute the interface of the module. Modules can be interconnected through interfaces. Customer-created computer programs and third-party applications, such as domain-specific analysis tools or CAD systems, can be embedded into a module. Modules interact with each other by exchanging information and services, reacting to each other's changes for man integrated system model. Modules can also be distributed over the network, collectively forming a distributed model for a collaborative, multidisciplinary, and concurrent design evaluation. Details can be found in Pahng et al. (1998a).

In the distributed design environment, each group of designers can define their own modules, loading them into their local work area and eventually connecting them to the other parts of the design problem through appropriate networked interfaces. Figure I.5.2(b) shows a DOME model involving two designers and three modules. Designer 1 defines modules A and B, while designer 2 defines module C. Two domains communicate through an Internet connection. Once the whole problem is loaded and interconnected, each group of designers typically has write access to local parts of the model, i.e., they can exert decisions within their local range of influence, and read

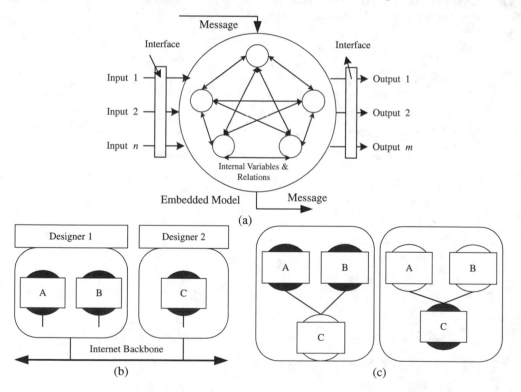

FIGURE I.5.2   (a) DOME module and (b) (c) DOME distributed modules.

access to relevant aspects of remote parts of the model. This allows them to see the remote effects of their local decisions. Figure I.5.2(c) shows that both designers 1 and 2 might see the complete problem, but with different access privileges. Designer 1 can see modules A and B as local and module C as remote. Conversely, module C is local to designer 2. The remote part seen by designer 2 could show modules A and B or just a single distributed object (AB) if designer 1 restricts their visibility/access.

The interactions between distributed modules can be achieved by publishing and subscribing services. The term publish refers to making the services of one's local model visible to other designers. The term subscribe refers to making use of published services. Such design problem models are mixed variables, where independent parameters within modules are set and catalog selections might be used to substitute entire modules. Design solutions can be assessed and compared with each other using decision-making tools embedded in the framework.

5.3.3 Web-Based Knowledge-Intensive Design Framework

5.3.3.1 Expert Systems as Knowledge Servers

The widespread use of the Internet and WWW provides an opportunity for making expert systems widely available. By implementing expert systems as knowledge servers that perform their tasks remotely, developers can publish expertise on the Web. Technologies and infrastructures that make this approach feasible are emerging. Simultaneously, interest in artificial intelligence support for network navigation services is growing. Wide-area networks and the Internet-based WWW allow developers to provide intelligent knowledge servers. Expert systems running on servers can support a large-scale group of users who communicate with the system over the network. In this approach, user interfaces based on web protocols provide access to the knowledge servers, and users do not need special hardware or software to consult these services with appropriate web browsers.

To make knowledge servers available, developers must distribute the software front ends that allow users to communicate with the servers. The web supports the Common Gateway Interface (CGI), which can provide form-based front ends to databases, and to expert systems with simple user interfaces. Common Gateway Interface is the protocol that interfaces web servers with external applications to generate dynamic knowledge in real-time. This approach, however, does not allow user interaction beyond form filling and the use of web documents as program output. Java programming language, conversely, provides a powerful basis for implementing user interfaces to knowledge servers. Developers can include programs written in Java (Java Applet) in HTML/XML documents; web browsers can download these programs over the Internet and run them locally. Java programs can then act as user interfaces to expert systems by opening network connections to knowledge servers. If the user-interface front end and the knowledge-server are object-oriented, protocols that support shared objects can be useful. Object communication models such as the Common Object Request Broker Architecture (CORBA) can be used as a standard protocol. Moreover, for user interface it is often better to use approaches that support remote procedure and method call, such as to implement callbacks. The remaining part of this chapter will concern how expert systems technology can be used to assist designers in navigating design knowledge present on the WWW, drawing on an alternative view of KBS, and making decisions in design process.

5.3.3.2 Architecture for Knowledge-Server-Based Design Framework

In this research, the proposed web-based design framework (WebDOME) adopts the design with objects, modules network, and knowledge-server paradigms, which are techniques by which knowledge-based systems utilize the connectivity provided by the Internet to increase the size of

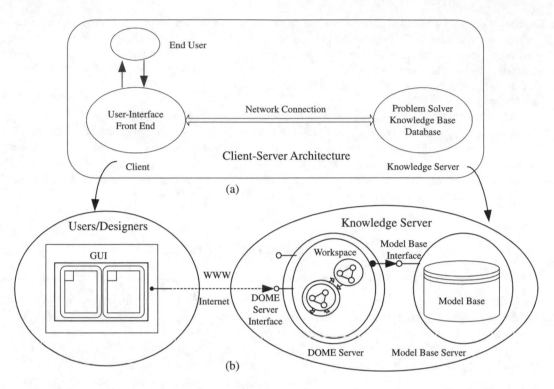

FIGURE I.5.3   (a) Client-knowledge server architecture, and (b) main components for WebDOME.

the user base while minimizing distribution and maintenance overheads. The knowledge-intensive system can thus exploit the modularity of knowledge-based systems, in that the inference engine and knowledge bases are located on a server computer and the user interface is exported on demand to client computers via network connections (e.g., Internet, WWW). Therefore, modules (objects) under WebDOME framework are connected together so that they can exchange services to form large integrated models. The module structure of WebDOME leads itself to a client (browser)/knowledge-server-oriented architecture using distributed object technology. Figure I.5.3 shows the main system components of the proposed client (browser) and knowledge-server architecture. Each of these components interact with one another using a communication protocol, CORBA, so that it is not required to maintain the elements on a single machine. As a gateway for providing services, the interface of a system component invokes the necessary actions to provide requested services. To request a service, a system component must have an interface pointer to the desired interface (Pahng et al., 1998a,b).

In the WebDOME architecture, the resultant service exchange network forms an integrated con-current system model if module services are connected, as shown in Figure I.5.4(a). As reviewed in Section 5.2, there are other architectures for network-centric collaborative design. These include the centralized multi-user system architecture (Figure I.5.4(b)) (e.g., the blackboard-based DICE (Sriram and Logcher, 1993), DIS (Bliznakov et al., 1995)), data and model exchange system (Figure I.5.4(c)) (e.g., the SHARE (Toye et al., 1993), EDN (Lewis and Singh, 1995), MADEFast (1999), NIIIP (1999), RaDEO(1998)), and multi-agent-based distributed system architectures (Zha et al., 1999;

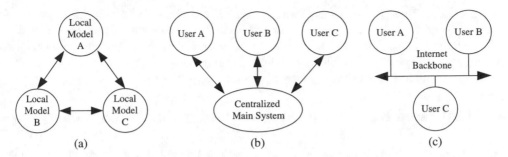

FIGURE I.5.4 Interactions between modules exchanging services: (a) Service exchanges in a module network; (b) Centralized multi-user system; (c) Data and model exchange system.

Zha and Du, 2001c). The comparisons between the WebDOME architecture and these architectures are described as follows (Pahng et al., 1998a):

1. For the WebDOME architecture, the distinct characteristic is that when module services are connected, the resultant service exchange network forms an integrated concurrent system model, as shown in Figure I.5.4(a).
2. For the centralized multi-user system, multiple users have access to the centralized main system, which stores and manages information such as product design models, design information, and design history. Although powerful, a central system is less suited for loose and flexible collaborations as it is not an open environment and does not allow for true knowledge encapsulation. However, such architecture could be supported within a module of a larger WebDOME network.
3. The data and model exchange system architecture tends to provide an "over-the-wall" sequential interaction between designers and models. When a designer receives a model or data from another designer, he/she works on the design and sends the result of design modification to others. Therefore, this architecture is not intended to provide concurrent system modeling functionality.
4. Multi-agent-based architectures are more appropriate for loosely coupled environments where mutual interactions between objects are not well defined. In the WebDOME architecture, the interactions between sub-problems are explicitly defined through design negotiation so that a communicating object paradigm is appropriate. However, within the WebDOME, agents will be useful when designers are not certain about what modules can provide the service they require. Agents could locate appropriate modules.

5.3.3.3 Module Interactions for Services Exchanging

The WebDOME architecture is designed to allow experts to publish and subscribe to design modeling and decision support services on the WWW. These services will operate when information is received from other clients or knowledge servers. When module services are connected, the resultant service exchange network forms an integrated concurrent system model. Any service request in the module network can invoke a chain of service requests if needed to provide correct information. When a design alternative is evaluated, the local model asks for the services of subscribed models. If the subscribed models themselves need services from other models in order to provide the request services, they will again request those services from their own network to remote models. Thus, the service requests are propagated through connected modules.

However, the complete system may not be visible to any given model. Since modules can only interact through services, it is possible for a module or local model to encapsulate its internal modules and hide intellectual property if desired. Before a designer publishes, they can assign access privileges for their services. Three levels of model access have been identified: owner, builder, and user. The owner is the original creator of the model and has access to all the services defined in the model and control over their publication. The builder can see the internal details of a model the owner chooses to make public and can add new modules. However, they cannot destroy modules created by the owner or other builders. The users can subscribe only to published services.

5.4   Web-Based Knowledge-Intensive Design Support System

In the previous section, a knowledge-server-based web distributed object modeling and evaluation framework (WebDOME) is discussed based on the design with objects scheme and the knowledge-server paradigm. This section concerns how expert systems technology can be used to assist in navigating knowledge present on the WWW, drawing on an alternative view of KBSs, and making decisions in design process. A web-based knowledge-intensive support system for product design will be built upon the WebDOME framework.

5.4.1   Knowledge-Intensive Design Support System on the Web

Initially, a stand-alone knowledge-intensive design support system (KIDSS) was developed to support top-down design for assembled products. It can also support bottom-up design process. Figure I.5.5 illustrates the activities that KIDSS can support. The backbone implementation of KIDSS was based on an assembly oriented design expert system (AODES) (Zha et al., 2001a,b) which originated from an even earlier version DFAES (Zha et al., 1999). Figure I.5.6 shows a knowledge-based framework for implementing KIDSS. The prototype of KIDSS was developed in close collaboration with several designers from small to medium-sized design enterprises, where the goal was to support designers during their generation, evaluation, and selection of assembly designs. It was designed in such a way to allow designers to generate, analyze, and/or modify the product assembly at any stage during the design process. It works in a full interactive mode and information associated with a particular subject such as product structure, number of components, component design, and component material, which enables designers to minimize the number of components of a product and select the most economic assembly technique for the specific product. All recommendations

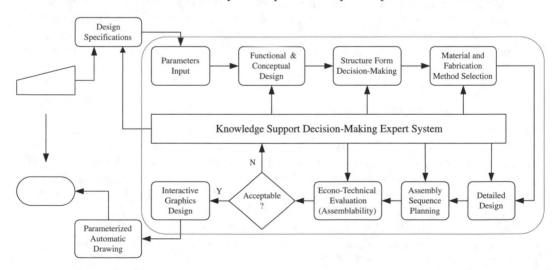

FIGURE I.5.5   KIDSS support in design.

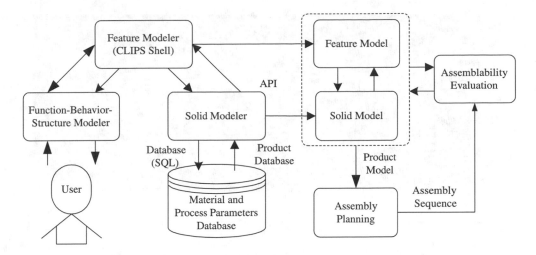

FIGURE I.5.6   Knowledge-based framework for KIDSS.

about product design for assembly, and assembly sequences, including optimal assembly sequence, can be generated and visualized by automatic reasoning, searching, computation, and simulation from a comprehensive assembly knowledge base when the user answers the questions presented by the system interactively.

The Internet-based WWW is seen as a natural medium for the establishment of a market in KBS design knowledge. The solution to providing distributed design support has been to extend the original stand-alone KIDSS system into WebKIDSS using WWW functionality. It is deployed on a Web server enabling access via the Internet. WebKIDSS adopts the knowledge-server paradigm. This is one technique by which knowledge-based systems can utilize the connectivity provided by the Internet to increase the size of the user base while minimizing distribution and maintenance overheads (Eriksson, 1996). WebKIDSS exploits the modularity of knowledge-based systems, in that the inference engine and knowledge base(s) are located on a server computer and the user interface is exported on demand to client computers via the Internet.

Currently, WebKIDSS is a Web-enabled prototype integrated intelligent design environment, consisting of design modeling and support software tools. It contains a powerful set of new modules that deliver capabilities previously unavailable in the existing design tools. WebKIDSS can also act as a repository for the accumulation and communication of design knowledge, which is in a rule-based form, similar to the stand-alone KIDSS' rule-based attribute models. Designers are able to browse this facility via the existing web browsers, and download the knowledge to their host machines. In turn, designers will be able to critique the KBS rules and to re-submit any modifications they have made to suit their own particular scenarios or tasks back to the design community via the Web, as suggested in Figure I.5.7. The comprehensive market of design knowledge, mainly in the form of KBS rules, will quickly be established. The philosophy underlying this research is a move from automated design using the expert systems approach to one embracing a more cooperative relationship between man and machine. It is believed that the most effective approach to computer-based design tools is for the computer to provide decision support and allow the human designer to supply the judgment.

5.4.2   WebKIDSS as a Navigation Aid

Originally, WebKIDSS was due to the need for a mechanism which would allow designers to access knowledge bases, their supporting references, and to permit these to be modified as required and subsequently made public for other users. However, with the development of design process on the

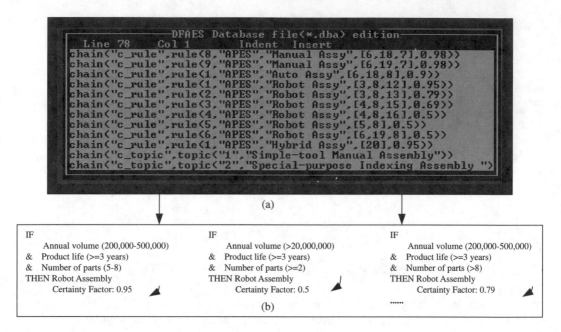

(a)

(b)

FIGURE I.5.7   KIDSS rule base and rule evolution.

Internet-based WWW, WebKIDSS creates the possibility of supporting designers in the navigation of on-line knowledge, in addition to other navigation and search tools available, such as search engines. This function is similar to that of the WebCADET (Rodgers et al., 1999). Thus, the following description in this section is based on the work about WebCADET.

The problem of navigation in design is actually to find and access design knowledge during design process. The access and retrieval of relevant knowledge are a combination of standard archive mechanisms and consulting with colleagues. There are a number of mechanisms already on the WWW, such as search engines (e.g., AltaVista) and catalog search engines (e.g., Yahoo!). However, these traditional WWW search engines merely serve to compound the problem of effective knowledge access and retrieval. Further problems with the search engine approach include the spasmodic availability of remote servers, out-of-date and repetitive knowledge, the possibility of overlooking an important document, and the time it takes to go through the results. The major problem is the lack of content awareness in traditional hypertext documents (Rodgers et al., 1999). Search engines must inspect all of the content on every page, and consequently return vast numbers of inappropriate hits. The advent of XML (Extensible Markup Language) on the mainstream Web provides a solution to this problem (Mace et al., 1998). XML describes the content of a document rather than the presentation, as is the case with HTML. Standard sets of XML tags have already appeared for a number of applications and others will follow suit. The greater accessibility and availability of meta-data describing the content of hypertext documents will substantially improve the relevancy of search engine results. Due to the time involved in checking through hundreds of document hits, only a percentage of the search engine hits were actually explored. The numbers of relevant hits from the initial search were found from the percentage of hits explored. Using this approach, there is still no guarantee that all relevant hits will be viewed.

In addition to taxonomies and word searches, according to Rodgers et al. (1999), another method is "social navigation" support, which relies on the knowledge and contribution of other users on the Internet. Social navigation is based on communication and interaction with other users (Dieberger, 1997). Actually, a number of other "social" mechanisms exist to support the WWW user in locating knowledge, such as questions posted to appropriate news groups or to specific persons identified from their web page interests. More recently, work on "shared bookmarks," in which a user can explore sites bookmarked by persons of similar interest, has also been developed. WebKIDSS can offer a similar functionality to these "social navigation" pages, by linking a number of similar pages together via task specific indices. "Shared bookmarks" are usually a very general set of links to related materials. There is no real structure other than simple clustering under headings to aid other users of these pointers.

Access to material would be made more effective if the pointers to on-line materials could be made appropriate to the specific problem at hand. WebKIDSS can support such task-dependent access by providing structured pointers through the design problem space, and allow designers to create task-based navigation links. That is, pointers to references from the rule bases or, more usefully, any explanation generated by WebKIDSS, create a "conceptual trail" in the WWW. If the reference is on-line, and it is expected that this will increasingly be the case, the designer can hyperlink from the rule or explanation to the complete associated reference material. From this new WWW resource, it is further likely that the designer will find links to related, additional, material.

Unlike the social navigation pages described above, however, the trail is derived from decisions made during the design process, and the list of pointers to references is dynamic. As rules are amended and updated, the links are continuously revised. Furthermore, as WebKIDSS takes in input data from the designer, these pointers are more appropriate to the current design task. Rules in the KBS are only triggered when the conditions match, and the pointers are triggered to support and associated materials when they are mostly likely to be of interest in the appropriate situation within the design process. This is very important in managing the knowledge given to the designer. WebKIDSS is an effective way of describing the current focus of interest of the designer, and thus coordinating the links to various on-line design knowledge resources.

The structure of the WebKIDSS referencing system can now be seen as an extension of other structuring mechanisms in print, such as checklists, tables of content, indices, etc. (Rodgers et al., 1999). The various WWW resources can now be linked by an additional layer of structure derived by their connection through the formal, inferential system. Figure I.5.8 illustrates how designers are able to select a number of KBS rules and supporting design knowledge on the Internet (via links from the explanation system) reflecting various design tasks, involving issues such as safety, reliability, and cost, as they explore the design space. Exploration of the design space is an iterative activity where designers consider many solutions and parts of the problem concurrently. WebKIDSS links to relevant knowledge and supplementary design resources are available for the designer to browse and utilize, and help cut down drastically the amount of time designers spend on fruitless searches.

5.5   WebKIDSS Design and Implementation

In this section, we will describe a prototype implementation of WebKIDSS with a Java-based user interface. However, the focus is on the description of the technologies employed in the design and development of WebKIDSS.

5.5.1   System Overview

The design and implementation of a prototype WebKIDSS system is actually a two-stage process. The first stage is to convert KIDSS implemented in C/C++/Visual Prolog into CGI executables that are then deployed on a standard web server, in terms of template web pages to contain dynamically generated input forms, the necessary code to extract knowledge from submitted forms, and display results. The second stage is to implement WebKIDSS front-end user interfaces using Java and CORBA technologies. WebKIDSS uses a Java program as a front end to an expert system and was implemented as a knowledge server to provide a web advisory and collaborative system for design process. WebKIDSS can run as an applet, freely available via the WWW and process CLIPS

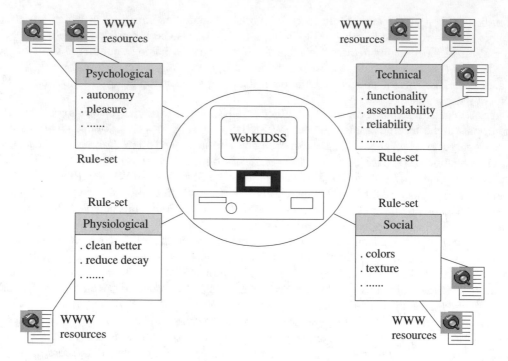

FIGURE I.5.8   Task-based links from WebKIDSS via KBS to WWW resources. *Source:* Rodgers et al., *Res. Engineer. Des.*, 11, 31–44, 1999. With permission.

(C Language Integrated Production System) or Jess rule bases that were modified slightly to work with WebKIDSS. It is interlinked with web pages of tutorials, and reference pages relating design process and explaining the facets and design results. WebKIDSS can perform its tasks using data from a remote web-based database that can be maintained and updated via the Internet (Smith, 1999; Zha and Du, 2001d). The details about the formal implementation of WebKIDSS will be discussed in Section 5.5.3 below.

5.5.2   Operation Scenarios

WebKIDSS may be accessed using any Java-enabled browser. The operation of WebKIDSS can be considered at both the system and application levels. At the system level, a WebKIDSS consultation will proceed until a point is reached where it is necessary to acquire input from, or send output to, the user. WebKIDSS determines the nature of the user's input or request and processes it accordingly. Processing continues until the next juncture for human–computer interaction occurs. At the application level, what is important is the process of design evaluation from selections of product type, assembly methods or systems, and processes and materials to resulting scores and associated explanations, etc. For example, in processes and materials selection, a frame is provided for a first-order cost estimation along with examples for selected processes, and provides for the generation of process chains using secondary processes to refine certain features on a part (Smith, 1999). While running, WebKIDSS generates a dialogue with the designer, inquiring about batch size, typical tolerances, size, overall shape, and cost requirements. After entering values for a set of facets, or attributes, for a conceptual part, the user is given real-time feedback regarding plausible fabrication methods. Once a process is selected, process chains or cost estimates can be explored. At each step along the way, the user is presented with an updated, ranked list of manufacturing possibilities. A similar method is used to define the attributes for material selection (yield strength,

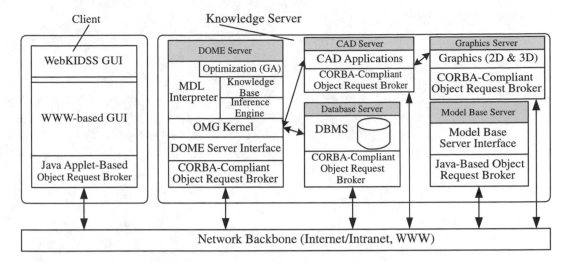

FIGURE I.5.9 Implementation of the open-design environment based on the WebDOME framework.

density, etc.), and generate material rankings. The final result is a ranked list of viable combinations, obtained through a process-material pair optimization.

5.5.3   System Implementation

Implementation of WebKIDSS uses the two-tier client-knowledge server architecture to support collaborative design interactions with a web-browser-based graphical user interface (GUI), as shown in Figure I.5.9. The underlying WebDOME framework and the knowledge engine were written in Java with integration into CLIPS or FuzzyJess (Java Expert System Shell with fuzzy extensions (Friedman-Hill, 1999)) (Zha and Du, 2000a,b; Zha and Du, 2001a–c). It was integrated with the existing application packages such as Java 3D and JDBC for 3D CAD and database applications. CORBA serves as an information and service-exchange infrastructure above the computer network layer, and provides the capability to interact with existing CAD applications and database management systems through other Object Request Brokers (ORB). In turn, the WebDOME framework provides the methods and interfaces needed for the interaction with other modules in the networked environment. These interactions are graphically depicted in Figure I.5.10. When a change is made by designer B, the service corresponding to the request from designer A will reflect the design change. The enumerated request shows the sequence for obtaining the service that is needed by designer A. The light gray module seen by designer A is a remote module published by designer B. The underlying collaboration mechanism is based on the board systems. Each modular system has two board systems, a blackboard and a whiteboard, as shown in Figure I.5.11. The blackboard system is used for the local modular system to store intermediate reasoning and calculation results. It dynamically flushes in running. The whiteboard system is used for collaboration, which is actually a bulletin board system.

5.5.3.1   WebKIDSS Graphics User Interface

The WebKIDSS graphics user interface (GUI) provides users with the ability to examine the configuration of design problem models, analyze tradeoffs by modifying design parameters within modules, and search for alternatives using knowledge-base systems and/or an optimization tool, e.g., genetic algorithms. The GUI is a pure client of the knowledge server, delegating all events to an associated DOME server. For wide accessibility and interoperability, the graphical user interface is implemented as a web browser-based client application (Figure I.5.12). The front-end side of

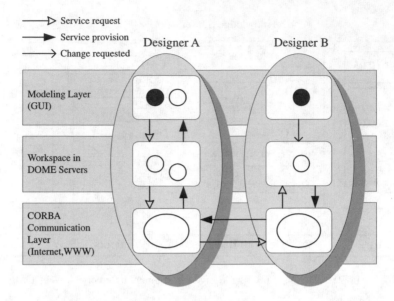

FIGURE I.5.10   Service exchanges between distributed modules.

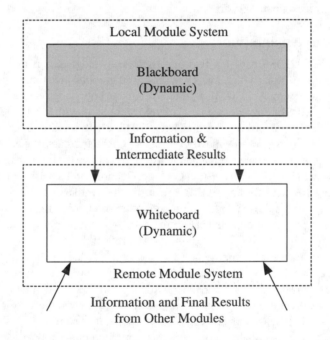

FIGURE I.5.11   Blackboard and whiteboard system for collaboration.

the application is implemented as a combination of HTML documents and Java applets. For the CORBA-based remote communication between the GUI Java applets and the back-end side system components, such as DOME server, model base server, graphics server, and database server, a commercial ORB implementation of Java applets (OrbixWeb) is employed (IONA, 1997).

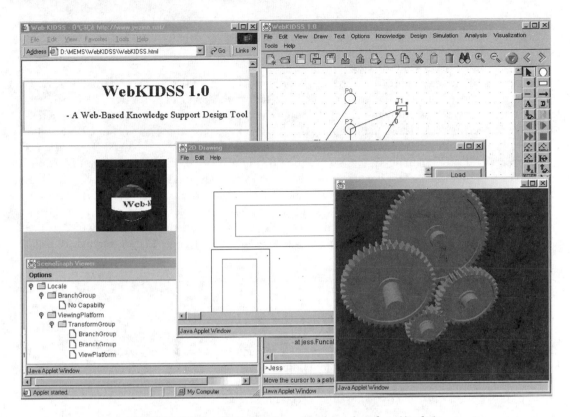

FIGURE I.5.12   WebKIDSS graphical user interface (Applet).

5.5.3.2   Knowledge Server

Based on the system architecture for implementation, the functionality of the knowledge server in WebKIDSS is achieved through implementing DOME servers, model base server, core knowledge engine, database server, and even knowledge-base assistant and interserver communications explanation facilities. The implementation details for the knowledge server are discussed below.

5.5.3.2.1   DOME Server

The WebKIDSS GUI interacts with designers and delegates user events and requests to a DOME server that provides the back-end implementation for the modeling of design problems. The core of the server is based upon object-oriented modeling and evaluation (OME) kernel (Pahng et al., 1997, 1998b; Zha and Du, 2000b, 2001c,d) written in Java, integrating FuzzyJess (Friedman-Hill, 1999). The back-end implementation of knowledge server including DOME server, model base server, graphics server, database server, and the front-end interface to the GUI are written in Java. The DOME server manages each design session in a workspace and can simultaneously maintain several workspaces. In addition, the workspace manages administrative aspects of a model, e.g., ownership, access privilege, links to other workspaces in different DOME servers. The DOME server itself is a CORBA-compliant distributed object and can communicate with other DOME servers.

5.5.3.2.2   Model Base Server

The model base server maintains persistent storage for models created by the DOME servers. The model repository stores a model in a Model Definition File (MDF) with two parts: meta definition and model definition (Pahng et al., 1997, 1998b). The meta definition contains the information such as model ID, ownership, and access privilege information. The model definition is based upon a Model Definition Language (MDL) used by the DOME server system.

5.5.3.2.3 *Core Knowledge Engine*

The core knowledge engine of WebKIDSS originated from KIDSS. The internal knowledge representation used in WebKIDSS has a more concise format than that of the original KIDSS, but is equally expressive. The problem-solving paradigm including inference engine used in WebKIDSS is more powerful than that of the original KIDSS. The knowledge base is built in CLIPS, which is a multi-paradigm programming language that provides support for rule-based, object oriented, and procedural programming system language (Riley et al., 1993). CLIPS is a forward-chaining, rule-based production system language, based on the Rete algorithm for pattern matching. CLIPS programs are expressed by means of commands, functions, and constructs. In CLIPS, a fact is presented as an ordered list of fields; the system also supports templates (or nonordered facts). Rules allow the user to specify a set of conditions to CLIPS, such that when the conditions in the left-hand side are satisfied, a set of actions in the right side is executed. A sample CLIPS rule that is used to select processes is shown as follows:

```
(defrule Rule XX
(goal (type identifyProcess) (value "yes"))
=>
(printout t "Is the process cost high? |explanatory | Answer the question"
    "by selecting one of the choices and then clicking on 'proceed'.
|yes|no|end")
(assert (attribute (type hasHighcost) (value (readline))))
)
```

As discussed above, the knowledge base in WebKIDSS can be classified into two categories: off-line (static) knowledge base and on-line (dynamic) knowledge base. The off-line knowledge base is referred to as the knowledge base without change during running, which mainly consists of the static text knowledge which is used for inquiring about the system overall structure and program executive flowchart, and the explanation for the results. The management of off-line knowledge base is accomplished by file operations. The on-line knowledge base is established and operated in the course of running (in design process), and used for storing intermediate results of designs and decisions. The off-line and on-line knowledge base can store a large number of facts and rules. The knowledge base is actually a rule base, which is constructed using CLIPS or Jess.

The problems solving paradigms include inference engine, search strategy, matching algorithms and conflict resolution, and meta-knowledge. Three main forms of control during the reasoning process are: forward reasoning (data-driven approach), backward reasoning (goal-driven approach), and hybrid reasoning, a combination of forward and backward reasoning. WebKIDSS employs mainly forward inference, a data- driven approach. The forward inference involves reasoning from the existing facts in the memory to conclusions, with the aim of establishing the final desired conclusions. Like many other rule-based systems, WebKIDSS works in a match, select, and act cycle. During matching, the facts in memory are matched against the conditions of the rules in the knowledge base. Those rules whose conditions are totally satisfied will be triggered. If more than one rule is triggered, a conflict resolution (or selection) technique must be used to determine which rule to fire (or activate) first. When a rule is fired, its conclusion or action part is added to the memory. The cycle is then continued with the modified memory. As WebKIDSS is a CLIPS-based system, the pattern-matching algorithm is the Rete algorithm. In order for the inference engine to decide which rules to use in a particular cycle, it has to match the rules with the data in the memory. A set of rules triggered in any particular rule-processing cycle has to be arranged in some order, so that only one is fired in any particular cycle. Conflict resolution strategies in common use can include specificity ordering, rule ordering, size ordering, recency ordering, and context limiting. In WebKIDSS, the following conflict resolution strategies are used:

1. If only one rule is triggered, then this rule is definitely a starting rule.
2. Knowledge base is organized in terms of the problem to be solved. Therefore, to solve a problem, only the knowledge related to this problem is chosen.
3. Knowledge in a knowledge base is sequenced in advance and used in order.

5.5.3.2.4 Web Database System

The most efficient way to store and operate the data, such as process capability and material properties data and curves, fabrication method, predefined assembly time and cost data and charts, equipment cost data, etc., is to use database techniques. To enhance the problem-solving efficiency and reduce the work needed in setting up an expert system, it is necessary to call directly the data organized by databases. These databases are open and can be searched by the interface between CLIPS and the database at any time upon the request of specific problems. They can be remotely updated by administrators or authorized users at any time if necessary. The fundamental information that forms these databases was obtained from our previous work (Zha et al., 1999) and Internet resources (e.g., http://cybercut.berkeley.edu/). One of the criteria used for the development of WebKIDSS databases is to create a repository of manufacturing data separate from the code for the main program.

In this research, the web database system software were developed by use of Microsoft Access databases to store the details of data and Java programs to access these databases through JDBC connections. Figure I.5.13 shows a pictorial view of Java database system scheme. For example, the manufacturing database shown in Figure I.5.14 supports many widely used manufacturing processes, such as: plastic injection molding, forging, sand casting, sheet metal forming, extrusion, milling, die casting, shell mold casting, investment casting, and EDM. This remote process capability database contains information about the component processes, materials, and vendors. Opening the database on the server brings up a menu of tools (Smith, 1999; Zha and Du, 2001c,d):

i. Vendor Editor is to provide the account management with companies that have processes.
ii. Material Editor is to edit the properties of the generic raw materials.
iii. File exporter is to generate data files and human readable reports.
iv. Process editor is to specify the performance of all processes, compatibilities with materials, and locations of on-line resources.

These tools can add new processes or materials to the WebKIDSS without any change to the compiled code.

5.5.3.3 Supporting and Explanation Facility in WebKIDSS

To help users, a step-by-step tutorial provides instructions to users who are not familiar with manufacturing terms. Samples are offered for the extensive on-line help "manual." Descriptions and sample values are given for each of the process and material requirements, which allow users to compare their tolerance values with common products. Each of the processes included in WebKIDSS has a set of descriptive web pages. The information includes production numbers, shape capabilities, design rules, sample parts, material usage notes, pros/cons, related processes, and links to equipment suppliers and fabrication sites. All of the documentation is linked through the applet itself. The designer may select any process, material, or requirement, and click the "Get Website Info" button to call up an informational web page. To begin an analysis of a design, the user should

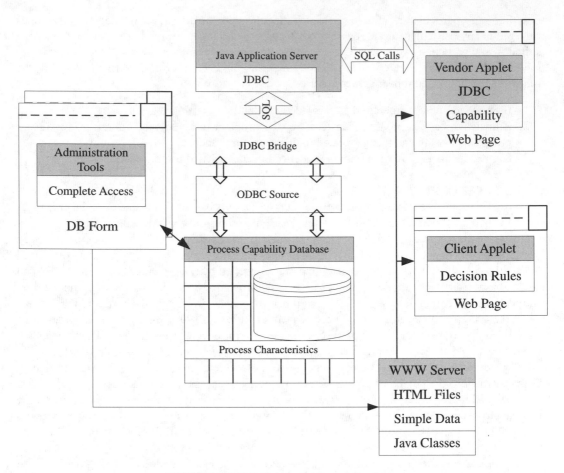

FIGURE I.5.13   Java database system scheme.

start the system; the instruction for the use of the system is very easy and key steps are shown in the startup screen.

5.5.3.4   Knowledge-Base Evolutionary Maintenance in WebKIDSS

To enhance the functionality and problem-solving ability continuously, WebKIDSS should have updating facility to upgrade its knowledge base evolutionarily, i.e., evolutionary maintenance. This is also a process for knowledge acquisition or learning. With this facility, the system will become smarter. In this research, this is achieved through providing a knowledge-base assistant and interserver communications facilities. The evolutionary maintenance, as a form of software maintenance, is used to extend and revise knowledge bases to include new knowledge and remove obsolescent knowledge (Watson et al., 1992). If the expert system is a sort of diagnostic expert system where chains of rules are essential to determine a probable cause for a set of symptoms, it is a troublesome problem. This is because in such systems, adding new knowledge almost always involves knowledge-base verification to ensure that the new knowledge does not contradict or interfere with established heuristics. In WebKIDSS, as rules are completely separated from each other, it is therefore less troublesome, owing to the autonomy of rules and rule-sets than it was in DFAES where knowledge base was organized in a chain list (Figure I.5.7a) (like diagnostic expert systems) (Zha et al., 1999). The effort in extending the knowledge base is confined to adding new rules and different versions of rules, and ensuring that such rules are appropriately linked to attribute categories and products, and that the necessary explanation facts are included.

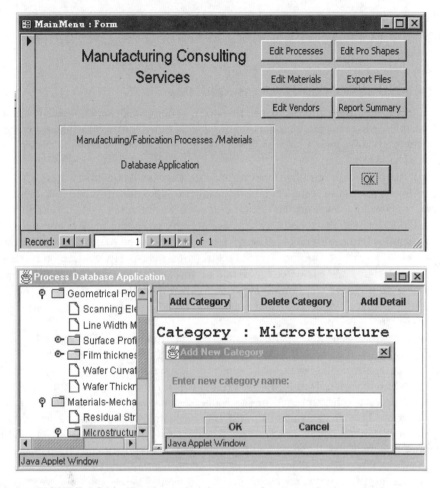

FIGURE I.5.14   Database administrator tools menu and process database.

To ease the addition of new knowledge, and to allow designers other than the developers to expand knowledge servers of WebKIDSS, an appropriate knowledge-base assistant was designed and implemented. During the collaboration, multiple but distinct instances of WebKIDSS are deployed and running at various remote sites. Each specific WebKIDSS server could thus grow to encompass the design knowledge of the designers at each site. By providing WebKIDSS servers with the capability to communicate with each other, it is possible to harness the community of design resources, so that when a local server is unable to evaluate a given design, it will be able to automatically interrogate other servers to utilize their knowledge. As rules are marked with both their applicable context and their authors or sources, quality control in the WebKIDSS community is in terms of actual usage of the design rules, with servers containing high-quality knowledge thriving and less well-respected servers falling into disuse.

5.6   WebKIDSS Application for Collaborative Design Process

In the previous section, the straightforward system WebKIDSS that uses a Java program as a front end to an expert system was implemented as a knowledge server which provides a web advisory and collaborative system for design process. In this section, to illustrate the design modeling and decision support of collaborative design process, the developed system WebKIDSS is used as a platform for very roughly collaborative design for a micro-robotic assembly system. Details are discussed below.

5.6.1 Collaborative Design Process Modeling

Suppose that designers from different teams, divisions, or companies in remote locations would like to participate in creating integrated models for the design of a micro-robotic assembly system. The robotic assembly system consists of three major components: the robot system, work platform, and components to be assembled.

The overall topology of the problem and design workspace is illustrated in Figure I.5.15, in which the designers from the robot and gripper manufacturing teams provide their models to the robot system design team, who in turn develops the technical models for the robot system. The robotic system manager collaborates with the robotic system designer by providing models and data for robotic operating conditions and requirements. The design models are used to develop cost evaluation and redesign models. The robot and gripper manufacturing teams respectively develop models for their products, so the designer and customer can obtain performance predictions for different parametric configurations and operating conditions. These individual models are constructed, published, and served by each party. If a single individual creates the model to provide these services, he/she will work in an individual workspace.

The design session creates the modules in the design workspaces. The designers can use any commercial web browser to access and work on these modules. As the robot system design and operation are tightly coupled, it would make sense for designers in these groups to share a common model. Thus, while designers from different groups are in remote locations, they can access into the same workspace, which is referred to as a shared workspace. Figure I.5.16 shows the design workspace as viewed by the designers from the robot system design team and the robot operation team. The robot system design team is connected to the robot and gripper manufacturing teams, so that their robot system design integrated with gripper and robot models can be tested.

Figure I.5.17 depicts a 2D-design model for the gripper in WebKIDSS. The users or operation team will be sharing their workspaces with the design team. The design team creates modules in the upper-left corner, while the robot system operation team makes the rest designed. In this case the design team owns the session and the operation team has joined as a builder. Although builders cannot modify the modules created by other builders or owners, they can add new modules and utilize all services. For example, the operation team can use a service from a design module to obtain robot accuracy and the open distance of the gripper, and can build new modules in the workspace that utilize this information.

Similarly, the design team can also use services from the models published by the robot and gripper manufacturing team. Utilizing models provided by other designers is referred to as subscribing to a model. It is the responsibility of the design team to provide these data or to locate other models that can provide these data as services. The robot system managers want to evaluate the design from a cost viewpoint. They link their models to the design module to obtain the information services needed by their models. The design team has published only cost-related aspects of their models. This means that the robot system managers can only observe elements of the design models that were published, as the designers wanted to protect their proprietary models.

The problems of robotic system simulation and design are also tightly coupled, so that the design and simulation teams should share a common model and access into the same workspace, although these teams may be in remote locations. The robot system can be operated by means of a virtual robot manipulation system constructed in the WWW scheme, in which 3D models of the components are manipulated virtually in a computer graphics. The robotic system simulator developed by the simulation team provides a new design tool for the design team to carry out the flexible assembly

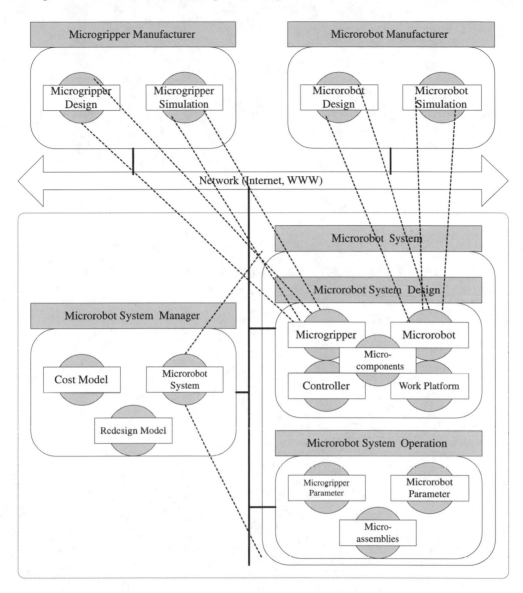

FIGURE I.5.15   Problem topology of the design model.

and the intuitive operations and simulations. This will help the designers to verify the design. When a simulation sequence is running, users or designers can control positions and orientations of the robot and the components, and open–close states of the gripper by clicking on them. The user interface graphically displays robot configurations, gripper states, and the component states. The simulation results will also help the designers in the design team to modify and redesign the design if necessary.

5.6.2   Decision Support in Collaborative Design

The consultation session in WebKIDSS for collaborative design was implemented through the application of advisory systems for design support, e.g., in assembly method selection (Zha et al., 1999), material and process selection (Zha and Du, 2001d), etc. To illustrate the use

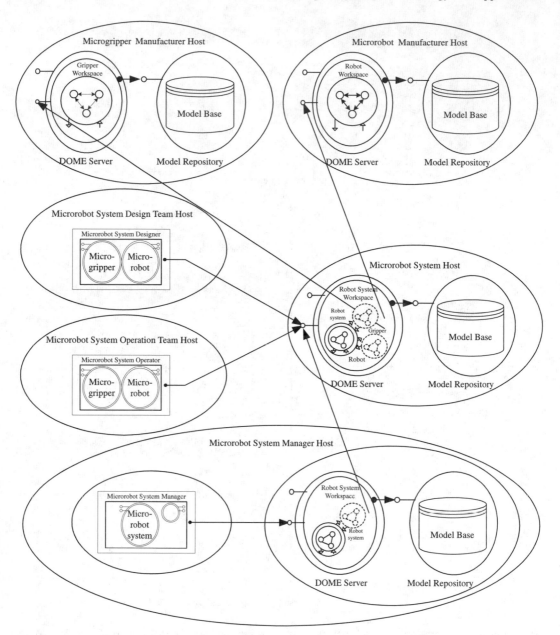

FIGURE I.5.16   Design workspace as viewed by the robotic system designers and operator; robot system manager model connected to the robot system design model.

of WebKIDSS for material and process selection, an example of a prototype gripper is explored to show the possibilities for making a full production run. Thus, it is necessary to use the process search, material search, and results survey mode (Smith, 1999; Zha and Du, 2001d). The specifications are made for the procedures of process search and material search. The process search is for the lowest possible cost over a long production cycle. At the end of the process search, the Electro-Dischargeable Machine (EDM) (rank 1.00) was ahead of the only other possibility, Etch (rank 0.96). Similarly, after material search, the system generated six viable materials, with carbon steel ranking the highest at 1.00 and aluminum and alloys ranking at 0.98. Furthermore, after process search and material search,

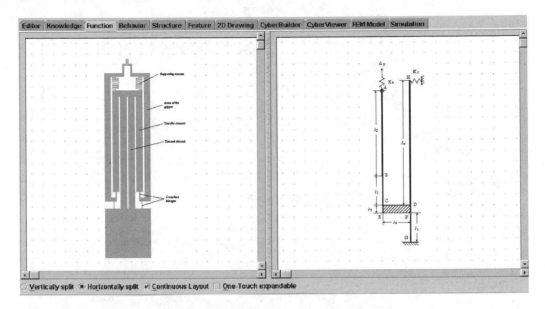

FIGURE I.5.17   2D design and simulation model for the gripper generated by WebKIDSS.

the "Final Result" button would be enabled and clicked to combine the results of both searches to find the best material/process combination. As shown in Figure I.5.18, the two boxes at the top are a summary of the viable materials and processes, and the final box lists all of the feasible combinations, taking into account a compatibility factor between each process and material. Thus, EDM with carbon steel is the best choice, with etched stainless steel second.

On completion of the reasoning process, WebKIDSS returns the score obtained for this design with respect to this attribute and indicates in the result page the Pass/Fail status of each design parameter. To request an explanation of the evaluation, the user can click on the button of the "Explain the Result" on the results page. The explanation may consist of the rationale for the score in terms of justifications and references, both to the design literature and various on-line resources. It also shows an example of how on-line expert design knowledge and resources can be accessed during the design process.

5.7   Summary and Future Work

This chapter presented a web-based knowledge-intensive collaborative environment for design modeling and decision support. Because of the heterogeneous structure, product design and simulation require different grades of abstraction, and need the cooperation and collaboration of different disciplines and resources. The proposed framework is intended to provide distributed designers with a tool for collaboratively building integrated system models. The advantage of the demonstrated modular concept consists in the flexibility of the program structure and the reduction of costly software support by integrating design tools and simulators. Large problems are decomposed into sub-problems with modules. Models or other software applications are encapsulated in modules. A module can provide information services through its interface, and the network of modules exchanging services form a concurrent design model. So the behavior of complex systems and the interactions of components can be analyzed and optimized during the design process, resulting in shorter manufacturing cycles.

The knowledge-server-based WebDOME framework is built upon to provide module network architecture for integrating modeling services available on the network. The framework can

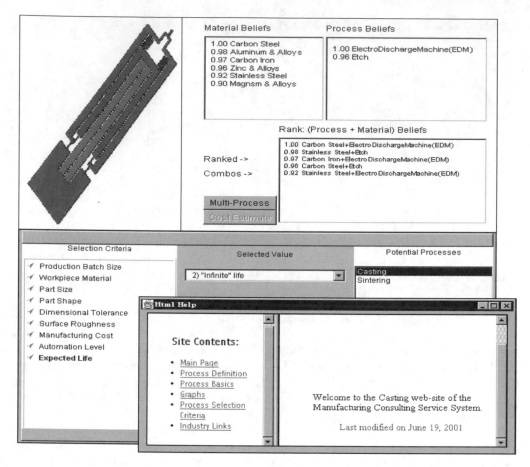

FIGURE I.5.18   WebKIDSS application for process and material search in design process and final results.

accommodate top-down and bottom-up approaches in the context of both traditional sequential design processes and concurrent design. In the module network, design resources, models, data, and activities are not centralized or concentrated in one location. They are distributed among many companies, designers, or design participants working together over the Internet/Intranet. Then, the module' network architecture is extended to a computer network environment, WebKIDSS, focusing on the web-based knowledge-intensive and collaborative design modeling and support. Design modules are created by fully implementing locally defined modules and subscribing to the services of remote modules. The software implementation hides the details of the remote interaction mechanism from the user, allowing the designer to model interactions between local and remote modules in a transparent manner. In turn, designers can selectively publish modeling services for use by others. The two-tiered client (browser)/server architecture was adopted to allow experts and designers to publish and subscribe modeling services on the WWW. In this architecture, when modules' services are connected, the resultant service exchange network creates an integrated concurrent system model or module network that invokes a chain of service requests if needed to provide correct information. The applications illustrate the concept and different models of collaboration supported by the prototype implementation.

For design knowledge modeling and design support in collaborative design, the efficiency and effectiveness of design knowledge retrieval are crucial in terms of their search times in use. The ability to determine relevant knowledge as well as to provide a browsable and searchable structure is also significant and of great value in operation. WebKIDSS meets these requirements by providing a highly practical means of finding relevant design knowledge efficiently and effectively. The structuring mechanism of WebKIDSS can act as an index to support material that is present on the WWW, and can also be used as a task-based navigation aid to the WWW-based content. As rule-bases, databases, and supporting multimedia materials are open and available on the WWW, the library of expertise will grow with user/designer access and use. Such dynamic libraries can significantly provide aid for the distributed and flexible collaborative design and development teams. The evolutionary maintenance facility in WebKIDSS can update and upgrade its knowledge base to enhance the functionality and problem-solving ability continuously.

Although the preliminary implementation illustrates the potentials of WebDOME framework for design modeling and decision support, there are some fundamental issues yet to be addressed, such as knowledge-base evolutionary maintenance, model interface standard, computational strategy for resolving circular dependencies, parallel service request invocation, etc. Also, there exists a large amount of work needed on the particular design paradigm or methodology for web-based collaborative design. For example, in a collaborative design environment, other aspects such as human interaction and knowledge sharing still require the integration of additional support tools with the framework, e.g., e-mail, video conferencing, hypertext documentation, etc. If integrated with other computer-based collaboration tools, such as Windchill project link (http://www.ptc.com/products/windchill/), WebKIDSS will provide designers with a powerful infrastructure for concurrent collaborative product design.

References

Bic, L.F., Fukuda, M., and Dillencourt, M.B., Distributed computing using autonomous objects, *IEEE Comp.*, August, pp. 55–61, 1995.

Blessing, L.T.M., A Process-Based Approach to Computer Supported Engineering Design, Ph.D. Thesis, University of Twente, 1993.

Blessing, L.T.M., Design process capture and support, in Tichem, M. et al. (Eds.), *Proc. 2nd WDK Workshop on Product Structuring*, Delft University of Technology, The Netherlands, pp. 109–121, 1996.

Bliznakov, P.I., Design Information Frame Work to Support Engineering Design Process, Dissertation, Arizona State University, 1996.

Bliznakov, P.I. et al., Design information system infrastructure to collaborative design in a large organization, *Proc. ASME DETC*, Boston, MA, vol. 1, pp. 1–8, 1995.

Bradley, S., Agogino, A., and Wood, W., Intelligent engineering component catalogs, in Gero, J.S. and Sudweeks, F. (Eds.), *AI in Design '94*, Kluwer Publishers, Dordrecht, The Netherlands, pp. 641–658, 1994.

Case, M.P. and Lu, S.C.-Y., Discourse model for collaborative design, *Computer-Aided Design*, vol. 28, no. 5, pp. 333–345, 1996.

Cave, P.R. and Noble, C.E.I., Engineering design data management, in Leech, D.J., Middleton, J., and Pande, G.N. (Eds.), *Proc. Engineering Management: Theory and Applications (EMTA '86)*, Swansea, UK, pp. 301–307, 1986.

Chappel, D., *Understanding ActiveX and OLE*, Redmond, WA, Microsoft Press, 1996.

Chen, H. et al., Toward intelligent meeting agents, *IEEE Comp.*, August, pp. 62–69, 1996.

CORNAFION, *Distributed Computing Systems*, Elsevier, Amsterdam, 1985.

Court, A.W., Culley, S.J., and McMahon, C.A., The information requirements of engineering designers, in Roozenberg, N.F.M. (Ed.), *Proc. ICED '93*, The Hague, pp. 1708–1716, 1993.

Court, A.W., Culley, S.J., and McMahon, C.A., A methodology for analyzing the information accessing methods of engineering designers, in Hubka, V. (Ed.), *Proc. ICED '95*, Praha, Czech Republic, pp. 523–528, 1995.

Court, A.W., Culley, S.J., and McMahon, C.A., The influence of information technology in new product development: Observations of an empirical study of the access of engineering design information, *Int. J. Inf. Manage.*, 17(5): 359–375, 1997.

Cross N., *Engineering Design Methods: Strategies for Product Design*, Wiley, Chichester, UK, 1994.

Dieberger, A., Supporting social navigation on the World Wide Web, *Int. J. Human Comp. Stud.*, 46(6): 805–825, 1997.

Dong, A. and Agogino, A.M., Text analysis for constructing design representations, in Gero, S.J. and Sudweeks, F. (Eds.), *AI in Design '96*, Kluwer Academic, Dordrecht, The Netherlands, pp. 21–38, 1996.

Dym, C.L., *Engineering Design: A Synthesis of Views*, Cambridge University Press, UK, 1994.

Eriksson, H., Expert systems as knowledge servers, *IEEE Expert*, 14(3): 14–19, 1996.

Ernest, J. and Friedman-Hill, J., The Java Expert System Shell, http://herzberg.ca. sandia.gov/jess, Sandia National Laboratories, 1999.

Evbuomwan, N.F.O., Sivaloganathan, S., and Jebb, A., State-of-the-art report on concurrent engineering, *Proc. Concurrent Engineering: Research and Applications*, Pittsburg, PA, pp. 35–44, 1994.

Friedman-Hill, E.J., Jess, the Java Expert System Shell, Sandia National Laboratories, 1999. http://herzberg.ca.sandia.gov/jess.

Frost, H.R. and Cutkosky, M.R., Design for manufacturability via agent interaction, *Proc. 1996 ASME DETC*, Irvine, CA, 1996.

Gero, J.S., *Knowledge-Based Design Systems*, Addison-Wesley, Reading, MA, 1990.

Gero, J.S., Design prototypes: a knowledge representation schema for design, *AI Magazine*, 11(4): 6–36, 1990.

Giachetti, R.E., A decision support system for material and manufacturing process selection, *J. Intell. Manufact.*, Volume 9, Issue 3, pp. 265–276, 1998.

Glaze, G., Johnson, J., and Cross N., Elicitation of rules for graphic design evaluation, in Gero, J.S., and Sudweeks, F. (Eds.), *AI in Design '96*, Kluwer Academic, Dordrecht, The Netherlands, pp. 527–540, 1996.

Gui, J.K., Methodology for Modeling Complete Product Assemblies, PH.D. dissertation, Helsinki University of Technology, 1993.

Hardwick, M. and Spooner, D., An information infrastructure for a virtual manufacturing enterprise, *Proc. Concurrent Engineering: A Global Perspective*, McLean, VA, pp. 417–429, 1995.

Huang, G.Q., (Ed.), Design for X: *Concurrent Engineering Imperatives*, Chapman & Hall, London, 1996.

Hubka, V. and Eder, W.E., *Design Science*, Springer-Verlag, London, 1996.

Huang, G.Q., Lee, S.W., and Mak, K.L., Web-based product and process data modeling in concurrent 'design for X', *Robotics and Computer-Integrated Manufact.*, vol. 15, no. 3, pp. 53–63, 1999.

Huang, G.Q. and Mak, K.L., Design for manufacture and assembly on the Internet, *Comp. Ind.*, vol. 38, no. 1, pp. 17–30, 1999.

IONA, Orbix2 Programming Guide: IONA Technologies Ltd., 1997.

Jones, J.C., *Design Methods*, Wiley, Chichester, UK, 1970.

Klein, M., Capturing geometry rationale for collaborative design, *J. Engineer. Appl. Sci.*, Proc. 1997 6th IEEE Workshops on Enabling Technologies: Infrastructure for Collaborative Enterprises, WET-ICE, p. 24–28, June 18–20, Cambridge, MA, 1997.

Konduri, G. and Chandrakasan, A., Framework for collaborative and distributed web-based design, *Proc. Design Automation Conf., Proc. 1999 36th Ann. Design Automation Conf. (DAC)*, p. 898–903, June 21–25, New Orleans, LA, 1999.

Kuffner, T.A. and Ullman, D.G., The information requests of mechanical design engineers, *Design Studies*, 12(1): 42–50, 1991.

Lander, S.E., Issues in multi-agent design systems, *IEEE Expert: Intell. Sys. Appl.*, vol. 12, no. 2, 1997.

Lawson, B., *How Designers Think—The Design Process Demystified*, Butterworth Architecture, London, 1990.

Lewis, J.W. and Singh, K.J., Electronic design notebooks (EDN): Technical issues, *Proc. Concurrent Engineering: A Global Perspective*, McLean, VA, pp. 431–436, 1995.

Lispscombe, B., Expert systems and computer-controlled decision making in medicine, *AI & Society*, 3: 184–197, 1989.

Mace, S., et al., Weaving a better web, *Byte*, 23(3): 58–68, 1998.

MADEFast, http:// madefast.stanford.edu/, 1999.

Madni, A.M., The role of human factors in expert systems design and acceptance, *Human Factors*, 30(4): 395–414, 1988.

Malinen P. and Salminen V., Product development in global networking environment, in Riitahuhta, A. (Ed.), *Proc. Int. Conf. Engineering Design '97*, Tampere, Finland, pp. 189–192, 1997.

NIIIP, http:// www.niiip.org/, 1999.

Norman, D.A., *Psychology of Everyday Things*, Basic Books, New York, 1989.

O'Grady, P. and Liang, W.Y., An object-oriented approach to design with modules, Iowa Internet Laboratory Technical Rep. TR98-04, 1998.

Orchard, R.A., FuzzyCLIPS Version 6.02A User's Guide, Knowledge Systems Laboratory, Institute of Information and Technology, National Research Council of Canada, Ottawa, Canada, 1994.

Owen, J., *STEP – An Introduction:* Winchester, 1993.

Pahl, G. and Beitz, W. *Engineering Design: A Systematic Approach*, Springer-Verlag, London, 1996.

Pahng, G.D.F., Bae, S.H., and Wallace, D., Distribution modeling and evaluation of product design problems, *Computer-Aided Design*, vol. 30, no. 6, pp. 411–423, 1998a.

Pahng, G.D.F., Bae, S.H., and Wallace, D., Web-based collaborative design modeling and decision support, *Proc. DETC'98*, Atlanta, GA, 1998b.

Pahng, F., Senin, N., and Wallace, D.R. Modeling an evaluation of product design problems in a distributed design environment, *CD ROM Proc. ASME DETC*, Sacramento, CA, 1997.

Pena-Mora, F., Sriram, R., and Logcher, R., Conflict mitigation system for collaborative engineering, *AI EDAM-Special Issue of Concurrent Engineering*, vol. 9, no. 2, pp. 101–123, 1995.

Pena-Mora, F., Sriram, D., and Logcher, R., *SHARED DRIMS: SHARED Design Recommendation intent Management System, Enabling Technologies: Infrastructure for Collaborative Enterprises*, IEEE Press, Morgantown, WV, pp. 213–221, 1993.

Petrie, C., Cutkosky, M., and Park, H., Design space navigation as a collaborative aid, *Proc. 3rd Int. Conf. Artif. Intelli. Design*, Lausanne, Switzerland, 1994.

Pugh, S., Total Design. Addison-Wesley, Wokingham, UK, 1991.

RaDEO, http://elib.cme.nist.gov/radeo/, 1998.

Riley, G., NASA, CLIPS Reference Manual (I, II and III), 1993.

Rodgers, P.A. and Huxor A.P., Knowledge-based design systems as texts: An emerging view, in Santo, H.P. (Ed.), *Advances in Computer-Aided Design: Pro CADEX '96 Int. Conf. Computer-Aided Design*, Hagenberg, Austria, pp. 162–170, 1996.

Rodgers, P.A. and Huxor, A.P., The role of artificial intelligence as 'text' within design, *Design Studies*, 19(2): 143–160, 1998.

Rodgers, P.A., Huxor, A.P., and Caldwell, N.H.M., Design support using distributed web-based AI tools, *Res. Engineer. Design*, vol. 11, pp. 31–44, 1999.

Roozenberg, N.F.M. and Eekels, J., *Product Design: Fundamentals and Methods*, Wiley, Chichester, 1995.

Rowe, P.G. *Design Thinking*, MIT Press, Cambridge, MA, 1987.

Roy, U. et al., Product development in a collaborative design environment, *Concurr. Engineer. Res. Appl.*, vol. 5, No. 4, pp. 347–365, 1997.

Saha, D. and Chandrakasan, A.R., Framework for distributed web-based microsystem design, *J. Engineer. Appl. Sci., Proc. 1997 6th IEEE Workshops on Enabling Technologies: Infrastructure for Collaborative Enterprises*, WET-ICE, pp. 69–74, June 18–20, 1997.

Salzberg, S. and Watkin, M., Managing information for concurrent engineering: Challenges and barriers, *Res. Engineer. Design*, no. 2, pp. 35–52, 1990.

Sanders, M.S. and McCormick, E.J., *Human Factors in Engineering and Design*, McGraw-Hill, London, 1992.

Schott, H., Buttner, K., and Birkhofer, H., Information resource management for design—illustrated by *hypermedial guidelines*, in Riitahuhta, A. (Ed.), *Proc. Int. Conf. Engineer. Design '97*, Tampere, Finland, pp. 179–184, 1997.

Senin, N., Borland, N., and Wallace, D.R., Distributed modeling of product design problems in a collaborative design environment, *CIRP International Design Seminar Proc., Multimedia* Technologies for Collaborative Design and Manufacturing, October 8–10, Los Angeles, CA, 1997.

Siegel, J., *CORBA: Fundamentals and Programming*, OMG, John Wiley, New York, 1996.

Singh, A.K., CONSENS – An IT solution for concurrent engineering, *Proc. Concurr. Engineer.: A Global Perspective*, McLean, VA, pp. 635–644, 1995.

Smith, C.S., Manufacturing Advisory Service: Web Based Process and Material Selection, Ph.D. Thesis, Univ. of California Berkeley, 1999.

Smith, P.G. and Reinertson, D.G., *Developing Products in Half the Time*, Van Nostrand Reinhold, New York, 1991.

Sobolewski, M.W. and Erkes, J., CAMnet architecture and applications, *Proc. Concurr. Engineer.: A Global Perspective*, McLean, VA, pp. 627–634, 1995.

Storath, E., Mogge, C., and Meerkam, H., Tele-engineering—product development in virtual design offices using *distributed engineering network services*, in Riitahuhta, A. (Ed.), *Proc. Int. Conf. Engineer. Design '97*, Tampere, Finland, pp. 173–178, 1997.

Sriram, D. and Logcher, R., The MITDICE project, *IEEE Comp.*, pp. 64–65, 1993.

Terry, J.C., Materials and design in Gillette razors, *Materials and Design*, 12(5): 277–281, 1991.

Toye, G. et al., SHARE: A methodology and environment for collaborative product development, *Proc. 2nd Workshop on Enabling Technologies: Infrastructure for Collaborative Enterprises*, Morgantown, WV, pp. 33–47, 1993.

Ullman D.G., Metrics for evaluation of the product design process, in: Riitahuhta, A. (Ed.), *Proc. Int. Conf. Engineering Design '97*, Tampere, Finland, pp. 563–566, 1997.

Watson, I., Basden, A., and Brandon, P., The client-centered approach: expert system maintenance, *Exp. Syst.*, 9(4): 189–196, 1992.

Westerberg, A.W., et al., Distributed and collaborative computer-aided environment in process engineering design, *Proc.*, 1995.

Wood III, W.H. and Agogino, A.M., Case-based conceptual design information server for concurrent engineering, *Computer-Aided Design*, 8(5): 361–369, 1996.

Yang, M.C. and Cutkosky, M.R., Automated indexing of design concepts for information management, in Riitahuhta, A. (Ed.), *Proc. Int. Conf. Engineer. Design '97*, Tampere, Finland, pp. 191–196, 1997.

Zha, X.F., Knowledge Intensive Methodology for Intelligent Design and Planning of Assemblies, Ph.D. Thesis, Nanyang Technological University, Republic of Singapore, 1999.

Zha, X.F. and Du, H., Knowledge intensive multi-agent framework for cooperative/collaborative assembly-oriented design modeling and decision support, *Proc. Int. Conf. Engineer. Tech. Sci. 2000*, Beijing, 2000a.

Zha, X.F. and Du, H., Web-based knowledge intensive collaborative design modeling and decision support for MEMS, *Proc. Int. Conf. Engineer. Tech. Sci.*, vol. I, pp. 80–92, Beijing, China, 2000b.

Zha, X.F. and Du, H., Web-based knowledge intensive collaborative design framework for MEMS, *Proc. Int. Workshop on MEMS 2001*, pp. 503–513, Singapore, 2001c.

Zha, X.F. and Du, H., A world wide web-based manufacturing consulting service system for processes/materials selection in concurrent design for manufacturing, *Proc. Int. Conf. Mat. Adv. Tech. (ICMAT 2001)*, Singapore, 2001d.

Zha, X.F., Du, H., and Qiu, J.H., Knowledge-based approach and system for assembly-oriented design, Part I: The approach, *Int. J. Engineer. Appl. Artif. Intell.*, vol. 14, no. 1, pp. 61–75, 2001a.

Zha, X.F., Du, H., and Qiu, J.H., Knowledge-based approach and system for assembly-oriented design, Part II: The system implementation, *Int. J. Engineer. Appl. of Artif. Intell.*, vol. 14, no. 2, pp. 239–254, 2001b.

Zha, X.F., Lim, S.Y.E., and Fok, S.C., Integrated knowledge-based product design for assembly, *Int. J. Comp. Integr. Manufact.*, vol. 12, no. 3, pp. 211–237, 1999.

6

A Multi-Agent Framework for Collaborative Reasoning

Chunyan Miao
Simon Fraser University

Angela Goh
Nanyang Technological University

Yuan Miao
Nanyang Technological University

Abstract. This chapter proposes a reusable multi-agent framework for supporting collaborative reasoning in multi-agent systems (MAS). It is based on the Agent Inference Model (AIM) and the knowledge combination/split capabilities of an extended Fuzzy Cognitive Map (FCM) theory. The proposed framework facilitates problem modeling, task decomposition, and task allocation among individual agents. The Dynamic Inference Agents (DIAs) underlying the framework are not only able to represent a vast amount of knowledge related to complex problems, but are also able to collaborate with each other and make inferences at a system level using numeric representations and computation instead of symbolic representation and deduction. The ensemble of collaborative DIAs has capabilities that exceed any singular agent in a multi-agent system. The construction of MAS, based on the framework, shows that each component of MAS can be developed as a reusable component. The whole MAS presents a plug-in /drag-off architecture.

Key words: Intelligent agent, multi-agent system, fuzzy cognitive map, software reuse.

6.1 Introduction

Single agents are independent entities with their own goals, behaviors, actions, and knowledge.[1] An agent in a single-agent system only models itself, the environment, and its interaction with the environment. In order to solve complex problems, a collection of agents that work cooperatively for a common goal in a heterogeneous environment and form a multi-agent system (MAS) is needed. Agents in a multi-agent system coordinate with each other, manifesting system behavior and achieving a higher-level goal that is beyond the capability of any individual agent within the MAS.

Multi-agent systems are increasingly playing more important roles in agent technology.[2] Nevertheless, due to some crucial problems,[3,4] some researchers argue that little "real progress" has been made in MAS. Therefore, the issues related to those problems are considered as research challenges in multi-agent systems.[5] Some of the major issues[4] are listed as follows:

- The ontology issue: Agents in MAS need to have common definitions of related concepts.
- The task decomposition issue: Complex problems need to be decomposed properly in order for different tasks to be allocated to individual agents in MAS.
- The reasoning and coordination issue: Agents in MAS need to carry out the reasoning and coordinate with each other.
- The construction, engineering, and integration issue: This relates to difficulties in constructing and implementing practical multi-agent systems and integrating them with existing software systems.
- The reusability issue: Almost all agent systems developed to date have been realized from scratch and achieved independently by different development teams.[6] Thus, the reusability of such agent systems is very low.

In this chapter, we propose a framework for multi-agent collaborative reasoning which is able to address the above issues. The proposed framework is a continuation of our previous work.[7] The theoretical foundation of the framework includes agent technology,[8,9] fuzzy cognitive map (FCM),[10] object-oriented methodology,[11] and software reuse.[12]

Object-Oriented Methodology (OOM) is considered to be mature and the most promising technology for MAS design and development. Several agent-oriented methodologies (AOM) based on OOM have been proposed.[13,14] These methodologies aim to promote OOM to AOM. In other words, they aim to promote conventional objects to active objects and then from active objects to autonomous agents. However, the current AOMs have limitations in properly representing MAS.[13] These limitations include the representation of the mental states of an agent and the collaboration among agents.[15]

In such a circumstance, it is necessary to take advantage of both the cognitive theory and OOM in order to handle the above-mentioned limitations. This chapter proposes an approach for representing the mental states of agents and supporting the collaborative reasoning among agents in MAS. This is achieved by extending the FCM theory and using OOM.

The structure of the chapter is as follows: Section 6.1 gives a brief introduction to some major issues and background related to multi-agent systems. The next section describes the overview of the framework, followed by details of how it supports the collaborative reasoning among multiple agents. Section 6.4 illustrates the construction of a multi-agent system based on the proposed framework. Finally, the conclusion and future work is presented in Section 6.5.

6.2 Overview of the Framework

The term "agent" is regarded as a computational entity that acts autonomously in order to accomplish tasks on behalf of its user. The life cycle of an agent in its environment is abstracted as a Perceive-Reason-Act (PRA) cycle.[16,17] An agent is able to perceive within the environment and represent the perception as its internal knowledge; to reason based on its knowledge; and to act in its environment. Agents continuously operate in an iterative PRA life cycle, which is shown in Figure I.6.1. Agents are proactive and goal oriented; they act autonomously toward their goals.

A multi-agent system consists of a number of decentralized, autonomous agents. Each agent has its own PRA life cycle. Nevertheless, each agent has incomplete capabilities to solve a complex problem. They perceive, reason, and collaborate with each other toward a common goal. Agents in a MAS environment need to have common sense, do collaborative reasoning, and perform decomposed tasks in order to reach a shared goal, which forms the challenges raised in Section 6.1.

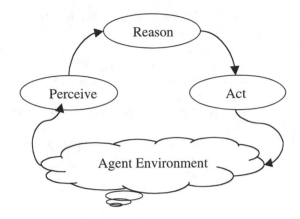

FIGURE I.6.1 An agent life cycle.

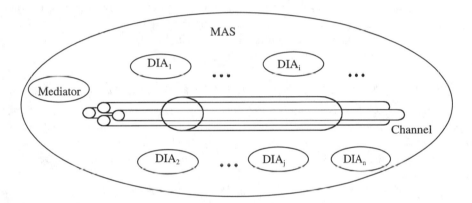

FIGURE I.6.2 A framework for multi-agent reasoning and collaboration.

In order to meet the challenges discussed in Section 6.1, this section describes a reusable framework for multi-agent reasoning and collaboration. As shown in Figure I.6.2, the multi-agent framework consists mainly of a number of dynamic inference agents (DIA), a mediator agent, and a set of communication channels.

[Definition 1] A multi-agent system is a 6-tuple (D, M, Ω, O, K, E), where:

a. $D = \{DIA_j | j = 1, 2, \ldots, n\}$ *is the set of dynamic inference agents*
b. M *is a mediator agent called coordinator*
c. Ω *is the set of the communication channels*
d. O *is an ontology server*
e. K *is a knowledge base*
f. E *is an agent environment*

The main component of the framework is the dynamic inference agents (DIAs). They are responsible for the environment perception, real-world problem modeling, knowledge representation, reasoning, and decision making on behalf of human beings. DIAs model real-world problems as a collection of factors and causal relationships between factors, based on the AIM model as described in the next section. A concept is the abstract description of a factor in the problem space. Each DIA represents some pieces of knowledge related to a complex problem. In other words, each DIA has the ability to represent a small collection of concepts/factors and their causal relationships related to a complex problem. They are able to collaborate with each other and present a system-level behavior for solving a complex problem in a MAS environment.

The DIAs communicate with each other via structured messages through the communication channels. The mediator agent coordinates the DIAs. It enables the ensemble of DIAs to take all related concepts/factors (knowledge) of the complex problem into account in order to carry out collaborative reasoning and decision making at a system level. The ontology server provides the common sense among DIAs. The knowledge base is the knowledge repository of DIAs.

With such a framework, each component plays its own role in the multi-agent system. They are independent of each other, and can be developed as reusable components.

6.3   How the Framework Works

In Section 6.2, the individual components of the framework and their roles have been described. This section illustrates how the framework supports task decomposition and collaborative reasoning among the multiple agents.

There are three major aspects for designing a multi-agent system: the mechanisms that define an individual agent's knowledge representation and behavior; the mechanisms that decompose complex problems; and the mechanisms for supporting reasoning and collaboration between agents. The following sub-sections address each aspect respectively.

6.3.1   Modeling, Knowledge Representation, and Reasoning

In the proposed framework, knowledge is represented by the DIAs. DIA perceives the environment and models real-world problems as a collection of factors and causal relationships between factors. It is constructed based on an agent inference model (AIM), which has the abilities to model, reason, make decisions, and act on behalf of human beings. The theoretical foundation of the AIM model includes FCM theory[10] and OOM.

The AIM model consists of two basic types of object: *factor*, which represents the factors that DIA has to take into account, and *impact*, which denotes the impacts (causal relationships) between the factors. A directed graph of an AIM model is shown in Figure I.6.3, in which factors are represented by nodes, and the impacts are indicated by directed and weighted arrows between factors. The direction of the impact arrow shows which factor "has impact on" the other; i.e., which factor "causes" the other. For example, in Figure I.6.3: $Factor_i$ has impact on or causes $Factor_j$ by *weight* w_{ji}.

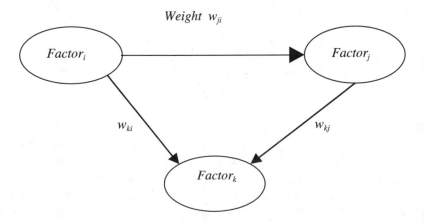

FIGURE I.6.3   Agent inference model.

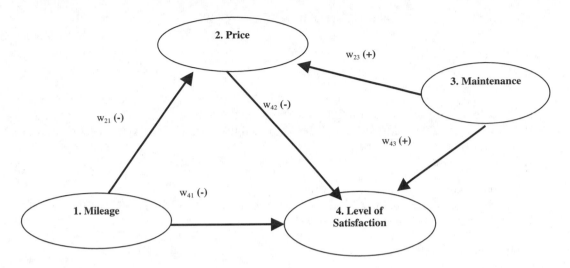

FIGURE I.6.4   The AIM model for agent Amy.

The status of the factor is described with a quantitative attribute value called *factor state value*. It shows the current status of the factor. It can be changed by the impacts from other factors. The degree of how strong a factor has impact on the other (i.e., one factor causes another) is illustrated by the *weight value* of the impact. As it has been shown, both the factor and the impact are associated with quantitative attributes, which form the value set of the AIM model.

[Definition 2] The value set of an AIM model V_{AIM}, is defined as a 2-element tuple:

$$V_{AIM} = \{F, I\}$$

where

- $F = \{x_i | x_i \in [-1, 1]; i = 1, 2, \ldots, n \text{ and } x \text{ is a real number}\}$ represents the state value of **factors**
- $I = \{w_{ij} | w_{ij} \in [-1, 1]; \ i = 1, 2, \ldots, n; \ j = 1, 2, \ldots, n \text{ and } w_{ij} \text{ is a real number}\}$ represents the weight value of **impacts**

Figure I.6.4 shows a simple example of a DIA based on the AIM model. Assume that a DIA named *Amy* acts as a buying agent for purchasing a second-hand car on behalf of its user from electronic markets over the World Wide Web (WWW). While communicating with its user, *Amy* has found that the user mainly takes into account the following factors: *Price, Mileage*, and the *Maintenance* of the car. Therefore, the *level of satisfaction* of the user can be determined by evaluating the above factors. *Amy* then models the relationships among these factors based on the AIM model, uses the model as a basis of its own knowledge, and sets its goal to recommend those cars with high *level of satisfaction*.

The directed graph of the AIM model that *Amy* built is shown in Figure I.6.4. Both the factor *Price* and the factor *Mileage* have negative impacts on the factor *Level of Satisfaction*. In another words, the higher the Price/Mileage is, the less the user will be satisfied. In contrast, the factor *Maintenance* positively affects the factor *Level of Satisfaction*. The better the car is maintained, the more the user will be satisfied. Besides the impacts on the factor *Level of Satisfaction*, the factor *Price* is negatively affected by the factor *Mileage* and positively affected by the factor *Maintenance*.

The value set of the AIM model of agent *Amy*, $V_{Ally} = \{C, W\}$ where

- $C = \{x_1, x_2, x_3, x_4\}$
- $W = \{w_{ij} \mid i = 1, 2, 3, 4; j = 1, 2, 3, 4\}$

where x_1, x_2, x_3, x_4 are the state values of factors *Mileage, Price, Maintenance,* and *Level of Satisfaction* respectively. w_{ij} is the weight value between the factors.

Besides the quantitative attributes, both the factor and the impact of the AIM model possess the following behaviors:

A **factor** object has:

- An input and output function to interact with its environment.
- A fuzzification mapping function to map a real causal activation value to the factor state value.
- A decision-making function to make decisions.

An **impact** object has:

- An input and output function to interact with its environment.
- A fuzzification mapping function to map a real causal activation value to the weight value of the impact.
- A dynamic relationship function to adjust the weight value of the impact dynamically by the state value of its causal factor. In the AIM model, the weight is not restricted to a fixed value of the given fuzzy set. It may also have a relationship to the state value of its causal factor. It can be computed using the dynamic relationship function based on the state value of its causal factor.

As has been illustrated, the knowledge can be easily represented by the AIM model. The following is a description of the inference algorithms that are carried out by the AIM model.

From the directed graph of the AIM model, it can be seen that the state of a factor node at *step* k can be computed by taking the sum of the impacts from its causal factors, i.e., the state values at *step* $k-1$ of its causal factor nodes multiplied by the corresponding weights. The detailed inference can be carried out as follows:

An *adjacency matrix* for the directed graph of an AIM model which consists of n factor nodes is an $n*n$ Weight Matrix (W) in which w_{ij} denotes a weighted arrow between nodes i and j.

$$W = \begin{bmatrix} w_{11} & w_{12} & w_{13} & \ldots & w_{1n} \\ w_{21} & w_{22} & w_{23} & \ldots & w_{2n} \\ w_{31} & w_{32} & w_{33} & \ldots & w_{3n} \\ \ldots\ldots\ldots\ldots\ldots\ldots\ldots \\ w_{n1} & w_{n2} & w_{n3} & \ldots, & w_{nn} \end{bmatrix} \quad (I.6.1)$$

The state values of the n factors at *step* $k-1$ forms a $1*n$ Factor Matrix (F_{k+1}):

$$F_{k-1} = [x_1, x_2, x_3, \ldots, x_n] \quad (I.6.2)$$

The state values of the n factors at *step* k can be obtained by multiplying Matrix (F_{k-1}) and Matrix (W):

$$[y_1, y_2, y_3, \ldots, y_n] = F_{k-1}*W \quad (I.6.3)$$

where

$$y_i = \sum_j w_{ji} x_j \tag{I.6.4}$$

y_i is the sum of the products of the state value of all the causal factors of *factor*$_i$ and the weight values between the two factor nodes.

Taking y_i as the input of the decision-making function of the *factor*$_i$, the state value of the *factor*$_i$ at *step k* can be further computed:

$$x_i = d_i(y_i) = d_i \left(\sum_j w_{ji} x_j \right) \tag{I.6.5}$$

where x_i is the new state value of the *factor*$_i$, w_{ij} is taken from the weight matrix W, and d_i is the decision-making function of *factor*$_i$.

Therefore, each step of the inference becomes a matrix multiplication, followed by the concept decision-making function (D):

$$F_k = D(F_{k-1}{}^*W) \tag{I.6.6}$$

Assuming that $x = (x_1, x_2, \ldots, x_n)^T$ is a n-tuple factor state value vector, $x(k-1)$ is the vector value at *step k*−1, and $x(k)$ is the vector value at step k, the inference from *step k*−1 to *step k* is shown as follows:

$$x(k) = d(W(x(k-1))) = [d_1, (y_1), d_2(y_2), \ldots, d_n(y_n)]^T \tag{I.6.7}$$

where W is the weight matrix and d_i is the concept decision-making function.

Based on the AIM model, the DIA is able to perceive the environment, model real problems in the read world using concepts and the causal relationships between the concepts, represent them as its own knowledge, and make inference and decisions/actions on behalf of its user. It performs PRA life cycle autonomously toward its goal. In particular, all the inferences are carried out using numeric computation with no symbolic deduction involved.

6.3.2 Advantages of the Agent Interference Model

As described, the individual agents in the proposed multi-agent framework are based on the AIM model. The theoretical foundation of the AIM model includes the fuzzy cognitive map (FCM) theory[10] and object-oriented (OO) methodology.[20] The concept of the fuzzy cognitive map was proposed by Kosko.[10] It is a combination of the cognitive map theory introduced by Alexrod[18] and the fuzzy sets theory introduced by Zadeh.[19] In a causal system, there are three basic elements; the cause, the causal relationship, and the effect. FCMs are signed, weighted, and directed graphs with feedback that model the world as a collection of concepts and causal relationships (weights) between concepts. It improves cognitive maps by describing the strength of the causal relationships through weights.

In the AIM model, we take advantage of both the cognitive modeling methods with numeric inference mechanism of FCM theory and the key concepts of object-oriented methodology. The directed graph of the AIM model can be viewed as an extended FCM. The components of the

AIM model can be viewed as active objects. Compared with the traditional FCM, the AIM model FCM extends in the following significant ways:

- The concept and the weight are extended to factors and impacts. Unlike FCM, the factor and impact are no longer of single values, but objects that encapsulate their quantitative attributes and behaviors.
- The state value of the concept is not restricted to a binary form as it is in FCM. Instead, it has been extended to any member data in a fuzzy set which enables the AIM model to have a richer ability of modeling the complex status of various factors in the real world.
- The factor has a mapping function to map a real causal activation value to a factor state value that enables the AIM model to represent various fuzzy information.
- The weight value of impact may have a dynamic relationship function with the state value of its causal factor. Therefore, the AIM model is able to describe a wider range of causal relationships between factors in the real world compared to traditional FCM, and dynamic inference can be carried out.

Above all, the AIM model provides sufficient mechanisms for defining an individual agent, DIA, in MAS. Unlike traditional intelligent agents, DIA is able to represent knowledge and carry out the inferences using numeric representation and computation instead of symbolic representation and deduction. Thus, the construction of DIAs is simplified and the implementation code is compact. DIA also has the ability to handle various types of fuzzy information and the ability to make intelligent decisions in complex situations in a heterogeneous environment.

6.3.3   Task Decomposition and Knowledge Combination/Split

Through the above illustration, it has been shown that each DIA in MAS is able to represent some pieces of knowledge related to a complex problem based on the AIM model. Moreover, the DIA is able to perceive-reason-act autonomously based on its knowledge. This section addresses the issue of task decomposition raised in Section 6.1. How is the entire knowledge of a complex problem represented by a collection of DIAs? How is the complex problem decomposed into different tasks, which can then be allocated to individual DIAs?

These questions can be answered by the knowledge combination/split capabilities of FCM. As described, the directed graph of the AIM model can be viewed as an extended FCM. One of the desirable features of FCM is that several FCMs can be combined into one by merging their adjacency matrix with different, weighted coefficients.[21] On the other hand, in Miao et al.,[22,23] we have proposed methods that make it possible to divide a complex FCM into several basic FCMs that are comparatively simple, and extend the FCM theory to dynamic cognitive net (DCN). Thus, it has been proved that both the knowledge combination and the knowledge split can be well performed by the FCM theory and its extension, DCN. However, the split can only be done when the FCM is composed of a few basic FCMs. Obviously, not every FCM satisfies the restriction. In the proposed multi-agent framework described in this chapter, restrictions are relaxed in order to apply the knowledge combination/split capabilities of FCM/DCN for decomposing complex problems into different tasks, which can then be allocated to individual DIAs in MAS.

Assuming a real-world complex problem Z is modeled by the AIM model, the corresponding directed graph of the AIM model is named as FCM G.

[**Definition 3**] An *FCM G* corresponding to a directed graph of an AIM model is a 4-tuple $\{V, A, W, F\}$ where:

a. $V = V(G) = \{v_i | i = 1, 2, \ldots, n; v_i$ *is a node in FCM G, n is the number of nodes in FCM G*$\}$ represents the nodes of FCM G
b. $A = A(G) = \{a_{ij} | a_{ij} \neq 0, i, j = 1, 2, \ldots, n, a_{ij}$ *is the arrow that starts from* v_i *and points to* $v_j\}$ represents the arrows of FCM G

  c. $W = W(G) = \{w_{ij}|w_{ij} \neq 0, i, j = 1, 2, \ldots, n, w_{ij} \text{ is the weight of } a_{ij}\}$ represents the weight matrix of FCM G

  d. $F = F(G) = \{f_i|i = 1, 2, \ldots, n, f_i \text{ is the vertex function of node } v_i\}$ represents vertex functions of FCM G

From the definition, we can get the following relationships:

$$w_{ij} = W(a_{ij}) = W(v_i, v_j) \tag{I.6.8}$$

Suppose we split FCM G into two sub-FCMs; G1, and G2:

$$G1 = \{V^1, A^1, W^1, F^1\}, \ G2 = \{V^2, A^2, W^2, F^2\} \tag{I.6.9}$$

While splitting, there is a basic restriction, which is:

$$A^1 \cap A^2 = \emptyset \tag{I.6.10}$$

This restriction is to make sure that different sub-FCMs do not have any conflicts. To facilitate the split, we need to introduce some interface nodes and channels for them to communicate, so that the inference pattern will be the same as the original one.

Here we define:

$$V' = \{v|v \in v^1 \cap V^2\}, \ A' = \{a_{ij}|v_i \in v^1, v_j \in v^1\} \tag{I.6.11}$$

According to the basic restriction mentioned above that the two sub-FCMs have two common arrows, each a_{ij} in set A' has to be split into either G1 or G2. If a_{ij} is split into FCM G2, we need to introduce two interfacing nodes, v_i^{out} and v_j^{in} for FCM G1, and two interfacing nodes, v_i^{in} and v_j^{out} for FCM G2.

$$\text{In G1, } W(v_i, v_i^{out}) = 1, \tag{I.6.12}$$

$$W(*, v_i^{out}) = 0, \tag{I.6.13}$$

$$W(v_i^{out}, *) = 0, \tag{I.6.14}$$

$$W(v_j^{in}, v_j) = 1, \tag{I.6.15}$$

$$W(*, v_j^{in}) = 0, \tag{I.6.16}$$

$$W(v_j^{in}, *) = 0, \tag{I.6.17}$$

where the symbol * stands for any other nodes in G1.

$$\text{In G2, } W(v_i^{in}, v_i) = 1, \tag{I.6.18}$$

$$W(*, v_i^{in}) = 0, \tag{I.6.19}$$

$$W(v_i^{in}, *) = 0, \tag{I.6.20}$$

$$W(v_j, v_j^{out}) = 1, \tag{I.6.21}$$

$$W(*, v_j^{out}) = 0, \tag{I.6.22}$$

$$W(v_j^{out}, *) = 0, \tag{I.6.23}$$

where * stands for any other nodes in G2.

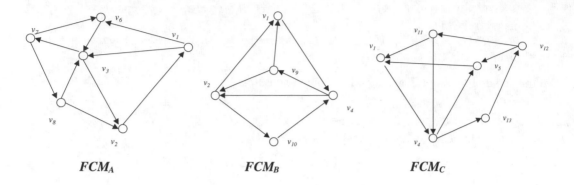

FIGURE I.6.5   Three direct graphs representing complex problem Z.

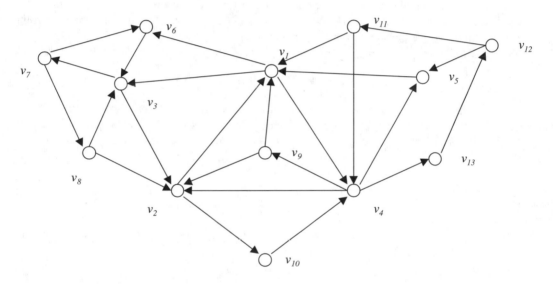

FIGURE I.6.6   The directed graph of AIM model for complex problem Z (FCM$_Z$).

The following is an illustration of how the split can take place. Assume that we are going to build up a multi-agent system for solving a complex problem Z, based on the proposed framework. First, Z is modeled using the AIM model Z. Since Z is very complex, it is not possible to take all the concepts into account at one time. Thus, modeling is undertaken step by step. At each step, we only take into account some factors related to the complex problem Z, and the causal relationships among those factors. Therefore, a corresponding directed graph of an AIM model will be generated at each step. Assuming that as a result, the following three directed graphs of the AIM model, shown in Figure I.6.5, are formed.

By composing the directed graphs of the three basic FCMs (Figure I.6.5), the directed graph of AIM model, FCM$_Z$ (Figure I.6.6) for the complex problem Z can be obtained.

Secondly, we need to decompose the complex problem Z into different tasks and assign them to a number of computation intelligent agents (DIAs). When decomposing, a basic restriction is defined based on the extension of FCM theory: one causal relationship is limited to one DIA only. That is to say, any two DIA agents may have the same concept, but they are not allowed to share the same causal relationship. Therefore, each causal relationship can be assigned to only one DIA.

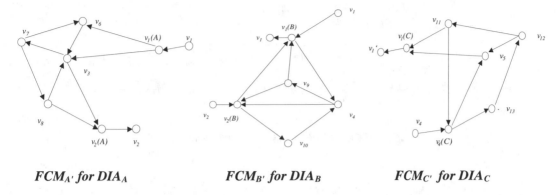

$FCM_{A'}$ *for* DIA_A $FCM_{B'}$ *for* DIA_B $FCM_{C'}$ *for* DIA_C

FIGURE I.6.7   FCMs for individual DIAs.

Thus, in order to maintain the original relationships and to ensure that the combination of the knowledge represented by each DIA is the same as the original AIM model, we cannot simply decompose FCM_z into FCM_A, FCM_B, and FCM_C. When decomposing, we have to follow the rules described above to adjust the FCM_A, FCM_B, and FCM_C to $FCM_{A'}$, $FCM_{B'}$, and $FCM_{C'}$ by introducing new concepts for the construction of the corresponding DIA_A, DIA_B, and DIA_C called interface concepts. This is illustrated in Figure I.6.7.

Based on the AIM model and the methods of knowledge combination/split of FCM, the proposed framework is able to model a complex problem, decompose the complex problem into different tasks, and allocate these tasks to individual agents.

6.3.4   Collaborative Reasoning and Coordination

After the complex problem is decomposed and different tasks are allocated to individual DIAs in MAS, we come to another important issue raised in Section 6.1: How do these DIAs reason collaboratively and coordinate with each other in MAS? For instance, since DIA_A, DIA_B, and DIA_C in AIM model z have common factors, they need to interact with each other in the multi-agent system. As mentioned, the ontology server supports interoperability by providing a common vocabulary among agents; the knowledge base is a repository of the knowledge for agents; and the mediator agent offers coordination services among DIAs. The following describes how the collaborative reasoning and coordination is done among the agents.

- Ontology management—Ontology management is accomplished through the ontology server. The ontology (or concept definitions) specifies the terms each party (agent) must understand and use during interaction with other parties (agents). It explicitly defines every concept to be represented by the individual agent. While collecting the factors/concepts and their descriptions through interaction with its user, the ontology server enables the DIAs to understand and interact with each other based on the common vocabulary sense.
- Knowledge base and agent communication language—The knowledge base is a knowledge repository of agents. The agents exchange messages and access the knowledge base through standard (Knowledge Query and Manipulation Language) KQML.[24] Therefore, the DIAs not only share a common vocabulary, but also speak the same language.
- Communication channel and sockets—The DIAs communicate with each other via communication channels. The communication channel is built by the coordinator for facilitating the agents to communicate with each other and for exchanging messages such as interfacing nodes in the graphs of the AIM model. As shown in Figure I.6.8, the channel between split FCMs has

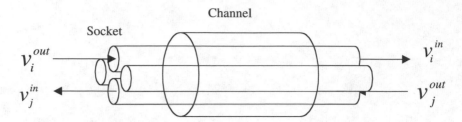

FIGURE I.6.8   Channel and sockets between split FCMs.

a number of sockets. We define socket i as v_i^{out} to v_i^{in}, and v_j^{out} should always pass its state to v_i^{in}. While implementing MAS based on the proposed framework, the socket can be realized as a buffer, where v_i^{out} always writes to and where v_i^{in} always reads from. A distributed MAS can also be implemented as a real network socket. A prototype using CORBA event channel as the communication channel in MAS has been developed.

- Interaction indexes—The coordinator maintains several interaction indexes in the knowledge base. Suppose the MAS contains n DIAs: $\{A_1, A_2, \ldots, A_n\}$. Every DIA has three sets of factors:

$$R_i, X_i, Y_i \in A_i \tag{I.6.24}$$

where R_i is the set containing all the factors that have inputs only from outside of the DIA, i.e., from other DIAs; X_i is the set containing all internal factors that do not have direct interaction with factors in other DIAs; and Y_i is the set containing all the factors that affect only the outside factors.

Taking the agent A in Figure I.6.7 as an example, the following is obtained:

$$R_i = \{v_1\}, \tag{I.6.25}$$

$$X_i = \{v_1(A), v_3, v_6, v_7, v_8, v_2(A)\}, \tag{I.6.26}$$

$$Y_i = \{v_2\}. \tag{I.6.27}$$

Defining $R = U_{i=1}^A R_i$, $Y = U_{i=1}^A Y_i$ for every $v \in R$, the coordinator maintains an index of all the DIA agents that has v as an input factor. Therefore, if an event causes the state of v to be changed, all these DIA agents can be notified so that they can take into account the latest state. For instance, during an inference, a DIA agent will call a routine task and send a message to the coordinator when a state change occurs in any of its output factors. The coordinator then checks the interaction index and informs all the other DIA agents who have the same factor as an input factor.

- Causal relationship index—The coordinator keeps an index for causal relationships, as no causal relationship is allowed to be repeated. This is used as a rule when a new DIA is added into the MAS.
- Adding new DIA—Adding a new DIA can be interpreted as combining a new portion of an FCM to an existing FCM. First, the coordinator will check the causal relationship index. If any of the causal relationships of the new FCM already exists in the index, some adjustments need to be done before the combination. If the causal relationship shown in Figure I.6.9(a) exists in the index, then an adjustment, as shown in Figure I.6.9(b), is made. The new FCM can be combined with the existing one.

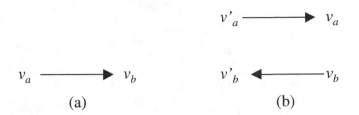

FIGURE I.6.9   Adding a new agent.

By introducing a new structure, Figure I.6.9(b), the coordinator can still maintain the original relationship. After combination, the coordinator will update the interaction index and the DIA can be plugged into the existing multi-agent system.

- Drag off a DIA—Drag off a DIA is straightforward and therefore the illustration is omitted.

Under the proposed framework, each DIA perceives its environment, represents partial factors and their causal relationships related to a complex problem as its knowledge, collaborates with other agents via communication channels and coordinators, and makes its own reasoning autonomously in a MAS environment. Through the collaboration and coordination, the ensemble of DIAs in MAS not only has the ability to represent the entire body of knowledge relating to complex problems, but can also make collaborate reasoning in a multiagent system.

Moreover, the proposed framework exhibits the reusable methods that define an individual agent's knowledge representation and behavior; reusable algorithms that decompose a complex problem and allocate sub-problems to individual agents; and reusable channels for supporting collaboration and coordination among agents. The proposed framework also presents a reusable plug-in/drag-out multi-agent system architecture. Every DIA need only infer on the concepts it is concerned with. The interactions of DIAs are carried out by the coordinators through communication channels. The details of how to use the proposed framework to facilitate code reuse and simplify agent construction in MAS will be discussed in Section 6.4.

6.4   Construction of Multi-Agent Systems

The design and development of multi-agent systems based on the proposed framework is comprised of the following levels: multi-agent system design, component design, component based development, and system implementation.

- Multi-agent system design—At this level, the reusable multi-agent system architecture will be defined based on the proposed framework. The definitions of basic agent types and the common attributes and behaviors of each basic agent type will be defined. The communication channel will also be specified. The MAS design is given as UML[25] diagrams in Figure I.6.10, which shows the class diagram of MAS design level. As shown in the figure, the MAS consists of a mediator agent, a number of DIAs, an environment, and communication channels.
- Component design—The component design includes the detailed design of the ontology server, the knowledge base, the abstract agents, and the abstract communication channel. An abstract agent is inherited from the basic agent type defined in the multi-agent system design level. Figure I.6.11 shows the class diagram of the DIA. As shown in the figure, DIA consists of a set of classes, where each class contains some important behaviors of DIA. A DIA has the ability to perceive within the environment. It possesses a goal as well as intelligence to act autonomously toward its goal. Therefore, it can reason and determine the actions to be taken

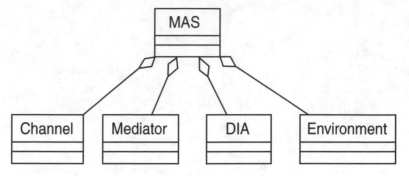

FIGURE I.6.10   Class diagram of MAS.

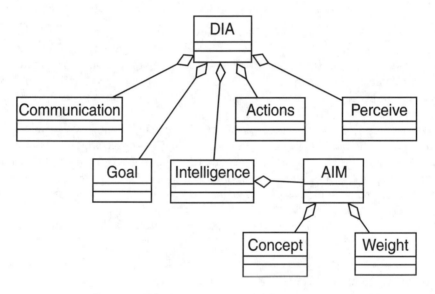

FIGURE I.6.11   Class diagram of DIA.

based on its goal and knowledge. The knowledge of DIA is represented by the AIM model. DIA is also able to communicate with other DIAs through structured messages.

The design of abstract classes includes the design of the most common behaviors of the abstract class and its sub-classes. Figure I.6.12 shows the aggregation relationship between class *concept*, *weight*, and *AIM*. The design of abstract classes promotes the reusability, since they can be transported from one MAS to another. The detailed class diagram of the described MAS design is shown in Figure I.6.13.

• Component-based development—Component-based development is to implement the real agents for construction of MAS. With the reusable abstract agent classes from the component design level, the skeleton code of real agents' implementation can be automatically generated by code generation functions. Subsequently, the agents in MAS can be developed into reusable black boxes, or white boxes. The black boxes imply that the developer does not need to know

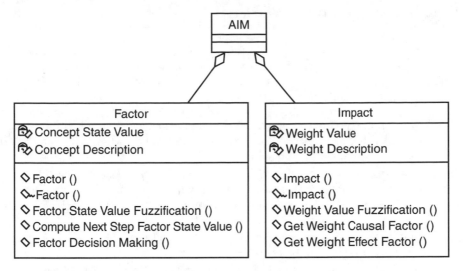

FIGURE I.6.12   Class diagram of AIM.

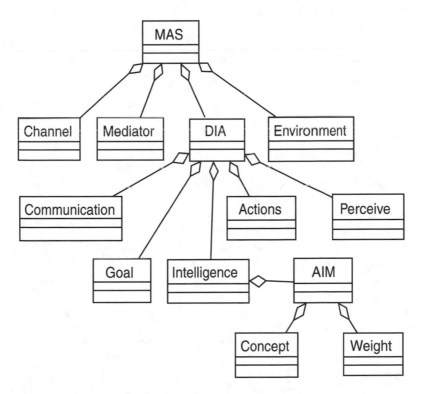

FIGURE I.6.13   Detailed class diagram of MAS.

how the agent is implemented when it is reused in the future. The white boxes indicate that some modifications might be needed before reuse.

- System implementation—The final level is system implementation. All the constructed agents are integrated in order to complete the construction of the multi-agent system based on the proposed framework.

6.5   Conclusion and Future Work

In this chapter, we have presented a reusable framework for supporting multi-agent collaborative reasoning and coordination. Unlike most of the current AOMs, which have limitations in representing the mental states of an agent and the collaboration among agents etc., the proposed framework has the following advantages:

- It is able to support problem modeling, task decomposition, and allocation of the decomposed tasks to the individual agents, based on the AIM model and the knowledge combination/split capabilities of the extended FCM theory.
- The DIAs proposed are not only able to represent the entire knowledge related to complex problems, but are also able to collaborate with each other and carry out the inferences using numeric representation and computation instead of symbolic representation and deduction.
- The mediator agent (or coordinator) not only enables a collection of DIAs to take into account all concepts/factors related to the complex problem, but also allows the individual agents to do collaborative reasoning in a MAS. This results in the collection of agents having a higher capability than any singular agent.
- The ontology server provides the common definitions of related concepts shared by each part (agents) in the MAS.
- A practical methodology for implementing and constructing a multi-agent system based on the proposed framework takes advantage of object-oriented concepts, software reuse, and component-based development technology.
- Each component of the framework is independent from each other and can be developed as a reusable component. Finally, the whole MAS presents high reusability with a plug-in/drag-off architecture.

With reference to the research issues in multi-agent system technology discussed in Section 6.1, the proposed framework has solutions that can be used to resolve those issues. Compared with existing designs of multi-agent systems, it presents a practical new approach to modeling, designing, and constructing multi-agent systems with precise knowledge representation, reasoning and coordination capability, easy plug in/drag off architecture, and reusable framework.

The investigation and case study for applying the proposed framework for building a wide range of decision support systems will be carried out in our future work. Decision making is vital to individuals, enterprises, and governments. The explosion of information makes decision making more complex and difficult. Human beings are limited in their decision-making capabilities when faced with a large number of interactive factors. As a result, computer-based decision support systems (DSS) are in great demand by different kinds of organizations, especially for large decision support systems (LDSS). An LDDS may contain thousands of interacting variables/concepts from different domains. Although a number of conventional decision support systems have been proposed, they are usually not designed for large complex problems that consist of a huge number of interactive factors. We believe the proposed framework will advance the application of building various multi-agent decision support systems.

References

1. Stone, P. and Veloso, M., Multi-agent systems: A survey from a machine learning perspective, *Autonomous Robots*, 8-3, 345, 2000.
2. Susan, E.L., Issues in multi-agent design systems, *IEEE Expert,* March–April, 18, 1997.

3. Lee, L.C. et al., The stability, scalability, and performance of multi-agent systems, *BT Tech. J.,* 16–3, 69, July 1998.

4. Nwana, H. and Ndumu, D., A perspective on software agents research, *Knowledge Engineer. Rev.,* 14–2, 1, 1999.

5. Syaima, K.P., Multi-agent systems, *AI Magazine,* 19, 79, 1998.

6. Kendall, E.A. and Krishna, M., Patterns of intelligent and mobile agents, *in Proc. Autonomous Agents,* 1998, 92.

7. Miao, C.Y. et al., Computational intelligent agent, *in Proc. 1st Asia Pacific Conf. Intell. Agent Tech.,* 1999, 299.

8. Wooldridge, M., Jennings, N., and Kinny, D., A methodology for agent-oriented analysis and design, in *Proc. Agent'99,* 1999.

9. Kendall, E.A., Agent roles and role models, in *Proc. Intell. Agents for Information and Process Management,* 1998.

10. Kosko, B., Fuzzy cognitive maps, *Int. J. Man-Machine Studies,* 24, 65, 1986.

11. Mellor, S. and Johnson, R., Why explore object methods, patterns, and architectures?, *IEEE Software,* 14–1, 27, 1997.

12. Krueger, C., Software reuse, *ACM Computing Survey,* 24–2, 131, 1992.

13. Wooldridge, M. and Dianaimini, P., Agent-oriented software engineering: the state of the art, *Agent-Oriented Software Engineering,* Lecture Notes in AI, 1957, Springer-Verlag, 2000.

14. Kostiadis, K., Hunter, M., and Huosheng, H., The use of design patterns for the development of multi-agent systems, *2000 IEEE Int. Conf. Systems, Man, and Cybernetics,* 1, 2000, 280.

15. Hongsoon, Y. et al., Architecture-centric object-oriented design method for multi-agent systems, in *Proc. 4th Int. Conf. Multi-Agent System,* 2000, 469.

16. Genesereth, M.R. and Nilsson, N., *Logical Foundations of Artificial Intelligence,* Morgan Kaufmann Publishers, San Mateo, CA, 1987.

17. Huhns, M.N. and Singh, M.P., *Readings in Agents,* Morgan Kaufmann Publishers, San Francisco, CA, 1998.

18. Axelrod, R., Structure of decision: The cognitive maps of political elites, *Princeton University Press,* Princeton, New Jersey, 1976.

19. Zadeh, L.A., Fuzzy Sets, *Information and Control,* 8, 338, 1965.

20. Rumbaugh, J. et al., *Object Oriented Modeling and Design,* Prentice-Hall, NJ, 1991.

21. Silva, P.C., New forms of combined matrices of fuzzy cognitive maps, in Proc. *IEEE Int. Conf. on Neural Networks,* 2, 1995, 771.

22. Miao, Y. and Liu, Z.Q., On causal inference in fuzzy cognitive map, *IEEE Trans. Fuzzy Systems,* 8–1, 107, 2000.

23. Miao, Y. et al., Dynamical cognitive net—an extension of fuzzy cognitive map, *IEEE Trans. Fuzzy Systems,* 9-5, 760, 2001.

24. Finin, T., Labrou, Y., and Mayfield, J., KQML as an agent communication language, in *Software Agents,* Bradshaw, J., Ed., AAAI/MIT Press, Cambridge, 1997.

25. Schumann, J. and Whittle, J., Automatic synthesis of agent designs in UML, in *Proc. First Intl. Workshop, Agent-Based Systems,* FAABS, Springer, Greenbelt, MD, 2000.

7

Architecting Multi-Agent Systems: A Middleware Perspective

Zhonghua Yang
Nanyang Technological University

Robert Gay
Nanyang Technological University

Chunyan Miao
Simon Fraser University

Abstract.   Multi-agent systems (MAS) are expected to become a fundamental enabling technology, especially in mutual interdependent, dynamic uncertainty environments that require sophisticated control and coordination. However, despite the compelling vision of ubiquitous multi-agent system technology and the significant progress in the field of intelligent agents and multi-agent systems, it is still a distance away to have any real experience building truly heterogeneous, realistically coordinated multi-agent systems that work together. MAS requires the fundamental support from the underlying infrastructure and middleware for multi-agent interaction, coordination, coalition formation, conflict resolution, dynamic organization, and reorganization, among others. As compared to the distributed objects middleware technology, there is no well-defined, well-architected middleware platform which provides the convenient environment to support the architecting and building of MAS systems. In this chapter, we draw on the advancement in distributed objects middleware and the recent standardization efforts in multi-agent systems and present a middleware-based MAS architecture. Examined and presented is a variety of middleware services and supports that are desirable for a mature MAS middleware. The focus is placed on agent communication, agent directory service, agent mobility support, and interoperable knowledge bases access and manipulation.

7.1   Introduction

In the past few years, we have witnessed a significant progress in the field of intelligent agents and multi-agent systems. An agent, as considered in this chapter, is a computational entity that can perceive and act upon its environment, and that is autonomous in a sense of depending on its own experience. *Intelligent agents* are agents with a certain degree of intelligence. Given the information they have and their perceptual and effectual capability, intelligent agents pursue their goals and execute their tasks flexibly and rationally in a variety of environments. *Multi-agent systems* (MAS) are the systems in which several intelligent agents cooperate with each other and coordinate their knowledge and activities, and reason about the processes of coordination to accomplish a common goal. Multi-agent systems are normally distributed systems in which agents are distributed across several computing platforms. Traditionally, multi-agent systems are one of two types of *Distributed artificial intelligence* (DAI): distributed problem solving and multi-agent systems (MAS). The distributed problem-solving systems are concerned with how a particular problem can be partitioned and solved by a number of nodes that cooperate in dividing and sharing knowledge about the problem and in developing solutions. The recent trend in DAI has blurred the boundary between the two. In fact, there seems no reason why a particular problem cannot be divided and solved by multiple agents residing in several nodes.

The multi-agent system has the following major characteristics:[1]

- Each agent has just incomplete information, and is restricted in its capabilities.
- A system control is distributed.
- A data is decentralized.
- A computation is asynchronous.

Multi-agent systems can differ in the agents themselves, the interactions among the agents, and the environments in which the agents act.

Despite the compelling vision of ubiquitous multi-agent system technology, it is still a distance away to have any real experience building truly heterogeneous, realistically coordinated multi-agent systems that work together.[2] Furthermore, the current significant demonstration of MAS and experimentation in MAS behavior and implementation remain very weak. MAS requires fundamental support from the underlying infrastructure and middleware for multiagent interaction, coordination, coalition formation, conflict resolution, dynamic organization and reorganization, among others. MAS is expected to become a fundamental enabling technology, especially in mutual interdependent, dynamic uncertain environments that require sophisticated control and coordination.[2] However, the widespread use of MAS won't occur until the stable, widely accessible MAS middleware takes hold that provides a sophisticated support for many commonly required services and facilities. As compared to the distributed objects middleware technology,[3] there is no well-defined, well-architected middleware platform that provides a convenient environment to support the architecting and building of the MAS systems. Worse yet, there is no tools or IDEs like that are available for distributed object system development.

Middleware is systems software that resides between the applications and the underlying operating systems, network protocol stacks, and hardware. The middleware solves typical, costly, and commonly-accepted technical problems in systematic and appropriate ways.[3,4] It allows much greater community attention to unique, domain-specific aspects. In a similar way, the multi-agent middleware isolates the agent- specific behavior from the underlying support which would be common to all of the agents in a multi-agent system.

In this chapter, we examine the MAS systems from a middleware perspective. For the convenience of exposition, we present a layered MAS architecture based on the multi-agent system middleware which itself is built on the distributed objects middleware. This discussion is drawn heavily on the mature distributed objects middleware and emerging agent standardization efforts (most noticeably, DARPA and FIPA).

7.2 Middleware for Multi-Agent Systems

As indicated earlier, middleware is interposed between applications and commonly available hardware and software infrastructure to make it easier, more feasible, and more cost effective to develop and evolve systems using reusable software. In their recent contribution, Schantz and Schmidt provide an excellent coverage of a state of the art of distributed objects middleware.[3]

Middleware provides more advanced and capable abstraction and support beyond simple lower-level connectivity to construct effective distributed systems. Similar to infrastructure, middleware has the following general characteristics:[2,5]

- It is embedded *inside* other structure and technologies.
- It is transparent to use, meaning that it does not have to be reinvented or assembled each time for each task.
- It has wide reach or scope.
- It tends to be taken for granted for artifacts and organizational arrangements.
- It is linked with conventions of practice.
- It embodies standards; and its takes on transparency by plugging into other tools or infrastructures in a standardized way.
- It is built on and shaped by an installed base of practice and technology.
- It is invisible when it works, yet highly visible upon breakdown. For examples: server is down, there is a power blackout.

The experience in distributed object middleware shows that the middleware is fundamentally important when constructing multi-agent systems. The requirements for a sophisticated multi-agent middleware focus on the identification, evolution, and expansion of our understanding of current middleware services, and then the definition of additional middleware layers and capabilities to meet the challenges associated with constructing sophisticated multi-agent systems.

The technological advancements in the past decade have provided a very solid base on which the new and shaping multi-agent system middleware can be defined and developed.[3,4] The hardware (such as CPUs and storage devices) and networking elements (such as IP routers) have become easily available commodities. The programming languages (such as Java and C++) and operating environments, such as POSIX, Windows, and Java Virtual Machines (JVMs), have reached maturity. More importantly, the enabling fundamental yet popular middleware with some critical mass is widely deployed, such as CORBA,[6] Enterprise Java Beans,[7] and .NET.[8] These middleware are helping to commoditize many software components and architectural layers for distributed object applications.

We have witnessed the success of distributed objects middleware as an advanced, mature, and field-tested technology. Distributed objects middleware defines higher-level distributed programming models based on an object-oriented paradigm. It encapsulates the native OS network and underlying platform details, and thus enables distributed applications to be developed much like stand-alone applications. In other words, it allows invoking operations on target objects without concern about or knowledge of their location, programming language, OS platform, communication protocols and interconnects, and hardware. Distributed objects applications built on this middleware are composed of relatively autonomous software objects that can be distributed throughout a wide range of networks and platforms. At the heart of distributed objects middleware are object request brokers (ORBs), which in essence are object buses on which the application components can plug. The mature distributed object buses include:

- The OMG's Common Object Request Broker Architecture (CORBA),[6] which is an open standard that allows objects to interoperate across networks regardless of the language in which they were written or the platform on which they are deployed. At the core of CORBA is the CORBA object bus *object request broker* (ORB). The transportation defined for ORB is IIOP for the Internet TCP/IP environments.

- Sun's Java Remote Method Invocation (RMI),[9] which is the object bus for the Java platform and enables developers to create distributed Java-to-Java applications. Using RMI, the methods of remote Java objects can be invoked from other JVMs, possibly on different hosts. RMI supports more sophisticated object interactions by using object serialization to marshal and unmarshal parameters, so the Java objects can be passed around by value across networks. This flexibility is made possible by Java virtual machine architecture and is greatly simplified by using a single language.

- Microsoft's Distributed Component Object Model (DCOM),[10] which is Microsoft's object bus that enables software components to communicate over a network via remote component instantiation and method invocations. Unlike CORBA and Java RMI, which run on many operating systems, DCOM is implemented primarily on Windows platforms.

- Simple Object Access Protocol (SOAP), which is an emerging distribution middleware technology based on a simple and lightweight protocol,[11] and is based on eXtensible Markup Language (XML), a universal format for structured documents and data on the Web.[12] SOAP allows applications to exchange structured and typed information on the Web.[13] SOAP is designed to enable automated Web services based on a shared and open Web infrastructure. SOAP applications can be written in a wide range of programming languages, used in combination with a variety of Internet protocols and formats, including the Hypertext Transfer Protocol (HTTP),[14] Simple Mail Transfer Protocol (SMTP),[15] and Multipurpose Internet Mail Extensions (MIME),[16] and can support a wide range of applications from messaging systems to Remote Procedure Call (RPC).[17]

Distributed objects middleware is often augmented by a set of common services and facilities that define higher-level domain-independent services. These services and facilities provide a further abstraction that hides lower-level middleware directly, and thus allows application developers to concentrate on programming business logic. These services provide, through a set of well-defined standard API, transactional behavior, security, database connection pooling or threading, and concurrency control. Examples of common middleware services include: the OMG's CORBA services,[18] Sun Microsystems Enterprise JavaBeans,[19] and Microsoft's .Net Web Services,[20] and they are summarized as follows.[3]

The OMG's *CORBA Common Object Services (CORBAservices)*[18] provides domain-independent interfaces and capabilities that can be used by distributed applications. The OMG CORBAservices specifications define a rich set of these services, including naming, event notification, life cycle, trading, security, transactions, persistence, multimedia streaming, persistence, global time, real-time scheduling, fault tolerance, and concurrency control. It also defines mobile agent facility.[21]

Sun Microsystems *Enterprise Java Beans (EJB)* technology[19] is the server-side component technology and allows developers to create n-tier distributed systems. With EJB and J2EE, the applications can be developed and assembled using a set of separated developed components (e.g., beans). The middleware services are provided by the component *container*. Since EJB is built on top of Java technology, EJB service components can only be implemented using the Java language. The CORBA Component Model (CCM)[22] defines a superset of EJB capabilities that can be implemented using all the programming languages supported by CORBA.

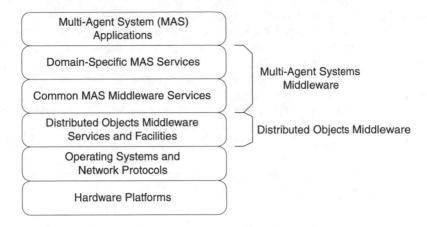

FIGURE I.7.1   The multi-agent system: layered architecture and MAS middleware.

Microsoft's *.NET Web services*[20,23] complements the lower-level middleware .NET capabilities and allows developers to package application logic into components. The .NET Web services combine aspects of component-based development and Web technologies, and thus the .NET applications can be accessed using standard higher-level Internet protocols, such as HTTP. Like components, .NET Web services provide black-box functionality that can be described and reused without concern for how a service is implemented. Unlike traditional component technologies, however, .NET Web services are accessed using Web protocols (e.g., HTTP) and data formats (XML).

The advancement of distributed object middleware provides the rich experience and solid technological base on which the future multi-agent systems and its middleware can be based (Figure I.7.1).

As shown in Figure I.7.1, the MAS middleware is layered on distributed objects middleware. The realization of MAS middleware will employ the mature and fundamental services and facilities available from the underlying distributed objects middleware, such as naming service for naming agents, directory (or repository) service for finding and locating agents, object bus for allowing agent components to be connected, and transport protocols for agent communication message transfer. The specific implementation of MAS middleware will depend upon the distributed objects middleware chosen, e.g., JAVA platform or CORBA.

The MAS middleware itself provides the common services and facilities that are used across all agent application domains which constitutes the *common MAS middleware services* layer in Figure I.7.1. These services include agent communication language (ACL) (communication acts, content languages, and interaction protocols), agent message transportation (message representation, message envelope representation, and transport protocols), agent mobility support, agent management services, and interoperable knowledge base access.

There are some services and facilities which are commonly required for specific agent application domains. These services and facilities constitute the *domain-specific MAS services* layer. The examples of domain specific services and facilities are domain-specific knowledge models, reasoning engines and inference mechanisms, and specific ontologies.

The overall physical infrastructure and middleware in which agents can be deployed is called the agent platform (AP).

The advantages of use of middleware and middleware-based architectures for faster development cycles, better quality, decreased effort, and greater software reuse are well documented.[3] The agent middleware shields application agent developers from low-level, tedious, and error-prone platform details, such as socket-level network programming; and thus functionally bridge the gap between agent applications and the lower-level hardware and software infrastructure. The agent middleware provides a consistent set of higher-level network-oriented abstractions that are much closer to application requirements, allowing the development of multi-agent systems to be simplified. A rich set of services offered by the middleware are proven necessary to operate effectively for multi-agent systems in a networked environment. Furthermore, it enables and simplifies the integration of components developed based on the same middleware by multiple technology suppliers; it helps leverage previous development expertise and capture implementations of key patterns in reusable frameworks.

In this chapter, we do not present the techniques of how a specific multi-agent system is constructed. We do not deal with the domain specific middleware concerns, either. Rather, we take a particular architectural view when building multi-agent systems: a middleware perspective by examining the common agent middleware services.

7.2.1   Architecting Multi-Agent Systems

When *architecting* a multi-agent system, we are concerned with fundamental aspects of the multi-agent system, i.e., we focus on the architecture of a multi-agent system. The architecture is defined as "the fundamental organization of a system embodied in its components, their relationships to each other and to the environment, and principles guiding its design and evolution."[24] It is well argued that several views contribute to defining an *architecture*.[24,25] Adopting this notion, we present a middleware perspective when architecting multi-agent systems.

7.3   Agent Communication

In a multi-agent system, an agent coordinates and cooperates to meet its goal, ultimately, by means of communicating with other agents, i.e., sending messages to them and receiving messages from them. Communication is involved in several aspects: *identification* of the counterpart you want to communicate with, the *content* of the communication, and *packaging* of the message in a way that makes communication clear. In the agent communication, it is important to separate the semantics of the communication languages (which must be application independent) from the semantics of the enclosed message (which tends to be application dependent). The communication language, therefore, must be universally shared by all agents. Agent communication support in terms of languages and protocols is considered as part of agent middleware.

An agent system is considered a *qualified agent system* only if the agent in the system is able to exchange knowledge using an agent communication language (ACL).[26] From a middleware perspective, agent communication language is at higher level than the remote procedure call (RPC), remote method invocation (RMI), and object request broker (e.g., CORBA). Agent communication languages handle propositions, rules, and actions instead of simple objects with no semantics associated with them. Moreover, a message of an agent communication language describes a desired state in a declarative language, rather than a procedure or method. Technically, the agent communication language itself defines the types of messages (and their meanings) that agents can exchange, and these messages are transported over networks using a lower-level protocol, e.g., SMTP,[15] TCP/IP, IIOP,[27,28] or HTTP.[14] At the same time, a higher-level conceptualization of the agent's behaviors drives the agent's communicative (and noncommunicative) behavior.[29]

When heterogeneous, autonomous agents exchange information, the meaning of the exchange is characterized by *communicative speech acts*. Communicative acts are illocutionary verbs that tell a receiving agent in which context to interpret the contents of the enclosed message. This method of communication is illocutionary in nature, not perlocutionary, since the sending agent expects the message to be understood, but has control over the effect of the message. For most computing scenarios, agent communicative acts fall into one of four categories:[30]

- *Constatives*: statements of fact, which include assertives, predictives, retrodictives, responsives, and suggestives.
- *Directives*, which include requestives, questions, requirements, prohibitives, permissives, and advisories.
- *Commissives*, which promise something and include promises and offers.
- *Acknowledgments*, which include apologize, condole, congratulate, greet, thanks, bid, accept, and reject.

When communicating, the agent selects communicative acts based on the relevance of the act's expected outcome or *rational effect* to its goal; the actual outcome, however, cannot be ensured. Consider the case that an agent plans to satisfy some *goal* (or objective) and adopt a specific *intention*. Assume that agent A cannot carry out the intention by itself. The question then becomes which message or set of messages should be sent to which agent, say B, to assist or cause its intention to be satisfied. If agent A behaves rationally, it will not send out a message whose effect will not satisfy the intention and hence achieve the goal. To make the point of *rational effect* clearer, let us consider that agent A wishes to buy an IBM share (G = "buy IBM share") and derives a goal to find out if the share will go up for it (G' = "know if it goes up"). Thus, its intention will be to find out the stock trend (I = "find out if it is up"). Obviously, agent A will be ill behaved to ask agent B "is it raining today?" From agent A's perspective, whatever agent B responds, it will not help him to determine whether the stock share will go up. Therefore, the meaningful communication requires agents to act rationally. In the example above, if agent A, acting more rationally, asks agent B "can you tell me if IBM share will go up?", he has assumed that agent B will know the answer, and then acted in a way he hopes will satisfy his intention and meet his goal. The reasoning behind agent A's behavior of asking agent B is that agent B would tell him, hence making the request fulfill his intention. However, simply on the basis of having asked, agent A cannot assume that agent B will act to tell him the stock trend. The important point is that in a multi-agent system, each agent is independent and autonomous, and thus agent communication does not ensure the *rational effect* from sending the messages.

The typical communication languages for agents to exchange information and knowledge are DARPA's KQML[31] and FIPA's ACL.[32] In the remainder of this section, we first briefly describe KQML, then focus on the agent communication as specified in the new FIPA specifications.

7.3.1 Agent Communication in KQML

Knowledge Query and Manipulation Language (KQML) is a Lisp-based language that was developed as part of the DARPA Knowledge Sharing Effort[31,33] and has been implemented in several agent environments. Agents speak KQML using pre-defined *communication acts library* (primitives) with the predefined meaning.

KQML is speech-act based, and the messages exchanged between agents are *performative*. KQML defines 42 basic performatives in its 1993 specification, but it can be extended.

Constatives (basic informatives)

- tell (share a piece of knowledge)
- deny or untell (negate a speech act)

Database performatives

- insert (ask recipient to add something to his knowledge base (KB))
- delete (ask recipient to delete a fact from his KB)

- delete-one (ask recipient to delete one of the facts that match X)
- delete-all (ask recipient to delete all facts that match X)

Responses from recipient

- error (what you said doesn't make sense)
- sorry (I can't do what you requested; also means "no (more) answers")

Query performatives

- evaluate (evaluate an expression; details depend on language)
- reply (I am sending you data to answer your query)
- ask-if (yes-no question)
- ask-about (tell me what you know about X, reply with 1 list)
- ask-one (send me one response that matches my query)
- ask-all (send me all responses that match my query)

Multi-response query performatives

- stream-about (like ask-about, but reply with a series of messages)
- stream-all (like ask-all, but reply with a series of messages)
- eos ("end of stream" marks end of series of messages)

Effector performatives

- achieve (change things to make X true)
- unachieve (you need not make X true)
- generator performatives
- standby (get ready to give me the answers to this question)
- ready (I am ready to give you the answers)
- next (give me the next answer)
- rest (give me all the remaining answers)
- discard (you need not give me any more answers)
- generator (like standby + stream-all)

Notification performatives

- subscribe (tell me about all future changes to this data item)
- monitor (like subscribe stream-all)

Capability-definition performative

- advertise (I hereby announce that I can handle such-and-such messages)

Networking performatives

- register (I hereby announce that I can deliver messages to X)
- unregister (I retract that announcement)
- forward (I am forwarding you this message from X; reply to him through me)
- broadcast (send this to everybody you know, unless it has looped)
- pipe (establish a communication path to X)
- break (remove the pipe communication path)
- transport-address (associate a name with a non-KQML address)

Facilitation performatives

- broker-one (get somebody to process this message, send me the result)
- broker-all (get everybody to process this message who can do so)

TABLE I.7.1   KQML Performative Parameters

:sender	symbol identifying the sender
:reply-with	identifier that must appear in the reply
:in-reply-to	symbol from reply-with field of the message being answered
:content	the content of the message
:language	language in which content is expressed
:ontology	ontology (knowledge base) used by content
:force	whether the sender will ever retract (deny) this message

- recruit-one (like broker-one but have him send result to me directly)
- recruit-all (like broker-all but have them send results to me directly)
- recommend-one (find me somebody who can process this message)
- recommend-all (find me everybody who can process this message)

Each KQML performative carries parameters that provide additional information. Parameters identify the sending agent and receiving agent and provide tags for pairing up messages with their replies. The application dependent content language and ontology can also be specified in the parameters. Note that parameter *force* can be used to mark a speech act as irrevocable. The basic set of parameters is shown in Table I.7.1.

Using these primitives (acts), agents exchange messages in the following form:

```
(performative Name
      :sender       A          :receiver    B
      :content      X          :language    L
      :ontology     N          :reply-with  W
      :in-reply-to  P
)
```

This is a message from agent *A* to agent *B* in reply to a previous message identified by *P*. Any message that is sent in response to this message should include *:in-reply-to W* as indicated by *reply-with W*. The content *X* has a syntax specified by the content language *L*, whose terms are taken from ontology *N*. Ontologics define the common concepts, attributes, and relationship for different subset of world knowledge. The definition of ontology terms give meaning to expressions expressed in language *L*. The content language commonly used in a KQML message is Prolog or KIF (Knowledge Interchange Format),[34] but other languages such as LISP, SQL, and XML can also be used. The message's meaning is determined by the combination of the performative *performativeName* and the content *X*. The performative has values such as *ask-if, tell,* and *insert*. The content of these messages provide the details of what is asked, told, or inserted. A more specific example of KQML is a message representing a query about the price of a share of IBM stock as follows:

```
(ask-one
      :sender   joe
      :content (PRICE IBM ?price)
      :receiver stock-server
      :reply-with IBM-Stock
      :language LPROLOG
      :ontology NYSE-TICKS)
```

In this message, the KQML performative is *ask-one*, the content is (PRICE IBM ?price), the ontology assumed by the query is identified by the token NYSE-TICKS, the receiver of the message is to be a server identified as stock-server and the query is written in a language called LPROLOG. A similar query could be conveyed using standard Prolog as the content language and the query requests the set of all answers:

```
(ask-all
        :content ''price(IBM,[price,time])''
        :receiver stock-server
        :language standard-prolog
        :ontology NYSE-TICKS)
```

The receiving agent can respond by sending the following KQML message:

```
(tell
        :sender         stock-server
        :content        (PRICE IBM 14)
        :receiver       joe
        :in-reply-to    IBM-Stock
        :language       LPROLOG
        :ontology       NYSE-TICKS)
```

In KQML, the semantics of the KQML primitives are defined only within a highly restricted communication environment, mostly through an *advertise* performative, which states that the sender is interested in receiving messages of a certain kind.[35] The advertise performative establishes the necessary precondition that the sender of the advertise wants to receive a certain type of message. This precondition enables the receiver of the *advertise* to send, at some appropriate time in the future, a message back to the original sender. For example, the server in the above example might have earlier announced:

```
(advertise
        :ontology NYSE-TICKS
        :language LPROLOG
        :content  (monitor
                        :content (PRICE ?x ?y)))
```

Agent communication in KQML consists of three levels, as illustrated in the above examples: *content level* identified by the value of *:content* keyword; *communication level* by the value of *:reply-with, :sender*, and *:receiver* keywords; and *message level* by the performatives *:language* and *:ontology*. Notice that the KQML expression can be nested.

KQML is possibly the first widely used agent communication language, and still a work in progress. Despite the wide interests and many years' effort since 1993, there is yet no official KQML specification that agent developers can rely on. An interesting alternative is a standard agent communication language from the Foundation for Intelligent Physical Agents (FIPA), hereafter called FIPA ACL.

7.3.2   Agent Communication in FIPA ACL

Agent communication in FIPA ACL is based on a model that is pre-defined, semantically rich, and well understood by agents. Similar to KQML, FIPA agent communication is concerned at the three different levels: *communicative acts, content languages*, and *interaction protocols*.

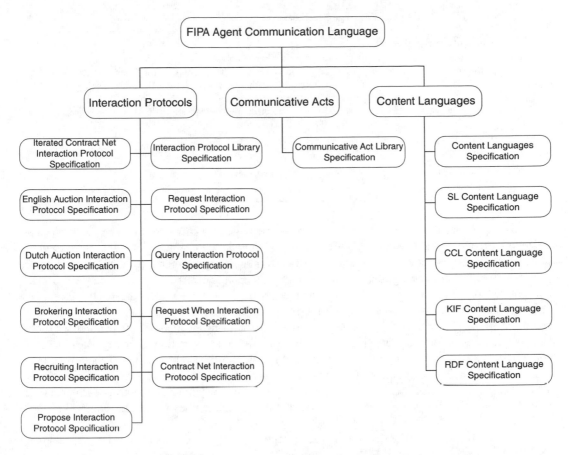

FIGURE I.7.2   FIPA agent communication language.

The basis of communication between FIPA agents is through the use of *communicative acts* (i.e., performatives in KQML) that are based on speech act theory.[36,37] FIPA specifies a number of communicative acts, such as *request, inform* and *refuse*[32] in a well-defined manner that is independent from the overall *content* of the message. The message that is supplied with a communicative act is itself wrapped in a well-specified message structure, called an agent communication language (FIPA ACL) message structure.[38] A FIPA ACL (message structure) provides mechanisms for adding context to the message *content* and the *communicative act*, such as identifying the sender and receiver, and the *ontology* and *interaction protocol* of the message. The actual content of a message is expressed in a *content language*, such as the FIPA semantic language (FIPA SL),[39] a constraint choice language (CCL),[40] KIF[41] or RDF.[42] Finally, the set of FIPA *interaction protocols* were based on work by Lux and Steiner.[43] The interaction protocols describe entire conversations between agents for the purpose of achieving some interaction or effect, such as auctioning, issuing a call for proposals, negotiating brokering services and the registration and deregistration of subscriptions. The FIPA specifications for agent communication are shown in Figure I.7.2.

7.3.3   FIPA Communicative Acts

FIPA communicative acts are superficially similar to that of KQML. The formal syntax are specified in FIPA Communicative Acts Library (CAL).[32] A brief overview of these performatives is presented in this section.

TABLE I.7.2   FIPA ACL Message Elements

Type of Message Elements	Elements
Performatives	Accept-Proposal, Agree, Cancel, cpf (Call-for-Proposal) Confirm, Disconfirm, Failure, Inform, Inform-If, Inform-Ref, Not-Understood, Propagate, Propose, Proxy, Query-If, Query-Ref, Refuse, Reject-Proposal, Request, Request-When, Request-Whenever, Subscribe
Participant in communication	sender, receiver, reply-to
Content of message	content
Description of content	language, encoding, ontology
Control of conversation	protocol, conversation-id, reply-with, in-reply-to, reply-by

An agent message contains a set of one or more message *elements* which can be categorized as: *communicative acts (performatives), participant in communication, content of message, description of content,* and *control of conversation.* The only mandatory element in an agent message is performative; others are optional. The meaning of each performative is straightforward and comprehensible from the act's name. For example, the act *accept-proposal* indicates the acceptance of a proposal that was previously submitted (typically through a *propose* act). An agent can *agree* to perform some action previously requested by another agent (using the *request* act). In other occasions, an agent can *request-when* a given precondition, expressed as a proposition, becomes true to perform some action, or an agent can use the *query-if* act to ask another agent whether (it believes that) a given proposition is true (for example, agent i asks agent j if j is registered with domain server ds). The act *request-whenever* can be used to request some action whenever some proposition becomes true (for example, *inform* me whenever the price of IBM share rises from less than 100 to more than 100).

In addition to the performatives, other types of agent message elements are used in a similar way to KQML. The complete list of FIPA communication acts is indicated in Table I.7.2.

Note that the FIPA compliant content language (as specified using the act *:content*) is defined in FIPA Content Language Library (CLL), which currently consists of FIPA SL (Semantic Language), FIPA CCL (Constraint Choice Language), FIPA KIF, and FIPA RDF (Resource Description Framework).

Furthermore, every FIPA-compliant content language must specify at least one canonical *encoding scheme* that can be used for this language within a FIPA ACL message. Most of the languages in the FIPA CLL include a *string-based encoding* scheme, but other encoding formats are allowed so that ACL messages can more efficiently represent the content of application specific data formats. The specific encoding scheme that is used in a particular FIPA ACL message must be specified in the *:language-encoding* parameter of the message.

The ontology in an ACL message is used in conjunction with the language element to support the interpretation of the content expression by the receiving agent. In many situations, the ontology will be commonly understood by the agent community, and so this message element may be omitted.

7.3.3.1   FIPA Agent Communication in Action

In this section, we use a set of examples to discuss the various aspects of FIPA agent communication. The reader will notice that the use of FIPA ACL elements in the agent message has similar syntax to KQML.

Consider a scenario where agent i informs agent j that it accepts an offer from j to stream a given multimedia title *rush-hour* to channel 30 when the customer is ready. Agent i will inform j of this fact when appropriate. In this scenario, agent *i* sends the following message, which indicates that FIPA-SL content language is used to interpret the message content.

```
(accept-proposal
   :sender (agent-identifier :name i)
   :receiver (set (agent-identifier :name j))
   :in-reply-to bid089
   :content
      ((action (agent-identifier :name j)
          (stream-content rush-hour 30))
      (B (agent-identifier :name j)
          (ready customer78)))
   :language FIPA-SL)
```

For the easy exposition, a few notes on FIPA-SL content language are in order. The content of an ACL message is expressed using FIPA SL content expressions, which can be one of three cases: (1) a *proposition*, which may be assigned a truth value in a given context. A proposition is used in the *inform* communicative act (AC) and other ACs derived from it; (2) an *action*, which can be performed. An action may be a single action or a composite action built using the sequencing and alternative operators. An action is used as a content expression when the act is *request* and other CAs derived from it; (3) an *identifying reference expression* (IRE), which identifies an object in the domain. This is the referential operator and is used in the *inform-ref* act and other CAs derived from it.

In the above example, the message content (:*content*) states that an action, which is introduced by the keyword *action*, is to be performed by agent *j*. The specified action is to "stream-content" the title "rush-hour" on channel 30 when agent *j* believes (denoted by *B*) customer78 is ready.

If agent *customer* requests agent *Pizza Hut* to deliver a box of pizza to a certain location, agent *Pizza Hut* answers that it agrees to the request, but it has low priority. In this message exchange, both agents employ the interaction protocol *fipa-request* (Figure I.7.3).

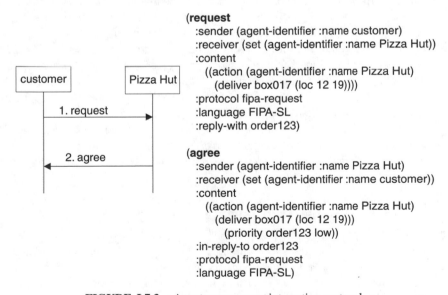

FIGURE I.7.3   Agents use *request* interaction protocol.

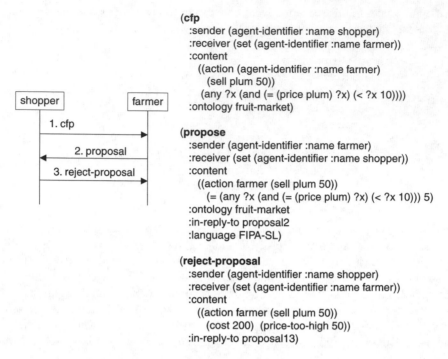

```
(cfp
  :sender (agent-identifier :name shopper)
  :receiver (set (agent-identifier :name farmer))
  :content
    ((action (agent-identifier :name farmer)
      (sell plum 50))
     (any ?x (and (= (price plum) ?x) (< ?x 10))))
  :ontology fruit-market)

(propose
  :sender (agent-identifier :name farmer)
  :receiver (set (agent-identifier :name shopper))
  :content
    ((action farmer (sell plum 50))
     (= (any ?x (and (= (price plum) ?x) (< ?x 10))) 5)
  :ontology fruit-market
  :in-reply-to proposal2
  :language FIPA-SL)

(reject-proposal
  :sender (agent-identifier :name shopper)
  :receiver (set (agent-identifier :name farmer))
  :content
    ((action farmer (sell plum 50))
     (cost 200)  (price-too-high 50))
  :in-reply-to proposal13)
```

FIGURE I.7.4   Agent negotiation using call for proposal.

Agents can use performative *cfp* (call for proposal) to initiate a negotiation process by making a call for proposals to perform the given action. The actual protocol under which the negotiation process is established is known either by prior agreement, or is explicitly stated in the *:protocol* parameter of the message. In normal usage, the agent responding to a *cfp* should answer with a proposition giving the value of the parameter in the original precondition expression. For example, the *cfp* might seek proposals for a journey from Beijing to Shanghai, with a condition that the mode of travel is by train. A compatible proposal in reply would be for the 18:30 express train. *cfp* can also be used to simply check the availability of an agent to perform some action.

A more complicated negotiation process can go through *call for proposal* by agent *i*, *propose* by agent *j*, but *reject it* by agent *i*. For example, as shown in Figure I.7.4, agent *shopper* negotiates with agent *farmer* to sell 50 boxes of plums at a certain price by sending *cfp* to *farmer* (step 1). Agent *farmer* submits the proposal to sell at the price of $5 (step 2), and agent *shopper* finds the price "too high" and so rejects the proposal (step 3).

It should be noted that agents can use appropriate performatives to acquire some knowledge about other agents' capabilities. For example, agent *traveler* asks agent *travel-service* for its available services using *query-ref*, and agent *travel-service* replies that it can reserve trains, planes, and automobiles using *inform* (Figure I.7.5).

7.3.4   FIPA Agent Message Transportation

The FIPA Agent Message Transport Specifications deal with the *delivery* and *representation* of messages across different network transport protocols, including wireline and wireless environments. At the message transport level, a message consists of a message envelope and a message body. The envelope contains specific transport requirements and information that is used by the Message Transport Service (MTS) on each agent platform to route and handle messages. The message body is the real payload and is usually expressed in FIPA ACL, but is opaque to the MTS since it may be

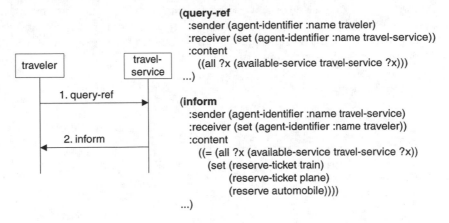

```
(query-ref
    :sender (agent-identifier :name traveler)
    :receiver (set (agent-identifier :name travel-service))
    :content
      ((all ?x (available-service travel-service ?x)))
  ...)

(inform
    :sender (agent-identifier :name travel-service)
    :receiver (set (agent-identifier :name traveler))
    :content
      ((= (all ?x (available-service travel-service ?x))
         (set (reserve-ticket train)
              (reserve-ticket plane)
              (reserve automobile))))
  ...)
```

FIGURE I.7.5　Agent updates its knowledge by querying other agents.

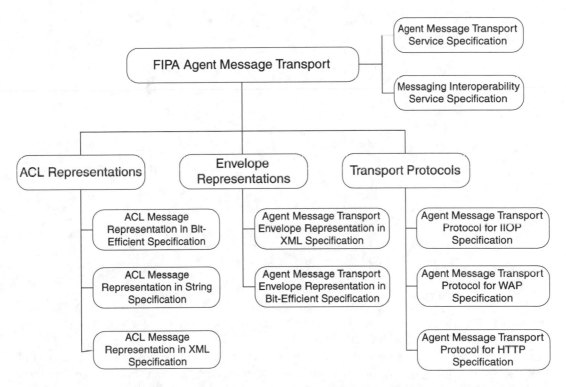

FIGURE I.7.6　FIPA agent message transport specifications.

compressed or encoded. At the representation levels, envelop representations and ACL representations are specified. Message envelope representations define the coding suitable for each transport such as an *XML encoding* for HTTP and a *bit-efficient encoding* for WAP.[44] The message envelope representation for IIOP is expressed in Interface Definition Language (IDL).[45] The coding for ACL messages can be a *string encoding*, an *XML encoding*, and a *bit-efficient encoding* (Figure I.7.6).

We would like to stress the potential significance of encoding ACL messages in XML. Using XML will facilitate the practical integration with various Web technologies. The link in XML messages can be used for specifying and sharing the ontologies used in a message and for other addressing purpose (linking *receiver* parameter to a network location that provides further information about the agent). In addition, the XML markup provides parsing information more directly. For an XML-based representation of ACL, the XML DTD is required.[46] The following is an example of using XML to encode an ACL *request* for delivering a box of pizza.

```
<?xml version="1.0"?>
<request>
    <sender>
        <agent-identifier>
            <name>customer</name>
        </agent-identifier>
    </sender>

    <receiver>
        <agent-identifier>
            <name>Pizza Hut</name>
        </agent-identifier>
    </receiver>

    <content>
        action (agent-identifier :name Pizza Hut)
            (deliver box017 (loc 12 19))
    </content>

    <protocol>
        (fipa-request)
    </protocol>

    <language>
        (FIPA-SL)
    </language>

    <reply-with>
        (order123)
    </reply-with>
</request>
```

The ACL message transport protocols can be the Internet Inter-Orb Protocol (IIOP),[27,28] the Wireless Application Protocols (WAP),[44] or HTTP over Web.

The messages transport protocol between agents using IIOP is based on the transfer of an OMG IDL structure containing the message envelope and an octet sequence representing the ACL message body. The envelope and the message body are transferred together within a single IIOP one-way invocation. Once the request has been received, the message envelope is used by the Agent Communication Channel (ACC) to obtain the instructions and information needed to correctly handle the message body. On the agent platform, the ACC provides the message transport service. In performing message transfer tasks, an ACC is required only to read the message envelope; the message transport behavior of an ACC is determined by the instructions expressed in the message envelope. For example, once a message has arrived at ACC, which can directly deliver it to the agent or agents named in the *:intended-receiver* parameter of the message envelope, this ACC should pass the message to the agents concerned.

The message transport protocol based on WAP transfers the entire agent message (including the message envelope) in a WAP message. Once the message has been received, the message envelope is parsed by the ACC and the message is handled according to the instructions and information given in the message envelope. The protocol requires that the transport addresses given must be complete, e.g., the complete URL such as *wap://foo.com/acc* for a WAP phone or a *http://bar.com/acc* for a WAP content server in a wireline network. In addition, the WAP content type for any data transfer must be set to *x-application/fipa-message*. There are two modes of interaction between wireless client devices and hosts in a wireline network: through a *WAP gateway* and to a *WAP server*; the agent message transport protocol does not distinguish between these. However, it should be noted that these two modes lead to different combinations of interfaces for the wireless and wireline environment hosts.

The agent message transport protocol using Web protocol HTTP transfers the entire agent message including the message envelope in a HTTP *request*. The HTTP uses the *request-response model*, and data transfer is a two-step process: the sender makes a HTTP request, and after receiving the data the receiver sends a HTTP response. The receiver then parses the message envelope and the message is handled according to the instructions and information given in the message envelope.

The agent communication requires the following system support:[29]

- A set of middleware (platform) services that assist agents with *naming, registration*, and basic *facilitation services* for finding other agents with required capability.
- A suite of API's that facilitate the composition, sending, and receiving of ACL messages.
- For a performative or communicative act that takes the action(s) prescribed by the semantics for the particular application, the ready-to-use code is available; this code will depend on the application language, the domain, and the details of the agent system using the ACL.

In the next section, we describe one of the middleware services that is used for agent registration.

7.4   Agent Naming

In a multi-agent system, an agent has to be named. The agent name is an external identity, which is normally human readable string. There is a corresponding internal agent identity, *agent-IDs*, which is a globally unique identifier (GUID). In a multi-agent system implemented in object-oriented environments (such as Java, CORBA), the agent-ID is represented as object reference (JaveObject or CORBA Interoperable Object Reference, IOR). Agent-ID issuing should occur in a way that tends to ensure global uniqueness across both space and time, with respect to the space of all GUIDs. There is a well-established algorithm for generating a GUID of 16 octets (128 bits),[47] which guarantees the uniqueness until the year 3400. In CORBA, the IOR contains information that is used for communication. The standard IOR format is flexible enough to support almost any inter-ORB protocol imaginable. For IIOP, an IOR contains a host name, a TCP/IP port number, and an object key that identifies the target object at the given host name and port combination.[18]

A useful role of naming an agent is to support the use of *belief desire intention* models within a multi-agent system. The agent name can be used to correlate propositional attitudes with the particular agents that are believed to hold those attitudes. Agents may also have "well-known" names by which they are popularly known. These names are often used to commonly identify an agent.[48]

The agent name service maps names to agent-IDs (e.g., object references). A name-to-reference association is called a *name binding*. The agent name service is a white-page service. The name service returns an agent object reference if supplying a registered agent name. The more sophisticated functionality for finding and locating an agent is provided by the agent directory service (or facilitator) as described in the next section.

7.5   Agent Directory Service

As indicated previously, agent communication is expected to be supported by middleware services which make it possible for agents to locate one another, find out each other's capability, and speak an agent communication language that both understand in distributed environments. The *agent directory service* provides a yellow-pages service to agents. As defined in FIPA00023,[49] an agent directory service is the trusted, benign custodian of the agent directory. It is trusted in the sense that it must strive to maintain an accurate, complete, and timely list of agents. It is benign in the sense that it must provide the most current information about agents in its directory on a nondiscriminatory basis to all authorized agents. In the agent community, the agent directory service is commonly called the directory facilitator (DF); in this chapter, both terminologies are used interchangeably.

In order to access the directory of agent descriptions managed by the directory service, each directory service must be able to perform at the least the following functions: *register, deregister, modify*, and *search*.

Every agent that wishes to publicize its services to other agents should find an appropriate agent directory and request for the *registration* of its agent description. Since the directory service as a middleware service does not place any restrictions on the information that can be registered with it, this advertisement does not imply future commitment or obligation. For example, an agent can refuse a request for a service which is advertised through a directory service. Additionally, the directory service cannot guarantee the validity or accuracy of the information that has been registered with it, nor can it control the life cycle of any agent.

The *deregistration* function has the consequence that there is no longer a commitment on behalf of the directory service to broker information relating to that agent. At any time, and for any reason, the agent may request the directory service to *modify* its agent description.

An agent may *search* for an agent that possesses the certain capability or properties from a directory service. As indicated above, the directory service does not guarantee the validity of the information provided in response to a search request.

The agent directory service is provided by the directory servers which operate in a request-response model. Directory clients submit service requests to directory servers and the directory servers handle the requests and provide responses to the directory clients. The implementation of agent directory service (DF) can be based on well-established Internet directory protocols such as the Lightweight Directory Access Protocol (LDAP).[50] The LDAP protocol operations for client request include: *bind, unbind, search, modify*, and *delete*. The protocol operations for server response include: *bind, searchResultEntry, searchResultReference, modifyResponse, deleteResponse*, and

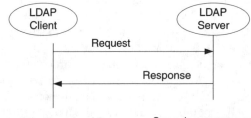

Client's request can be one of these:

> *Bind, Unbind, Search, Add,*
> *Modify, Delete, Compare,*
> *Abandon, Extended Request*

Server's response can be one of these:

> *Bind, SesearchResultEntry,*
> *SearchResultReference,*
> *SearchDone, Add, Modify,*
> *Delete, Compare,*
> *Extended Response.*

FIGURE I.7.7   The LDAP client server model.

CompareResponse (Figure I.7.7). For the purposes of LDAP protocol exchanges, all protocol operations are encapsulated in a common envelope, the *LDAPMessage*, which is defined in ASN.1 as follows:

```
LDAPMessage ::= SEQUENCE {
    messageID    MessageID,
    protocolOp   CHOICE {
                         bindRequest      BindRequest,
                         bindResponse     BindResponse,
                         unbindRequest    UnbindRequest,
                         searchRequest    SearchRequest,
                         searchResEntry   SearchResultEntry,
                         searchResDone    SearchResultDone,
                         searchResRef     SearchResultReference,
                         modifyRequest    ModifyRequest,
                         modifyResponse   ModifyResponse,
                         addRequest       AddRequest,
                         addResponse      AddResponse,
                         delRequest       DelRequest,
                         delResponse      DelResponse,
                         modDNRequest     ModifyDNRequest,
                         modDNResponse    ModifyDNResponse,
                         compareRequest   CompareRequest,
                         comparcResponse  CompareResponse,
                         abandonRequest   AbandonRequest,
                         extendedReq      ExtendedRequest,
                         extendedResp     ExtendedResponse },
    controls        [0] Controls OPTIONAL }
```

The directory service available in the agent platform will provide a set of APIs that agents can use for registering and locating other agents. For example, in an Java Agent Service environment,[51] Java API for LDAP-based agent directory service is shown in Table I.7.3.

Traditionally, LDAP directories have been used to store data. Users and programmers think of the directory as a hierarchy of directory *entries*, each containing a set of *attributes*. You look up an entry from the directory and extract the attribute(s) of interest. For example, you can look up a person's telephone number from the directory. Alternatively, you can search the directory for entries with a particular set of attributes. For example, you can search for all persons in the directory with the surname "Clinton," or search for agents that *know* about IBM share price.

For agents written and executed in the distributed objects environments, such as Java platform and CORBA, a kind of data that is typically shared are Java/CORBA objects themselves. In this context, the directory can be used as a repository for Java objects[52] or CORBA object references.[53] The directory provides a centrally administered and possibly replicated service for use by object-based agent applications distributed across the network.

A Java object is stored in the LDAP directory by using the object class *javaObject*. Three types of Java objects are stored in the LDAP directory: a Java serialized object, a Java marshalled object, and a JNDI reference. A Java remote object is stored as either a Java marshalled object or a JNDI reference. The *javaObject* is the base class from which other Java object-related classes derive: *javaSerialized-Object, javaMarshalledObject*, and *javaNamingReference*. More specifically, the javaObject is an abstract object class, which means that a javaObject cannot exist by itself in the directory; only auxiliary or structural subclasses of it can exist in the directory.

When agents are developed and deployed in CORBA environments with a LDAP-based directory, CORBA objects-based agent applications use the LDAP directory as a repository for CORBA object

TABLE I.7.3   LDAP as Agent Directory Service

	Java Agent Service LDAP APIs	
Classes	Methods	Remarks
LdapDirectory	void register(DirEntry entry)	Registers the entry and returns an id of the registered entry
	Enumeration search(DirEntry pattern)	Searches entries that satisfy the specified pattern
	void deregister (String id)	Deregisters an entry bound to the specified id
	DirEntry getNewEntry()	Creates a new DirEntry instance
LdapDirEntry	Object getAttribute(String name)	
	Name getName()	Gets the name of the entry
	Object getObject()	
	void setAttribute(String name, Object obj)	
	void setName(Name name)	Sets the name of the entry
	void setObject(Object obj)	
LdapDirectoryFactory	LdapDirectoryFactory()	Constructor
	Directory getInstance()	Creates a new directory instance
	Directory getInstance(Hashtable hash)	Stores additional information into hashtable

references. Thus, an agent (developed in CORBA) may use the directory for "registering" CORBA objects representing the services and capabilities that it manages or possesses, so that other agents can later search the directory to locate those services and agents as they need.

7.6   Middleware Support for Mobile Agents

The environment of mobile computing imposes different requirements for supporting agent applications from traditional distributed (stationary) systems in many aspects: bandwidth, latency, delay, error rate, interference, interoperability, computing power, quality of display, etc. An agent middleware is expected to provide the functionality and services to allow the agent application to adapt the mobile environments. In this section, we discuss the agent middleware support based on the recent FIPA specification.[54]

7.6.1   Monitoring and Controlling Quality of Service

The services and functionality required for agent middleware are categorized into monitoring and controlling support of an agent middleware, and often realized by middleware entities *monitor agent (MA)* and *control agent (CA)*. The MAs and CAs are responsible for:

- Providing information about expected performance
- Observing and measuring the quality of service of message transport protocol (MTP) and message transport connection (MTC)

- Collecting information from the observing and measuring sources
- Analyzing the information
- Controlling an MTC and selecting an MTP

In a heterogeneous mobile agent environment, the agent middleware can use the following approach to providing these functionalities, especially to determining the MTP, message representation, and content language to be used between communicating agents:

- Communicating agents *know* a peer agent's preferences of the protocol, representation, and content language beforehand and use them.
- The activating agent *tries* to use a method and if the peer agent is not capable of using the suggested method, then the activating entity may try another one (and so on).
- The communicating agents reach agreement through *negotiation*.

In the stationary agent environments, when agents interact or communicate with each other, it is assumed that the message transport connection has been established before the agent message exchange and that it is reliable. However, this is not always the case within a mobile environment. The common approach in this case is negotiation, which can be used for reaching agreement on the use of message transport protocols and message representations, as illustrated below.

An agent in a one-agent platform can propose a message transport protocol (MTP) to its middleware entity (e.g., *control agent*). It is the responsibility of the CA to either accept or reject the proposal, based on whether it is possible to use the proposed MTP. If accepted, CA negotiates with peer middleware CAs to use proposed MTPs. The whole negotiation process is shown in Figure I.7.8. Suppose that an application agent issues a request to the CA to use the WAP transport protocol (step 1). Based on the internal state of the CA (i.e., the CA knows whether a requested MTP can be activated or not), the CA agrees to activate an MTP and negotiate with its peers (step 2). This is also the point where the CA can reject the request for some reasons, e.g., the middleware cannot support

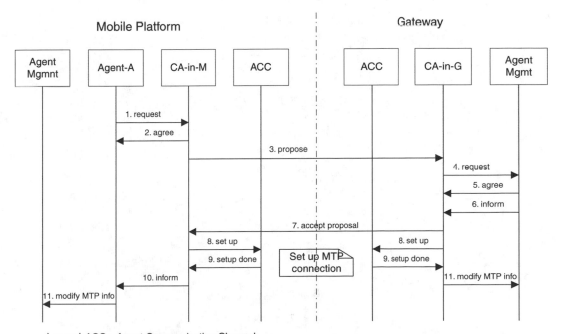

Legend: ACC---Agent Communication Channel

FIGURE I.7.8   Message transport protocol negotiation.

the requested transport protocol. Now suppose that the CA in the mobile host (*CA-in-M*) agrees to propose to its peer CA in the gateway host (*CA-in-G*) that the WAP transport protocol should be used in communication between the two agent platforms (step 3). The CA in the gateway host requests from the local agent middleware to determine whether the WAP MTPs are supported (step 4). The local middleware service agrees and informs the CA that the requested MTP is supported, and the CA decides to use the protocol WAP based on the current quality of service requirements of the message transport connection (steps 5, 6). Thus, the CA in the gateway host accepts the proposal to use the protocol WAP and sends the response to the CA in the mobile host, informing it of the accepted MTP (step 7). The CAs request their respective middleware entity, *agent communication channels* (ACCs), to set up the protocol WAP (step 8). The ACCs inform their respective CAs that the protocol WAP has been established between the mobile host and the gateway host (step 9). At this point, the CA can inform the application agent that the message transport connection (MTC) is established, and the middleware service modifies the related agent platform description to show that the WAP is now active and in use (step 11).

In other cases, a CA can activate the selection of an MTP without the application agent's request. In other words, the CAs can, without the application agent's request, initiate the negotiation with peers to obtain the knowledge of the transport protocols supported.

The negotiation messages use a chosen ACL, and the message envelope can be encoded using a chosen representation language, e.g., as an XML message to transport messages over a Message Transport Protocol. The following is an example of the XML encoded envelope used for the negotiation:

```xml
<?xml version="1.0"?>
<envelope>
  <params index="1">
    <to>
      <agent-identifier>
        <name>CA-in-G@ntu.edu.sg</name>
      </agent-identifier>
    </to>
    <from>
      <agent-identifier>
        <name>CA-in-M@singnet.com</name>
      </agent-identifier>
    </from>
    <acl-representation>fipa.acl.rep.string.std</acl-represen
      tation>
    <date>20010928T100900000</date>
  </params>
</envelope>
```

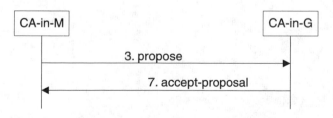

FIGURE I.7.9   ACL messages of negotiation.

During the negotiation, the control agent *CA-in-M* in the mobile environment sends out the *propose* message on the behalf of *Agent-A* (Figure I.7.9). The *propose* message is encoded in FIPA ACL as follows:

```
(propose
  :sender
    (agent-identifier
      :name CA-in-M@singnet.com)
  :receiver (set
    (agent-identifier
      :name CA-in-G@ntu.edu.sg))
  :ontology FIPA-Nomadic-Application
  :language FIPA-SL0
  :protocol FIPA-Propose
  :content
    ((action
      (agent-identifier
        :name CA-in-M@singnet.com)
      (use
        (transports
          :send (sequence
            (transport-protocol
              :name fipa.mts.mtp.wap.std))
          :recv (sequence
            (transport-protocol
              :name fipa.mts.mtp.wap.std))))
    true)))
```

It states the *sender* (*CA-in-M@singnet.com*) and *receiver* (*CA-in-G@ntu.edu.sg*). It also indicates the use of FIPA-Nomadic-Application as its ontology and FIPA-Propose as its interaction protocol. The message content proposes the transport protocol (WAP) to be used for both "send" and "receive" application agent communication. The accepted transport protocol is the FIPA WAP protocol as confirmed in the following *accept-proposal* message:

```
(accept-proposal
  :sender
    (agent-identifier
      :name CA-in-G@ntu.edu.sg)
  :receiver (set
    (agent-identifier
      :name CA-in-M@singnet.com))
  :ontology FIPA-Nomadic-Application
  :language FIPA-SL0
  :protocol FIPA-Propose
  :content
    (action
      (agent-identifier
        :name CA-in-M@singnet.com)
      (use
        (transports
          :send (sequence
            (transport-protocol
              :name fipa.mts.mtp.wap.std))
```

```
                         :recv (sequence
                            (transport-protocol
                               :name fipa.mts.mtp.wap.std))))
               true))
```

Similar support is provided by the middleware service to negotiate message representations. For example, an application agent in a mobile host (*Agent-A-in-M*) negotiates with its peer application agent in a fixed host (*Agent-B-in-F*) by proposing the use of the *FIPA ACL Message Representation in Bit-Efficient format* (fipa.acl.rep.bitefficient.std)[55] for their communication (Figure I.7.10).

The message exchanges for negotiating message representation also use the communication acts *propose* and *accept-proposal* as follows. First, an agent *proposes*:

```
(propose
  :sender
    (agent-identifier
      :name Agent-A-in-M@singnet.com)
  :receiver (set
    (agent-identifier
      :name Agent-B-in-F@fixed.com))
  :ontology FIPA-Message-Representation
  :language FIPA-SL0
  :protocol FIPA-Propose
  :content
    ((action
      (agent-identifier
        :name Agent-A-in-M@singnet.com)
      (use
        (msg-rep-selection
          :send (sequence
            (msg-representation
              :name fipa.acl.rep.bitefficient.std))
          :recv (sequence
            (msg-representation
              :name fipa.acl.rep.bitefficient.std)))))
      true))
```

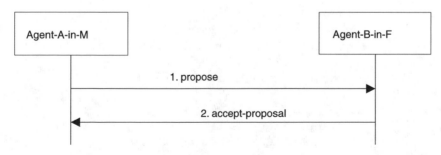

FIGURE I.7.10 Negotiation of message representation.

and the counterpart agent then accepts the proposal by sending the following *accept-proposal* message:

```
(accept-proposal
  :sender
    (agent-identifier
      :name Agent-B-in-F@fixed.com)
  :receiver (set
    (agent-identifier
      :name Agent-A-in-M@singnet.com))
  :ontology FIPA-Message-Representation
  :language FIPA-SL0
  :protocol FIPA-Propose
  :content
    (action
      (agent-identifier
        :name Agent-A-in-M@singnet.com)
      (use
        (msg-rep-selection
          :send (sequence
            (msg-representation
              :name fipa.acl.rep.bitefficient.std))
          :recv (sequence
            (msg-representation
              :name fipa.acl.rep.bitefficient.std))))
      true))
```

7.6.2 Message Exchange over a WAP Message Transport Protocol

We have described in the last section how the application agents agree through negotiation to use the WAP protocol for their communication. Now we describe the message exchange details over a WAP Message Transport Protocol (WAP MTP) to illustrate the necessary support from the agent middleware and platform. When using Wireless Application Protocol to transport messages between agents, there are two modes of interaction between wireless client devices and hosts in a wireline network: through a WAP gateway, and to a WAP server. Messages between the mobile host and gateway host are delivered mainly using the FIPA WAP protocol (denoted as fipa.mts.mtp.wap.std), and messages between gateway host and other agent platforms in the fixed network are delivered using the FIPA IIOP protocol (denoted as fipa.mts.mtp.iiop.std) or HTTP protocol.

In the gateway-based mobile agent applications, three specific agent platforms (agent middleware) are involved in agent interactions: one running in a mobile host, one running in a gateway host, and the last one running in a host, situated in a fixed network which represents the rest of the network. Figure I.7.11 shows the agent message exchanges using WAP and IIOP. In this setting, the agent in a mobile host has a WAP address and the agent in a stationary host has an IIOP address. In order to be reachable from an agent platform (AP) operating in a fixed network environment, an agent in the mobile host must register with the AP running in the gateway host. Subsequently, the ACC in the gateway host AP can forward messages intended for the agent operating in the mobile host to the ACC in the mobile platform. This registration can be achieved by message exchange between the agent (in mobile host) and gateway agent middleware using the acts *request, agree,* and *inform* (steps 1–3, Figure I.7.12).

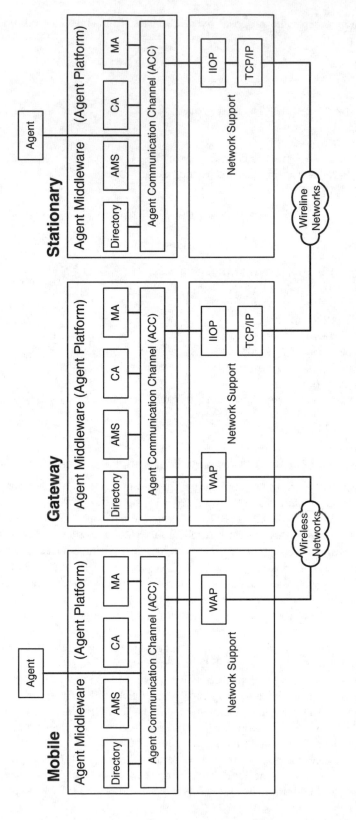

FIGURE I.7.11   Mobile agent communication using WAP and gateway.

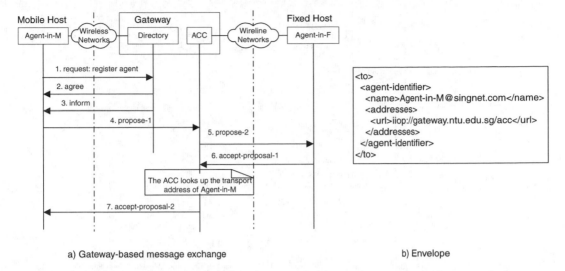

FIGURE I.7.12   Gateway-based message exchange in a mobile environment.

After successfully completing the registration (i.e., *Agent-in-M* received *inform* from the gateway directory service), *Agent-in-M* can be reached via the gateway host using the *to* envelope parameter as in Figure I.7.12b). The ACC in the mobile host forwards the message to the ACC in the gateway host using fipa.mts.mtp.wap.std MTP as previously negotiated. In the *from* envelope parameter, *Agent-in-M* can inform *Agent-in-F,* using *propose*, that its primary return address is its address in the gateway host (step 4). This proposal is forwarded by the ACC in the gateway host to *Agent-in-F*, also using fipa.mts.mtp.iiop.std MTP (step 5). *Agent-in-F* accepts *Agent-in-M*'s proposal by sending an *accept-proposal* message to *Agent-in-M* using its gateway host address (step 6). This *accept-proposal* message is again forwarded by the ACC in the gateway host to the ACC in the mobile host using the fipa.mts.mtp.wap.std MTP, which is agreed upon by both parties in the previous negotiation.

7.6.3   Mobile Agent Handoff Support

It is often required that agent middleware provides a seamless mobility support by maintaining the connectivity while agents move from one wireless cell to another. At the higher level, the maintenance of connectivity is achieved by the *handoff* control. The handoff control uses the monitor agents (MAs) to monitor the quality of the communication service and the control agents (CAs) to manage the establishment, teardown, suspension, and activation of the connection between the communicating application agents. The MA informs application agents about the status and changes of the network services. For the following discussion, we assume that the fixed host is connected to the ATM LAN. Thus, when the mobile host is connected to the wireless LAN, the IIOP-based message transport protocol can be used directly between the agents. When the mobile host is connected to the wireless WAN (e.g., GPRS), all agent message communication takes place through the gateway host. The WAP-based message transport protocol is primarily used between the mobile agent platform and the gateway platform.

When the monitor agent in a mobile host (MA-in-M) senses the disconnection, it *informs* the control agent (CA-in-M). Of course, the monitor agent in a fixed agent platform (MA-in-F) also detects the disconnected communication channel, and sends out the notification (Figure I.7.13, step 1, 1'). Receiving the notification of broken connection, the control agent in the fixed agent platform (CA-in-F) requests the control agent in the gateway host (CA-in-G) to open a wireless wide-area message transport connection (MTC) (step 2). CA-in-G *agrees* that it will try to open the

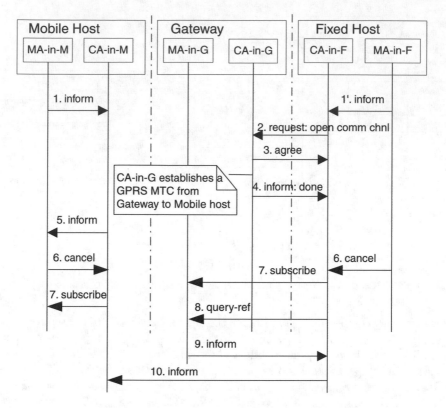

FIGURE I.7.13   Handoff control to maintain connectivity.

GPRS connection (step 3), and after successful establishment, CA-in-G informs the CA-in-F of the outcome (step 4). CA-in-M *informs* the MA-in-M that a new MTC has been established. At this point, both monitor agents in the fixed and mobile host have to *cancel* the subscription about the previous MTC (step 6), and subscribe for the new GPRS connections (step 7). After establishing a new connection, the agents can inquire about the quality of service by exchanging appropriate messages (steps 8–10). The handoff protocol as shown in Figure I.7.13 illustrates the seamless roaming of a mobile host (and its agent) from the wireless LAN to the UMTS network. All the message exchanges use agent communication language and its communicative acts.

7.7 Agent Security

Agent systems may be vulnerable to security threats due to weaknesses in the communications infrastructure and programming languages. The best support for security at the programming language level is Java.[56] This chapter is concerned with agent middleware, and thus we focus on security threats in communications, which include:[21]

- Denial of service, which is a reduction of the availability of an agent or agent system to legitimate users.
- Unauthorized access or use that occurs when an unauthorized person or computer program invokes operations of an agent or agent system.
- Unauthorized modification or corruption of data that occurs when an agent's or agent system's data is altered or destroyed, or false data is added.

An attacker can use a number of techniques to attack agent and agent system communications. Some possible attack techniques are:

- Spamming, which is flooding a service with illegitimate (or even legitimate) requests.
- Spoofing or masquerade, which is an agent or agent system falsifying its identity to get access to information and services.
- Trojan horse, which is an agent or agent system posing as a legitimate agent or agent system that can potentially receive private information from unsuspecting clients.
- Replay, which is recording and replaying a communications session. If a replay attack is not detected, and the repeated operations are cumulative, the attack can have a disastrous effect on the provided service.
- Eavesdropping, which is monitoring communications to obtain private information.

There are various strategies for countering threats and attacks. To ensure that agents act responsibly, security policies are defined. These policies are a set of rules used to govern an agent's activities. The security service provided in an agent middleware enforces the rules. The particular policy to enforce is determined based on the authenticity of the communicating parties' credentials, agent class, agent authority, and/or other factors. Security policies are used for restricting or granting agent capabilities, for setting agent resource consumption limits, and for restricting or granting access.

The agent platform should allow the quality of network communication security to be specified when an agent invokes an operation or wishes to travel (for mobile agents). The quality of network communication security refers to:

- Confidentiality—information is disclosed only to users authorized to access it. It is concerned with to what extent the communications channel is secure from eavesdropping, and the strength of the encryption.
- Integrity—there is no corruption or unauthorized modification of data during network communications, and information is modified only by users who have the right to do so, and only in authorized ways. It is transferred only between intended users and in intended ways.
- Authentication—communication takes place only with agent systems that it has authenticated.
- Replay detection—replay detection algorithm is in place to prevent duplication of the agent during a communications session.
- Accountability—users are accountable for their security-relevant actions. A particular case of this is nonrepudiation, where responsibility for an action cannot be denied.

Agent middleware provides the security service that protects an agent system from unauthorized attempts to access information or interfere with its operation, and meets the agent's requirement for secure mobile agent communications. Noticeably, the CORBA middleware security service[57] provides adequate service for secure agent communication, including mutual authentication of agent systems, agent system access to authentication results and credentials, agent authentication and delegation, setting and control of security policies for agent and agent system, and honoring agent's request for integrity, confidentiality, replay detection, and authentication. However, as far as the agent middleware is concerned, the security service is still a very weak area in multi-agent systems development.

7.8 Agent Management Service

An agent management service (AMS) is an important service provided by the agent middleware and agent platform (AP), but it is also an area that has received much less attention. The AMS is

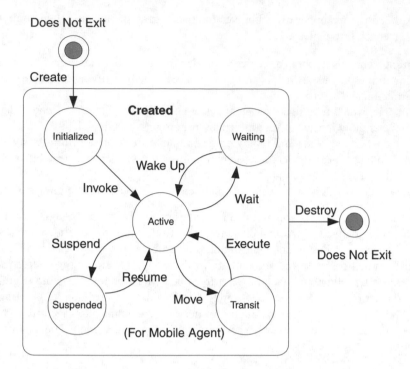

FIGURE I.7.14   Agent life cycle.

responsible for managing the life cycle of agents executed on the platform, such as the creation of agents, the deletion of agents, and, if agent mobility is supported by the agent middleware, overseeing the migration of agents to and from the AP. A life cycle is associated with each agent on the AP, which is maintained by the AMS. An agent comes into existence when it is created and instantiated, and then it can enter one of the states (Figure I.7.14): *active, suspended,* or *waiting*. For the mobile agent, the agent can also enter in *transit* state when the agent is moving from one agent platform to another.[49] When the agent is in the active state, the transport system delivers messages to the agent as normal. When the agent is in *initiated/waiting/suspended* state, the transport system either buffers messages until the agent returns to the active state, or forwards messages to a new location (if a *forward* is set for the agent).

In a sense, the agent management service represents the managing authority of an AP. An AMS can request that an agent performs a specific management function, i.e., terminate all execution on its AP, and has the authority to forcibly enforce the function if such a request is ignored.

The AMS maintains the knowledge about all the agents that currently reside on its platform. Residency of an agent on the agent platform implies that the agent has been registered with the agent management. Each agent must register with the agent management when it is created in order to be managed. Registration with the agent management grants the agent permission to access the platform's message transport service in order to send or receive messages.

For the mobile agent which travels from one platform (one transport system) to another, the message delivery is more complicated. When a mobile agent travels, it enters into the *transit* state. The transport system either buffers messages until the agent becomes active (e.g., the agent was successfully started on the destination agent platform), or forwards messages to a new location (if a *forward* is set for the agent).

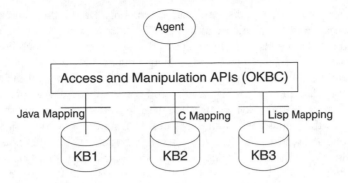

FIGURE I.7.15   Interoperable KB access and manipulation.

In a sophisticated multi-agent system, there might be different agent middleware and platforms that have different capabilities, e.g., between a mobile agent platform and a stationary agent platform. Thus, it is necessary for the AMS to be queried to obtain a description of its AP.

7.9   Supporting Knowledge-Base Interoperability

In the multi-agent systems, each agent platform may have different knowledge bases (KBs) and knowledge representation systems (KRS). These platforms even support different ontologies. It is often desirable that the multi-agent system middleware provides an interoperability support among these knowledge bases. The role of common middleware KB access and manipulation is to shield the difference in KBs and KRS, and to create a generic access and manipulation API layer as an interoperable knowledge-base service, thus allowing application agents to use more than one KB/KRS during the life of an application. The rational of common knowledge base access layer is that although there are significant differences among KRS implementations, there are enough common properties that we can describe a common knowledge model and an API for KRSs. This is the approach taken by Open Knowledge Base Connectivity (OKBC).[58] The common API layer allows an application some independence from the idiosyncrasies of specific KRS software, and enables the applications that operate on many KRSs to be interoperable and portable.

An application written to use these API operations has the potential of portability over a variety of KRSs and knowledge bases. This common knowledge base access API supports operations that are efficiently supported by most KBs and KRSs. The API is language-independent, and there is language binding for a specific implementation language, such as Java, Lisp, or C, to access and manipulate the knowledge bases implemented in these languages. This is shown in Figure I.7.15.

7.10   Conclusion

Despite the compelling vision of ubiquitous multi-agent system technology and the significant progress in the field of intelligent agents and multi-agent systems, the widespread development and deployment of MASs will not occur until the stable, widely accessible MAS middleware takes hold that provides a sophisticated support for many commonly required services and facilities. In this chapter, we examined the desired MAS middleware supports by drawing on the distributed objects middleware and standardization efforts of agent technology.

The cooperation and coordination via agent communication are the essential features of any reasonable MAS. At the core of agent middleware is the support for agent communication in terms of ACL, content representations, and transportation protocols. Before the communication can take place, agents need support for naming and locating or finding each other. The agent directory is the facilitator that enables agents to find peers that possess the required capabilities and properties. The middleware support for agent mobility allows agents to cope with various issues encountered in the dynamic mobile environments; the middleware entities monitor and maintain the communication connectivity and the quality of service, and carry out the negotiation for the (wireless) transportation protocols and representations when the application agents request it. In addition, the middleware handles the handoff when an agent moves from one environment to another. Agent management service and security are the areas that receive much less attention; our discussions have also merely touched the surface. The middleware support for the agent's knowledge and intelligence is another very weak area. The important issue is that middleware needs to support the interoperable knowledge bases and knowledge representation systems, while allowing different knowledge models and knowledge representation to co-exist in a reasonably scaled multi-agent system.

Over the past few years, we have seen the emergence of a multitude of multi-agent systems and other intelligent agent systems built around the standard infrastructure or middleware specifications. Among these, the most noticeable are JADE (Java Agent DEvelopment Framework), claimed to be FIPA-compliant[59] and Java Agent Service.[51] The reader is referred to Labrov et al.[29] for the more detailed survey.

References

1. Jennings, N.R., Sycara, K.P., and Wooldridge, M., A roadmap of agent research and development, *J. Auton. Agents and Multi-Agent Syst.,* 1998, 1, 7–36.
2. Gasser, L., MAS Infrastructure definitions, needs, and prospects, in *Infrastructure for Agents, Multi-Agent Systems, and Scalable Multi-Agent Systems*, Wagner, T. and Rana, O., Eds., LNCS 1887, Springer-Verlag, 2001.
3. Schantz, R.E. and Schmidt, D.C., Middleware for Distributed Systems: Evolving the common structure for network-centric applications, in *Encyc. Software Engineer.*, Marciniak, J. and Telecki, G., Eds., John Wiley & Sons Ltd., 2002.
4. Bernstein, P.A., Middleware: A model for distributed system services, *Commun. ACM*, 1996, 39, 86–98.
5. Star, S.L. and Ruhleder, K., Steps towards an ecology of infrastructure: Design and access for large information spaces, *Inf. Syst. Res.,* 1996, 7, 111–134.
6. Object Management Group, The Common Object Request Broker: Architecture and Specification (Revision 2.4.1), OMG Technical Document formal/00–11–07, 2000.
7. Sun Microsystems, Java 2 Platform Enterprise Edition Specification V 1.2 Sun Microsystems, 1999.
8. Microsoft, Overview of .Net Framework, 2000.
9. Sun Microsystems (White Paper), Java Remote Method Invocation—Distributed Computing for Java, Sun Microsystems Inc., 1999.
10. Sessions, R., *COM and DCOM: Microsoft's Vision for Distributed Objects*, John Wiley & Sons. Inc., New York, 1998.
11. W3C, Simple Object Access Protocol (SOAP) 1.1, W3C Notes 08, 2000.
12. W3C, Extensible Markup Language (XML) 1.0, Bray, T., Paoli, J., Sperberg-McQueen, C.M., and Maler, E. (Eds.), W3C Recommendation, 2nd ed., 2000.
13. Snell, J., MacLeod, K., and Kulchenko, P., *Programming Web Applications with SOAP*, O'Reilly & Associates, New York, 2001.
14. Fielding, R. et al., Hypertext Transfer Protocol—HTTP/1.1, IETF RFC 2616, 1999.

15. Postel, J., Simple Mail Transfer Protocol, IETF RFC 821, 1982.
16. Freed, N. and Borenstein, N., Multipurpose Internet Mail Extensions (MIME) Part One: Format of Internet Message Bodies, IETF RFC 2045, 1996.
17. Birrel, A.D. and Nelson, B.J., Implementing remote procedure calls, *ACM Trans. Comp. Syst.*, 1984, 2, 39–59.
18. Object Management Group, CORBAservices: Common Object Service Specification, OMG Technical Document formal/98–12–09, 1998.
19. Sun Microsystems, Enterprise JavaBeans Specification V.2.0 (Final Draft), Sun Microsystems, Inc., 2000.
20. Lam, H. and Thai, T.L., *.Net Framework Essentials*, 1st ed., O'Reilly & Associates, Combridge, 2001.
21. Object Management Group, Mobile Agent Facility Specification, Object Management Group, OMG Document formal/00–01–02, 2000.
22. Object Management Group, CORBA Component—Volume I (Joint Revised Submission), OMG Document orbos/99–07–01, 1999.
23. Meyer, B., .NET Is Coming, *IEEE Computer*, 2001, 34, 92–97.
24. IEEE Standard, IEEE Recommended Practice for Architectural Description of Software Incentive Systems, The Institute of Electrical and Electronics Engineers, Inc., 2000.
25. Maier, M.W., Emery, D., and Hilliard, R., Software architecture: Introducing IEEE standard 1471, *IEEE Computer*, 2001, 34, 107–109.
26. Genesereth, M.R. and Ketchpel, S.P., Software agents, *Commun. Assoc. Comput. Machin.*, 1994, 37, 48–53.
27. Object Management Group, General inter-ORB protocol, in *The Common Object Request Broker: Architecture and Specification (Revision 2.4.2)*, Object Management Group Document formal/ 01–02–01, 2001.
28. Ruh, W., Herron, T., and Klinker, P., *IIOP Complete: Understanding CORBA and Middleware Interoperability*, Addison-Wesley, Reading, MA, 2000.
29. Labrou, Y., Finin, T., and Peng, Y., Agent communication languages: The current landscape, *Intell. Syst.*, 1999, 14, 45–52.
30. Bach, K. and Harnish, R. M., *Linguistic Communications and Speech Acts*, MIT Press, Cambridge, MA, 1979.
31. Finin, T. et al., Draft Specification of the KQML Agent Communication Language, External Interfaces Working Group, The DARPA Knowledge Sharing Initiative, 1993.
32. FIPA00037, FIPA Communicative Act Library Specification, Foundation for Intelligent Physical Agents, 2001.
33. Finin, T. et al., KQML as an agent communication language, in *Proc. 3rd Int. Conf. Inf. Knowledge Manage. (CIKM'94)*, ACM Press, 1994.
34. Genesereth, M.R. and Fikes, R.E., Knowledge Interchange Format Version 3.0 Reference Manual, Logic Group, Computer Science Department, Stanford University, 1992.
35. Labrou, Y., *Semantics for an Agent Communication Language*, Thesis, University of Maryland, Baltimore County, 1996.
36. Austin, J.L., *How To Do Things with Words*, Oxford University Press, London, 2nd ed., 1975.
37. Searle, J.R., *Speech Acts*, Cambridge University Press, Cambridge, 1969.
38. FIPA00061, FIPA ACL Message Structure Specification, Foundation for Intelligent Physical Agents, 2000.
39. FIPA00008, FIPA SL Content Language Specification, Foundation for Intelligent Physical Agents, 2000.
40. FIPA00009, FIPA CCL Content Language Specification, Foundation for Intelligent Physical Agents, 2001.

41. FIPA00010, FIPA KIF Content Language Specification, Foundation for Intelligent Physical Agents, 2001.

42. FIPA00011, FIPA RDF Content Language Specification, Foundation for Intelligent Physical Agents, 2001.

43. Lux, A. and Steiner, D., Understanding cooperation: An agent's perspective, in *Readings in Agents*, Huhns, M.N. and Singh, M.P., Eds., Morgan Kaufmann, San Francisco, 1998.

44. WAP Forum, Wireless Application Protocol Specification Version 2.0, WAP Forum, http://www.wapforum.org/what/technical.htm, 1999.

45. Object Management Group, OMG IDL syntax and semantics, in *The Common Object Request Broker: Architecture and Specification (Revision 2.4.2)*, Object Management Group Document formal/01–02–01, 2001.

46. FIPA00071, FIPA ACL Message Representation in XML Specification, Foundation for Intelligent Physical Agents, 2001.

47. Open Group CAE Specification, DCE 1.1: Remote Procedure Call, Open Group, Document C706, 1997.

48. FIPA00001, FIPA Abstract Architecture Specification, Foundation for Intelligent Physical Agents. FIPA Specification 2000, 2001.

49. FIPA00023, FIPA Agent Management Specification, FIPA Specification 2000, 2001.

50. Wahl, M., Howes, T., and Kille, S., Lightweight Directory Access Protocol (v3), RFC 2251, IETF, 1997.

51. Sun Microsystems, Java Agent Service Specification, http://www.java-agent.org/, 2001.

52. Ryan, V., Seligman, S., and Lee, R., Schema for Representing Java Objects in an LDAP Directory, IETF RFC 2713, 1999.

53. Ryan, V., Seligman, S., and Lee, R., Schema for Representing CORBA Object References in an LDAP Directory, IETF RFC 2714, 1999.

54. FIPA00014, FIPA Nomadic Application Support Specification, Foundation for Intelligent Physical Agents, 2001.

55. FIPA00069, FIPA ACL Message Representation in Bit-Efficient Specification, Foundation for Intelligent Physical Agents, FIPA Specification 2000, 2001.

56. Gong, L., Java Security Architecture (JDK1.2), Sun Microsystems, Inc., 1998.

57. Object Management Group, Security Service Specification (v1.8), CORBAServices Ch. 15, Object Management Group, OMG Doc security/00–09–01, 2000.

58. Chaudhri, V.K. et al., Open Knowledge Base Connectivity (2.0.3), SRI International and Stanford University, 1998.

59. Bellifemine, F. and Trucco, T., Java Agent Development Framework, The latest official version is JADE 2.4. http://sharon.cselt.it/projects/jade/, 2001.

8

Applied Intelligent Techniques in Assisted Surgery Planning

Chua Chee-Kai
Nanyang Technological University

Chcah Chi-Mun
Nanyang Technological University

8.1 Introduction

8.1.1 Historical Background of Computers in Medicine

In the last three decades, advancement in both computer software and hardware technology has raised the profile and application of computer techniques in almost every discipline of science and technology. The steep rise in the popularity of computer techniques is enhanced by the significant reduction and ongoing decline in the cost of computing equipment. The built-in obsolescence of computing equipment caused by rapid advances in the development of microprocessors, data storage devices, and modern manufacturing techniques has made high computing power, memory, and data storage capabilities affordable to academic institutions, industry, and the general public. With improved processing capabilities and newly enabled technologies for the manipulation, transfer, and storage of large quantities of data, computing techniques have become standard tools and are essential prerequisites in research and academic institutions, and industry.

In the past, progress in medical science was highly constrained due to the limited variety and quality of scientific instruments capable of isolating simple general principles from the very complex behavior and nature of biological matter. Although computer technology has been around since the 1950s, its introduction in the field of medicine is late compared to significant break-throughs achieved in applications such as computer-aided design, manufacturing, and engineering (CAD/CAM/CAE) under engineering sciences.[1] The setback in the development and utilization of computer-based medical applications is mainly due to the lack of computing power, memory, and data storage capabilities available to medical practitioners during those days. Such high-performance computing resources are critical for the handling, characterization, and manipulation of overwhelming quantities of complex data sets generated from the investigation of morphology and metabolic behavior of the human anatomy typical in the area of computer-aided surgery planning and simulation. While early CAD applications in engineering entail the modeling of solid or rigid constructions, the modeling of human organs or soft tissues is highly complex due to their intricate morphology, their continuously deforming behavior, and the lack of bio-image data available during those days. The modeling and visualization of anatomical structures, which form the basis and are essential building blocks for computer-aided surgical planning, was deemed fictional until the development of computer-assisted 2- and 3-dimensional biomedical imaging modalities during the mid-70s.[2-4] The use of biomedical imaging allows the biological organs to be visualized in their true form and modeled using reverse engineering techniques, whereby representative 3-dimensional computer-generated models of the organs are derived from topological and volumetric data generated by the imaging systems.

The rapid technological advancements in computer techniques, and the entry of enabling tech-nologies such as digital imaging, virtual reality (VR), the Internet, artificial intelligence (AI), and computer-controlled robotics have created a profound and remarkable impact on the medical commu-nity. Such advances in scientific technology have dictated a steep growth in the number of complex surgical procedures, which were regarded as impossible a few decades ago, to be routinely performed on a daily basis around the world. There is currently a remarkable progress in and a heavy reliance of modern medicine on computing techniques to aid in diagnostic procedures, pretreatment planning and prognosis, treatment or therapy, procedural training simulation, and monitoring and assessment of medical cases.

8.1.2 Computer-Assisted Surgery Planning

Increasing health cost is a significant factor that has played a very important role in instigating immense demands for the development and implementation of safe and cost-effective approaches and solutions in the delivery of health care. Surgery planning is one of the implemented approaches that has become an important discipline in modern medicine, and is transcending the gulf between computer techniques and medical sciences. The application of computers to provide elaborate infor-mation regarding the patients' condition has increased diagnostic accuracy and laid foundations for effective delivery of treatment. As such, computer-assisted surgery planning is gaining widespread acceptance and popularity within the medical community and the general public. Since its emergence, it is not uncommon to find reports on the advances and successful applications of computer-aided surgery planning appearing in public newspapers and magazines. The field of computer-assisted surgery planning is multi-disciplinary and encompasses the full integration of the latest advances in computer graphics, computer simulation systems, manufacturing techniques, imaging modalities, and virtual reality (VR) environment. This pre-surgical process has been implemented with great success to a host of complicated surgeries, ranging from the excision of tumors from body organs,[5] neurosurgery,[6] and cardiac surgery,[7,8] to the separation of joined twins.[9]

In computer-aided surgery planning, surgical steps and procedures are planned and rehearsed based on accurate 3-dimensional model representations of human anatomical structures created through computer modeling or image reconstruction techniques using data input from medical imaging systems. Modern computer-aided manufacturing techniques such as rapid prototyping (RP) and computer numerical control (CNC) machining are capable of fabricating physical 3-dimensional plastic or metal replicas of soft and hard tissue structures using computerized 3-dimensional solid or volumetric data inputs. The wide variety of representative anatomical and biological models are important, as their utilization will lead to enhancement of interpretation, communication, and visual and physical evaluation of medical cases, and will serve as valuable and effective tools to aid surgery planning, rehearsal, and education.

8.1.2.1 Advantages of Surgery Planning Procedures

With either computer-generated 3-dimensional representations or physical anatomical models, the advantages derivable through their application in surgery planning prior to surgical intervention are many. The more tangible benefits of the process are as follows:

- **Effective Treatments:** Effective treatments can be secured and implemented as optimal procedures can be achieved through accurate diagnosis, careful planning, and repetitive rehearsals. Besides increasing the precision and success rates of operations, as well as reducing morbidity and risks imposed on the life of the patient, optimized surgical procedures can avoid otherwise complicated or costly procedures. With properly planned surgical steps and trajectory of surgical instruments, the technique for incision and invasion of any anatomical organ can be perfected, leading to increased surgical precision and minimal damage to healthy tissues. High surgical precision is vital for the segregation of diseased tissues or the repair of damaged tissues, to ensure preservation of sufficient post-operative organ functionality.

- **Rehearsal of Surgical Procedures:** Rehearsal of complex surgeries can be accomplished following planned surgical procedures, allowing enhanced interpretation and familiarization of medical problems, and improved levels of understanding and coordination between surgical team members. Surgery rehearsals will instill higher levels of awareness, preparation, and readiness in the event of intraoperative complications, since such complications can be simulated and smoothened with repeated rehearsals. Also, with better pre-surgical assessment of the patient's condition, minimally invasive surgery techniques can be adopted to improve healing, and to lessen the pain and discomfort caused by unnecessarily large incisions.

- **Invention and Discovery of New Equipment/Techniques:** The knowledge and experience acquired through the planning process and rehearsals may lead to the discovery, invention, and assessment of new forms of treatment, surgical tools, and techniques. The utilization of such newly created tools and equipment can be assessed during rehearsals.

- **Surgery Time:** Familiarity with the patient's case attained through surgical planning and rehearsals can minimize the duration required for the operation, making the procedure safer for the patient while maximizing the productivity of medical facilities and utilization of resources.

- **Improved Health Cost:** The exploitation of state-of-the-art technologies in modern medicine incurs high procedural costs due to the large amounts of investments poured into the purchase of new biomedical equipment and systems, and the training of personnel to man and operate such systems. However, the use of properly planned surgical procedures can help offset increased healthcare cost through improved intraoperative procedures and postoperative prognosis, which are achievable with accurate diagnosis and delivery of treatment. Other cost-reduction measures include shortened hospitalization and patient recovery time.

- **Education and Training:** Surgery planning process and rehearsals can be applied effectively as simulation platforms or educational tools for the training of surgeons and medical personnel.

8.1.2.2 Review of Surgery Planning Techniques

This chapter serves to review the aspirations behind the development of intelligent or computer-assisted surgery planning procedures, the technologies involved, and their applications. The methods reviewed range from traditional methods of using 2-dimensional X-ray radiology interpretations and the construction and modeling of human organs from cardboard slices cut from traces created using computer tomography (CT) scans, to advanced methods using computer-aided manufacturing (CAM) technology, 3-dimensional modeling, and VR. A brief history and description of the technology and application of computer-aided systems such as biomedical imaging modalities, biomedical visualization, CAM techniques, and computer simulation in modern-day surgery planning procedures are discussed in the following sections.

8.2   Biomedical Imaging and Visualization

8.2.1   Brief Historical Development of Biomedical Imaging

Until the beginning of the last century, studies on the anatomy and metabolic activities of visually discernable human body organs have mainly been conducted through the physical dissection and careful examination of dead bodies. The accumulation of knowledge on the specific metabolic functions of various human internal organs is acquired through close monitoring and observation of functional deficits or malfunction of the organs due to injuries sustained or diseases inflicted on the living. The constraints faced due to the lack of proper medical investigative tools and instruments available during that era limits the amount and the depth of knowledge acquired.

For centuries, mankind has been intrigued by the prospects of being able to "see" and comprehend the true form of living human internal organs. It was not until 1895 that the ability to conduct noninvasive examination of the interior of a living human body became a reality. The discovery of X-ray by W.K. Roetgen during that year started a revolution in both medical imaging and diagnosis. The discovery created considerable public interest leading to the commercialization of home X-ray kits and the first X-ray-guided operation, which was performed by an English doctor within a year of the discovery. For the next couple of decades, contributions to radiography, such as intensifying screens, tomography, and the rotating anode tube were invented. Major technological breakthroughs and discoveries by physicians and engineers during the beginning of the 1950s to the mid-70s led to a second revolutionary era for imaging and diagnostic systems. The introduction of the X-ray CT, Ultrasound, MRI modalities, and new data display tools during this period of time eliminated the need for physical dissection, and provided newer and more powerful opportunities for medical diagnosis and biological investigations.[10] Other modern-day 3- and 4-dimensional imaging modalities include Positron Emission Tomography (PET), Single Photon Emission Computed Tomography (SPECT), and 3D tomographic microscopy.

8.2.2   Biomedical Imaging Modalities

For the purpose of securing an understanding on the use of medical imaging equipment to assist in surgery planning, only certain primary tomographic imaging modalities, namely, CT, MRI, PET, and SPECT, will be discussed. The display and reconstruction of image data attained through the use of the various imaging modalities are described in Section 8.2.3.

8.2.2.1   Computed Tomography (CT)

The introduction of an X-ray tomographic method in 1963 by A.M. Cormack made possible the production of 2-dimensional cross-sectional images of the human anatomy. The images are produced through the utilization of a large collection of lateral sections acquired using X-ray projection measurements. The concept introduced by Cormack[11,12] provided the operating principles for the development of Computed Tomography (CT). Although the underlying mathematical principle such as "Fourier Filtered Backprojection" employed in CT was first reported by an astronomer, J. Radon, in 1917, the implementation of CT was not possible until the invention of digital computing. The combination of the digital computer, back projection or inversion algorithms, and X-ray tomography

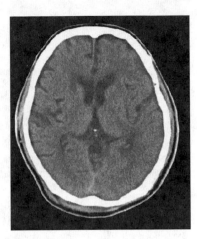

FIGURE I.8.1   Single slice of a CT head tomogram. Courtesy of the Department of Diagnostics Radiology, Singapore General Hospital.

method led to the development of the first CT scanner. The invention made its first appearance in clinical medicine in 1972 with the first CT application carried out by G.N. Hounsfield et al.[13–16] The entry of CT is a remarkable achievement in clinical medicine and created a revolutionary impact on the practice of medical radiology, superseding all classical tomographic X-ray methods involving the use of moving X-ray sources. G.N. Hounsfield and A.M. Cormack shared a Nobel Prize in 1972, the former for inventing the CT scanner and the latter for developing some of the underlying algorithms.

Among the main components of a CT scanner are the X-ray tube, a circular array of radiation detectors, the data acquisition system, and the operating console, which contains the system computer. Typical setups for modern-day CT scanners consist of a tilting gantry capable of achieving tilt angles of up to $+/-25°$, which houses the X-ray source, detectors and data acquisition system, and a patient couch that can be translated through the measurement field of the gantry. Cross-sectional images of the scanned object are generated through algebraic computation of projection measurements obtained from the attenuation of a large number of precisely controlled X-ray radiation paths emitted at different directions and passing through a thin planar section of the object. The generated images are a series of axial 2-dimensional transverse tomographs (Figure I.8.1) taken perpendicular to the longitudinal axis of the human body with fixed predetermined interspacing as the patient is advanced through the gantry in a discrete/incremental manner.

Although only transverse images are captured by the CT scanner, both sagittal and coronal sections of the anatomy can be derived from these axial slices with the use of 3-D reconstruction techniques together with surface rendering. The incorporation of tilting gantry designs has alleviated some limitations encountered in the measurement of coronary sections and allowed the free projection of organs. With CT, superposition effects inherent with the use of classical X-ray radiography due to the attenuation of radiation by overlapping soft and hard tissues are eliminated. The sharp and high contrast of reconstructed CT images allow the different types of soft and hard tissue in the human body to be easily recognized and observed with unprecedented precision, thereby increasing the accuracy in radiographic image interpretation.

Since its introduction, CT scanners have gone through five generations of design and setup improvements aimed at reducing scanning times and the quality of acquired images. Implemented changes in techniques range from the original single and parallel beam configurations to the current spiral scanning technique. The developments of the helical or spiral CT scanner have enabled highly detailed and sharper multiple images to be produced in much shorter scanning times. Using spiral scanning technique, the patient is translated continuously through a rotating X-ray source and radiation detector system, resulting in a continuous scan of the entire volume of the patient's body, compared to conventional discrete slice-by-slice acquisition. Despite the usefulness of CT, several major drawbacks exist with this modality, which include the risks involved in exposing patients to relatively large doses of harmful ionizing radiation during scanning, and poor spatial resolution encountered in the imaging of soft tissues.

8.2.2.2 Magnetic Resonance Imaging (MRI)

Imaging of the human anatomy using Magnetic Resonance Imaging (MRI) was pioneered by P. Lauterbur in 1973,[17] with images generated for a human finger, and R. Damadian, who demonstrated whole body imaging in 1977 with the invention of the first MR scanner. As the term "magnetic" in MRI suggests, the technique quantifies the responses of the human body when subjected to various magnetic fields, and the manner in which it relaxes upon the removal of the fields, for the acquisition of image data. Figure I.8.2 is a sample MR image showing the lumbar spine.

The key components of an MRI scanner are the main magnetic field assembly, gradient magnetic field coils, shim coil system, radio frequency (RF) coil, patient couch, operating console, and system computer. In whole body MRI systems, the main magnet assembly forms the horizontal bore, where scanning is carried out on a patient lying on the patient couch. Modern-day MRI scanners employ superconducting magnets in the main magnet assembly, as the use of such magnets allow the generation of reliable, static, and homogeneous main magnetic fields with high strengths of 1.5–2.0 Tesla. The stability and high strength of the main magnetic field are necessary for the polarization of atomic nuclei within the patient's body and critical for achieving high-quality images. The use of superconducting magnets carries low running costs due to low electrical requirements. Spatial homogeneity of the main magnetic field within the horizontal bore is guaranteed and maintained with the use of shim coils. The cylindrical gradient magnetic field coils consist of three windings located concentric with the bore and are used to provide the short pulses of linear gradient fields, which are proportional to the superimposed current, in the direction of the Cartesian coordinate axes. The linear gradient fields alter the main magnetic field at specific targeted regions of interest on the anatomy to be imaged.

At about the same instance when the gradient fields are applied, the RF coil is used to direct and transmit RF energy necessary to cause the protons at the targeted tissue region to resonate. When the RF transmission is switched off, the coil acts as a receiver to pick up MR signals generated by the relaxation of the protons. To increase the efficiency of transmission and reception of energies, and to reduce unwanted electromagnetic noises from the surroundings, different RF coil geometry to suit different regions of the human anatomy are engineered. By altering the local magnetic field strength in the region of interest through the application of gradient fields, different tissue types at the region

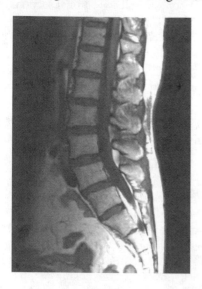

FIGURE I.8.2   MR image of lumbar spine. Courtesy of the Department of Diagnostic Radiology, Singapore General Hospital.

of interest will return differing MR signals in response to the changing field strengths. The differing signals acquired from the different tissues are computed and reproduced as images with exceptional soft-tissue spatial resolution and unparalleled level of detail.

Although MRI technology has been around for the past two decades, the technology is still in its infancy as compared to the well-matured X-ray CT method. Unlike X-ray CT, MRI is not associated with the use of harmful ionizing radiation, and produces no side effects following its application. Another major advantage of MRI over CT is that while CT images are confined to the axial plane, MR images can be obtained in three different planes (i.e., axial, sagittal, and coronal). In Magnetic Resonance Angiography (MRA), MRI is used to produce exquisite images of the flow of blood through the body. The advent of functional-MRI (fMRI) for functional brain mapping, customized MRI systems for the imaging of specific parts of the anatomy, and the evolution of open magnet system for intraoperative imaging and guidance are some of the new developments being explored and commercialized in MRI. Despite its great potential, MRI has several major drawbacks associated with its application. Due to the use of powerful magnetic fields, MRI is not safe for certain groups of people with implanted ferromagnetic devices (e.g., pacemakers, aneurysm clips). The tight bore confinements of earlier systems are a cause for concern with obese patients or patients who are claustrophobic.

8.2.2.3   Positron Emission Tomography (PET)

In most radiology methods, externally emitted radiation such as X-ray, electromagnetic, or sound waves are attenuated, reflected, absorbed, or released to provide the main contrast mechanism required to differentiate between differing tissue types. However, in nuclear diagnostics imaging, the region under study forms the active source of radiation, and contrast between differing tissues is provided by the 3-dimensional radioactive content distribution in the region. The use of radioactive isotopes for tomographic imaging purposes was pioneered by D.E. Kuhl and R.Q. Edwards in 1963.[18] Positron Emission Tomography (PET) is based on the detection of gamma radiation emitted from the annihilation of positron and electron pairs upon collision within the electron-rich tissues of the region under study. The positrons are emitted from the decay of short-lived radioisotope tracers, such as Carbon-11, Fluorine-18, Nitrogen-13, and Oxygen-15 bound to suitable metabolically active carriers. The radioactive substances are administered into the patient's body through intravenous injections and carried selectively to specific targeted regions of the anatomy, where they are preferentially accumulated.

Individual annihilation process, which occurs near the site of each emitted positron, is accompanied by the release of two gamma photons, each having energies equivalent to 511 keV and traveling in opposite directions. The trajectory of each pair of gamma rays defines a straight line, which is positioned close to the site of positron emission. The gamma rays emerging from the body of the patient are detected by a PET camera, which consists of a series of detector rings, each segmented by a finite array of gamma detectors, as the patient is moved past the detector rings in increments to produce a series of tomograms. The gamma detectors are made of bismuth germanate scintillation crystals, which convert gamma radiation into light photons before they are transferred into photomultiplier tubes and output as amplified electrical signals.

The detection of a pair of gamma rays reaching the detector ring at the same time (i.e., coincidence detection) gives rise to an event whereby a "line of sight" connecting the two detectors upon which the rays are incident is assigned. The density of the assigned lines passing through a region on the anatomy is dependent upon its concentration of radioactive tracers. By collecting and linking the many events leading to the assignment of line of sights, tomographic reconstruction procedures are used to produce the 2-dimensional image slices indicating the concentration of radioactive tracers within the region of study.

PET is useful in detecting and locating tumor growths and aneurysms, and can provide useful information regarding regional blood flow and the metabolism and physiology of tissues and organs.

It should also be noted that the radiation levels encountered with the use of minute quantities of radioactive substance in PET is much lower than the doses of radiation required in radiographic contrast studies. However, despite the advantages discussed, the spatial resolution of PET images is low and in the range of 2–5 mm. The requirements for short-lived radioisotopes also mean that the setup has to be located near a cyclotron.

8.2.2.4   Single Photon Emission Computed Tomography (SPECT)

Single Photon Emission Computed Tomography (SPECT) is based on the same principles of operation as PET. The main differences between SPECT and PET are in the choice of radioactive tracers used and the application of a collimator in SPECT for photon detection. In SPECT, the radioactive tracers employed, such as Iodine-123, Technetium-99, and Xenon-133, have much longer decay times and emit only single gamma photons compared to double emission in PET. A gamma camera based on the Anger principles is used to detect the randomly emitted gamma photons within the anatomical region under study. The camera is rotated through a wide range of angles relative to the patient and a lead collimator with a large array of parallel holes is used to capture the gamma photons. Only photon trajectories coincident with or close to the axis of the holes on the collimator are detected by the scintillator crystal and amplified by the photomultiplier tubes located behind the collimator. For each detected gamma photon, its line of sight or trajectory can be determined. The reconstruction of SPECT images follows the tomographic procedures used for CT or PET.

Similar to PET, SPECT can be used to provide valuable functional or metabolic information for soft tissues and information about blood flow. The use of longer-lived radioactive isotopes in SPECT implies that SPECT imaging facilities can be remotely located from cyclotrons. However, for SPECT, the use of a collimator and data filter algorithms to improve signal-to-noise ratios limits the spatial resolution achievable with SPECT. Current SPECT systems have a spatial resolution which is much poorer compared to PET, CT, and MRI.

8.2.3   Biomedical Visualization

Biomedical visualization is another important discipline made possible by advancements in computing techniques, graphics hardware, software, and medical imaging modalities that is becoming heavily relied on in surgery planning. The term biomedical visualization collectively represents the generation of realistic displays for representing multidimensional image data acquired through medical imaging or CAD systems, processing and manipulation of such data through specifically developed interactive or automated tools, and design, analysis, and validation of digital models that influence the decision-making processes in medicine.[19,20] Among the major objectives of biomedical visualization is to maximize the quality and detail of scientific or biological information that can be faithfully mapped onto and subsequently extracted from biomedical images. The data output of most modern biomedical imaging systems is in the form of 2-dimensional data slices (pixel arrays) or 3-dimensional volumetric data (voxel based). For the 3-dimensional volumetric data, a variety of image processing techniques is utilized to transform and project the 3-dimensional information onto a 2-dimensional image screen display that effectively portrays and preserves the 3-dimensional nature of the imaged structure.

Based on the characteristics of the different types of image data, their representation, and the user interface they support, image data in medicine can be classified under four different classes, as presented in the pyramid in Figure I.8.3. The complexity in data processing and the extent to which visualization technology is embraced for each class of image data increases toward the tip of the pyramid. The four classes of image data and their representation can also be further categorized according to their roles in medical visualization; namely, illustrative, investigative, and imitative visualization.[21] Each class of image data and their representation is briefly described in the following sections.

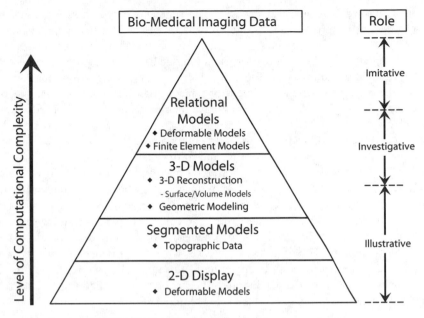

FIGURE I.8.3   Classes of image data representation.

8.2.3.1   Two-Dimensional Image Slices and Display

Two-dimensional cross-section slice images displayed on the viewing monitors of medical imaging systems are generated from the reconstruction of physical parameters (i.e., linear attenuation coefficient, MR energies, level of radioactivity) measured based on the properties of tissues within the anatomical region of interest. Tomographic reconstruction processes generally rely on the use of mathematical algorithms (e.g., inverse or backprojection algorithms) to determine the distribution of the physical parameters within, transmitted through, or emitted from the region of interest. The information is transferred to 2-dimensional image data composed using a light intensity image function, $f(x, y)$, containing a value, (w) representing the physical parameter. The two variables in the function, x and y, are real, non-negative bounded values denoting the spatial coordinates of the data point. The image data is displayed as an array of picture elements or pixels where the value of the image function (w) is mapped onto each pixel.

For monochrome displays, mapping of the pixels is carried out using a gray scale level setting consisting of 256, 1024, and 4096 distinct gray levels for 8-bit, 10-bit, and 12-bit data representations, respectively. For color displays, the RGB (red, green, and blue) value of each pixel is composed by the individual color elements, each represented by an 8-bit value having 256 settings. The resulting monochrome or color digital images are hence mappings of the image function discretized in both spatial coordinates and brightness. The images can also be considered as matrices with the rows and columns identifying the spatial position of the pixel and the individual matrix elements denoting the gray- or color-level settings. High-resolution images from medical imaging systems are usually displayed using 128×128 to 512×512 matrix arrays, depending on the type of imaging modality. To enlarge the images to proper display sizes on viewing monitors, interpolation techniques such as bilinear interpolation are applied.

8.2.3.2   Segmentation of Image Data

Image segmentation refers to the decomposition of an image into its natural primitive units and classification of the units into object classes based on its attributes in order to distinguish homogeneous image regions for the purpose of selecting anatomical structures of interest to carry out further visualization and measurements.[22,23] The segmentation process separates anatomical regions of interest from the rest of the image, allowing users to gain better understanding of the geometrical relationships of the various structures, and to facilitate the quantification of volumes for the structures being studied. The process forms an intermediate step required for the definition of anatomical regions and boundaries in 2-dimensional tomographic slices or 3-dimensional voxelized volumetric data before progressing to 3-dimensional reconstruction of the slice data, registration of multiple image data sets, atlas-matching, quantitative measurement of morphological descriptions, and motion detection.

In the case of 2-dimensional slice data, the segmentation process can be carried out with respect to visual elements such as the intensity, texture, and color of the pixels, whereas for 3-dimensional volumetric data, segmentation is carried out based on the attributes of the voxels. Many different approaches have been used for the segmentation of image data, which can be manual or through the use of semi- or fully automated segmentation algorithms.[24] In manual segmentation, the process is tedious, time consuming, and subjective due to the large volumes of generated image data. The use of semi- or fully automated algorithms (i.e., region growing) allows standardized results to be achieved in a much shorter time frame. However, there are cases when the presence of artifacts, arising from superposition, causes the detection and generation of false boundaries during the process. Under such circumstances, visual information does not suffice to allow proper definition of boundaries, and the experience and anatomical insight of the user is required to produce accurate delineation of the structure.

Commonly employed semi- or fully automated 2- and 3-dimensional segmentation techniques include thresholding,[25] region growing,[26] k-nearest-neighbor (kNN) classifier, etc. Newer and more recently developed algorithms for image segmentation include adaptive Bayesian and adaptive fuzzy c-means. Despite the wide variety of segmentation algorithms available, it should be noted that no single segmentation technique is capable of producing satisfactory results for all visualization applications. In the following paragraph, only two of the above-mentioned techniques, thresholding and region growing, will be briefly described.

The simplest of these techniques is threshold-based segmentation, where a pixel or voxel is considered to belong to an object class if a certain feature value assigned to it is above or below a certain fixed value, called the threshold. Thresholding techniques can be grouped under two different categories, depending on the choice of the threshold value; namely, global thresholding and local (adaptive) thresholding.[27] Global thresholding is the most intuitive approach of the two and involves the selection of only one threshold value for separating structural data of interest from background data during the processing of the entire image. The technique is simple to implement on computers and possesses fast processing speed. However, to achieve accurate segmentation results, the use of images with high signal-to-noise ratios, good contrast, and uniformity in the intensity of both structural and background data is critical. In local (adaptive) thresholding, the threshold value is selected based on the average, standard deviation or gradient of certain local properties within a sub-region on the image. As such, different threshold values are selected and used for different groups of pixels or voxels. Such selection procedures are necessary especially when dealing with images with varying structural and background contrast or background data with varying intensities, and also for the segmentation of small and sparsely spaced regions. The selection process can be conducted by subdividing the image data into a collection of smaller sub-images, which are sized to accommodate both object and background elements. The threshold value for each sub-image is then calculated based on the distribution of pixel or voxel intensities within each sub-image. A second method for the calculation of local threshold values involves the evaluation of the intensity distributions of neighboring elements for each pixel or voxel considered. Compared to global thresholding, local thresholding techniques are more complex to implement and involve significantly larger amounts of computing resources.

While threshold-based algorithms focus on the differences in the assigned attributes of the elements, region-growing algorithms are based on groups or regions of these elements having similar attributes. Using the region-growing segmentation technique, the initial step is for the user or an automatic seed-finding procedure to place an initialization seed or group of seeds within the region of interest. Based on the initialization seed(s), the algorithm scans and examines each neighboring pixel or voxel element one at a time. If a neighboring element meets the similarity or homogeneity criteria set based on a specific attribute of the initialization seed, the element is added into the growing region of interest. Neighboring elements to the newly added element are then selected and examined. The process is terminated when none of the neighboring elements tested passes the similarity criteria or the region of interest has grown to a pre-determined size.

8.2.3.3 Three-Dimensional Volume and Surface Reconstruction

Three-dimensional model reconstruction algorithms are employed to generate 3-dimensional surface or volume models from stacks of segmented image slices for visualization and transformation purposes. An early attempt at 3-dimensional reconstruction of anatomical structures, known as contour modeling, arranges entire sets of manually traced contour lines delineating the anatomical structure of interest from stacks of medical image slices to provide 3-dimensional topographic views of its boundaries. Further improvements to this technique involve the generation of wire-frame models from the contour line sets,[28] and the formulation of contour stitching techniques. Modern 3-dimensional reconstruction techniques allow highly accurate digital models to be derived from 2-dimensional tomographic image slices. Biological structures that can be modeled and visualized range from microstructures such as cells, proteins, and viruses[29] to visually discernable human organs and bones. The following paragraphs serve to provide a brief description of several commonly employed techniques in 3-dimensional reconstruction such as triangulation, voxel-based modeling, and NURBS-based modeling.

The triangulation approach is one of the simplest reconstruction techniques that can be utilized for producing 3-dimensional polygonal models from slice image data sets. In this approach, tiling algorithms are used to form triangular isosurface patches by connecting sample points from boundaries or contours lying on adjacent slices. The objective is to create a mosaic of joined triangular patches that envelope the entire set of contours. Optimized connections between sample points can be achieved using global,[30] local,[31] and heuristic[32] optimization methods. The use of triangles or polygons to represent the surfaces of the structure of interest allows only an approximation of the actual surface to be obtained. To reduce the extent of approximation error, more polygons can be utilized to bring about a closer piecewise linear approximation. However, an increase in the number of polygons used will definitely lead to increased data file size and longer computation times. Other major drawbacks encountered with the use of triangulation approach are correspondence and tiling problems due to the large spacing between individual image slices, whereby incorrect connections are made between the sample points on the two boundaries.

The second method, which is more commonly employed in biomedical applications for volumetric data representation, is voxel-based modeling, also referred to as spatial-occupancy enumeration. Using this method, structural information contained within the stack of image slices are represented using a large collection of adjoining, nonintersecting, identical spatial elements called voxels or volume elements. The voxels, which are primitive cubic cells, are arranged in a fixed regular grid to occupy the space containing the anatomical structure. Each voxel is assigned a value or density based on the magnitude of the physical parameter it represents. Under this representation method, the structure is encoded into a unique list of voxels that occupies its volume. Interpolation algorithms are employed in voxel-based modeling to enhance the resolution and accuracy of the 3-dimensional model by reducing the slice thickness between adjacent slices of the original data set. In voxel-based modeling, no constraints are faced in the type and complexity of the structure to be represented, making it a useful method to be used for the representation of natural and artificial structures. Surface information can be extracted from voxel-based models through the use of the surface tracking or marching cubes method. The surface tracking or marching cubes method is a thresholding algorithm which is capable of generating detailed representative surfaces of a model using a face of voxels.

The third method employed in 3-dimensional reconstruction, NURBS-based modeling, uses a set of Non-Uniform Rational B-Spline (NURBS) curves to interpolate points on adjacent slice boundaries. The splines are then patched via a skinning or sweeping process to produce the surfaces of the reconstructed model. NURBS volumes are defined by adding a third parameter to the 2-parameter NURBS representation. The 3-parameter NURBS representation represents not only the surface of the model but also its inner points. This method of 3-dimensional reconstruction is capable of producing 3-dimensional models with smooth freeform surfaces and, as such, provides better and more precise representations of the actual structure. To exploit the advantages of voxel-based modeling in NURBS modeling, the NURBS volume can be descretized through voxelization and reassembled using Boolean operations.

8.2.3.4 Geometric Modeling of Anatomical Structures via Constructive Solid Geometry (CSG)

Geometric modeling of anatomical structures differs from 3-dimensional reconstruction of 2-dimensional image slices in the manner the required 3-dimensional structures are constructed. 3-dimensional reconstruction methods apply reverse engineering principles in the construction of the anatomical model, whereby the actual structures are scanned using the various imaging modalities and digitized to serve as input points for the reconstruction process. Geometric modeling using constructive solid geometry (CSG) techniques assembles the anatomical structures directly from a combination of generalized geometric primitives by means of regularized Boolean set operators in a CAD environment.

Geometric modeling benefits from the fact that it obviates the use of data segmentation and 3-dimensional reconstruction process algorithms. The representation of anatomical structures by geometric models allows modifications on the size, shape, and relative locations of the structures and manipulation of data to remove overlapping structures, to reveal hidden features to be accomplished with ease. Further to this, accurate and efficient rendering processes can be carried out to improve the depth perception of the representation. However, the use of geometric models to represent anatomical structures is limited, due to the intricacy of some biological organs and tissues, and the inability of currently available geometric primitives to conform to and replicate such features.

8.2.3.5 Display of 3-Dimensional Reconstructed Models

The arrival of 3-dimensional images in medicine prompted the development of new methods for the display and visualization of 3-dimensional volumetric data. To date, numerous techniques[33,34] have been proposed and developed for projecting 3-dimensionl volumetric data to form 2-dimensional on-screen or pseudo-3-dimensional representations. The techniques can be classified under two major categories; surface rendering[35,36] and volume[37,38] rendering. Although either of the two methods can be used to generate displays, the methods are quite different, and the choice of method depends on the characteristics of the original input data and the kind of results required for the final application of the generated display.

Under surface rendering, the visualization and analysis of multidimensional image data are based on the surfaces or boundaries of the anatomical structures, whereby the surfaces are depicted on the screen of the display device with the aim of producing an image, which closely resembles a photographic image of the structure. The initial step taken by the process is to form or extract the surface information of the structure using the methods listed in Section 8.2.3.3 before proceeding with the rendering process. The rendering process is carried out by illuminating the structure with a set of parallel rays which are coincident with the viewing direction. Assuming the absence of perspective effect, the image formed on the viewing plane or display screen will be an orthogonal projection of the surfaces of the structure. The use of projection techniques reduces the original 3-dimensional data structure of the surfaces to 2-dimensional computer display data. To preserve the 3-dimensional form of the structure, illusions have to be created in the display data using hidden surface or part removal, shading, dynamic rotation of surfaces, and stereo projection techniques.

The advantages of surface rendering techniques include higher rendering speeds resulting from the processing of relatively smaller amounts of surface and boundary data, and the ease of transforming the patched surface representations into formats that can be processed and manipulated using CAD systems or standard computer graphics software and hardware. Moreover, lighting and shading algorithms common in computer graphic applications can be exploited to enhance the display of surfaces. However, the setting of threshold values and the use of interpolation techniques during surface extraction may result in the loss of image data and lead to the introduction of false surfaces, respectively. Figure I.8.4 shows a computer-generated skull model reconstructed from CT data and displayed using surface rendering and shading techniques.

Volume rendering techniques[39] are among the most powerful and versatile types of image display techniques available. The principles of volume rendering differ from that of surface rendering, as the technique does not require the prior extraction of surface or boundary information.

Similar to X-ray radiography, volume rendering techniques portray images reflecting the pseudo-surfaces and interfaces of different types of overlapping tissues in a more controlled and sophisticated manner. Volume rendering techniques are capable of producing 3-dimensional image renditions directly from gray-level scene data, hence eliminating the need for any intermediate geometric primitive representations (i.e., segmented images). This is advantageous, as problems such as the introduction of artifacts and data loss arising from the shortcomings of segmentation, surface, or boundary extraction, and other intermediate processes can be avoided, leading to preservation of the quality of original image data sets.

Under volume rendering, algorithms for ray-casting techniques (object-order, image-order, or domain-based) are used to directly project 3-dimensional volume data onto a user-defined viewing plane to generate the 2-dimensional displays. Ray-casting techniques project certain selected attributes (i.e., density value) of voxels within the anatomical structure of interest to form the required displays. To display the anatomical structures of interests, weighted factors are used during the rendering process to enhance data collected from the regions of interest and to suppress all other unimportant data. The projection is carried out using a series of straight rays emitted from a pre-determined source, which are then cast through the body of voxels that form the structure of interest. Each ray emerging from the body of interest accumulates the attribute values of the voxels directly or through interpolation techniques, depending on the manner in which the rays pass each

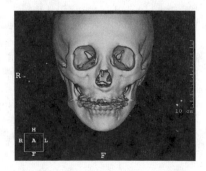

FIGURE I.8.4 Surface rendering of skull model. Courtesy of the Department of Diagnostic Radiology, Singapore General Hospital.

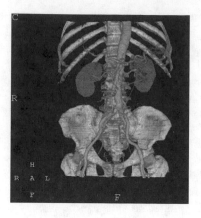

FIGURE I.8.5   Volume rendering of abdominal aorta. Courtesy of the Department of Diagnostic Radiology, Singapore General Hospital.

voxel along its line of sight and assigns appropriate weights for each data point collected. The screen displays are then generated using pixel values computed based on the aggregates calculated from the projected data of each ray. For volume rendering, any changes in the viewing plane will entail the re-calculation and re-projection of the rays, making the technique computationally demanding as compared to surface rendering.

One major advantage of volume rendering is that the rendered images produced are capable of being sectioned to show the original image data, and of carrying out voxel-based measurements since the rendering process involves the contribution of the entire volume data set. Figure I.8.5 illustrates the abdominal aorta together with the spinal column reconstructed from CT data and displayed using volume rendering technique.

8.2.4   Application of Medical Imaging and Computer Modeling in Surgery Planning

Visual information provided by medical imaging devices and reconstruction algorithms are aimed at providing noninvasive diagnosis, systematic planning of pre- and intra-operative procedures, custom design and fitting of implants and prosthesis, and quantitative measurements of the anatomical structure's morphology descriptions. In certain complex surgical procedures, such information is critical to equip the surgeons with precise knowledge of the anatomical structures, their dimensions, locations, and local geometric variations before an optimal surgical plan can be devised.

The routine use of 2-dimensional radiographs to provide such information and to aid in the planning of surgical procedures has several drawbacks. The main drawback is the limited quantity of diagnostic information carried by the 2-dimensional radiographs. Projection methods adopted in the production of the radiographs are not able to completely capture the 3-dimensional morphologic descriptions of anatomical structures that are being imaged. The use of 2-dimensional projection methods to cast images of 3-dimensional structures would imply that certain information relating to the geometry and volume of the structures would be missing due to the loss of dimensionality relating to depth perception. The resulting 2-dimensional image is blurred and contains significant amounts of unwanted artifacts due to superposition effects, whereby shadows of the various overlapping tissues or organs are superimposed on one another. With such limitations, radiographs cannot provide sufficient information necessary to establish the extent of complex injuries, deformities,[40] or spread of disease. In conjunction with the use of radiographs, it is not unusual for surgeons to heavily rely on personal insight, knowledge, and experience to understand and interpret the 2-dimensional information, and to construct a mental 3-dimensional portrayal of the physical shape of the tissues or organs.

3-dimensional digital models generated through computer reconstruction of medical image data or geometric modeling systems are capable of alleviating the restrictions imposed by the use of X-ray radiographs, as they provide an intuitive way for the accurate display and extraction of information relating to real-life objects. The need for such capabilities has led to continued advances in medical imaging and computer modeling technology, which are both widely recognized as invaluable and powerful analytical tools in the planning of complex surgical procedures. To counteract the loss of dimensionality in the display of 3-dimensional image data, visual illusions to recreate depth perception are achieved using perspective transformation, rendering, and shading techniques. The digital models can be scaled, rotated, and translated using dedicated software packages and can provide surgeons with necessary information relating to the geometry, relative sizes, locations, and geometrical interrelations of the various organs within the anatomy.

In the resection of tumors, such information is necessary to establish the spread of the tumor and for post-surgical monitoring. A surgeon could choose to address various internally located or hidden anatomical features, tissues, or organs by creating cut-away sectional views of the object under study. The ability to manipulate data associated with the digital models will significantly enhance visualization, diagnosis, and interpretation of a patient's condition. Using current technology, multi-modality imaging, whereby data of the same anatomical structure belonging to the same patient are acquired using different imaging modalities (e.g., CT, MRI) are registered, fused, and accurately aligned with one another to produce a morphological correlation of the structure of interest, can be accomplished.[41] Using multimodality imaging, the information available to the user is much more complete, as it is presented as a combination of individual information sets extracted from the various imaging modalities.

The importance of 3-dimensional models as visual tools should not be underestimated as studies[42] carried out have proven the effectiveness of the models for locating complex fractures that are missed most of the time using conventional radiology. Besides surgery planning, accurate displays of the anatomical models can be deployed during surgery to provide surgeons with additional visual information beyond the normal field of view. Preoperative models can be registered with intraoperative real-time data acquired from the patient on the operating table during actual surgery to provide on-line guidance to the surgeon. Besides informing the surgeon on the spatial locations of certain tissues and organs, images obtained through fMRI, PET, and SPECT modalities can provide information related to the physiology of the anatomical structures.

8.2.5 Virtual Reality and Computer Simulation in Surgery Planning

Virtual reality (VR) is a branch of computer science dealing with the generation of highly realistic virtual "worlds" or environments using a range of sophisticated 3-dimensional computer graphic programs. Such simulation fidelity coupled with high levels of visual and kinematics interaction can be exploited to imitate or provide the user with sensorial experiences similar to that in real life. VR offers exciting promises and many potentially novel applications as it unifies and bridges the gulf between computer animation and real-world, real-time patient data. To achieve sufficient levels of realism in VR, an essential and critical prerequisite is the availability of high-performance computing systems capable of processing and displaying information with reasonably high levels of detail and speed. Among the information or data that need to be acquired and processed include 3-dimensional digital models of real-life objects, their physical and physiologic properties, their locations and interactivity, and sensory inputs.[43] Compared with the other forms of visualization techniques discussed in the previous sections, VR is considered an infant science with limited capabilities, as the current state of computer technology is unable to match the resource requirements associated with proposed VR applications.

In medicine, the capability of VR in providing a highly realistic and interactive working environment holds great potential as a powerful educational tool in providing effective surgical training programs and education. Other important applications of VR in medicine include the practice of preventive medicine,[44] prototyping and testing of medical equipment and instruments,[45] and real-time surgery planning and rehearsals. For surgical planning and rehearsals, the advantages attainable with the use of VR simulators are unparalleled. The first VR surgical simulator incorporating a head-mounted display system and virtual surgical instruments for surgery planning and rehearsals on the abdomen region containing simplified computer-generated organs was created in 1991.[47] Despite the simplicity of the anatomical representations in such early simulators, they provided sufficient levels of realism for surgeons to perform effective procedural rehearsals and to plan and optimize surgical procedures on specific patient data. Latter developments in VR simulators employ texture-mapping techniques using photographs taken of actual anatomical structures to enhance the realism of the computer-generated models.

One of the most common current applications for VR in medicine is in the field of virtual endoscopy,[46] where the detection of exact locations of epithelial lesions within anatomical structures can bring forth much information to aid in the planning of surgical procedures. Virtual endoscopy is opening up new frontiers in biomedical imaging and carries the potential of realizing fully noninvasive screening where unobstructed observations can be conducted by "flying through" the internal organs[47] using sophisticated flight-tracking programs. In virtual endoscopy, reconstructions of 3-dimensional image data obtained through medical imaging modalities are utilized to provide simulated visualizations of the internals of hollow anatomical structures[48] similar to results attainable using conventional endoscopy. Such simulations can provide the user with thorough screening and exploration of the various internal organs without being visually occluded by the actual structures. The use of virtual endoscopy avoids the necessary incision procedure inherent with conventional endoscopy, therefore allowing visualization of anatomical regions which were previously considered impossible or too risky to be carried out using conventional endoscopy. A major advantage in the use of virtual endoscopy compared to conventional endoscopy is the elimination of any risks of infection, damage to tissues, and discomfort and pain to the patient. Despite the immense benefits that can be reaped through the use of this technology, the current state of the art of virtual endoscopy is very simple, with a level of realism comparable to cartoon graphics.

8.3   Rapid Prototyping

8.3.1   Introduction to Rapid Prototyping

Despite the increasing popularity in the use of soft prototyping methods (i.e., CAD, Virtual Reality) for simulating 3-dimensional objects in product design and development, physical prototyping is still an essential key element in the quest for superior product designs.[49] In product design as well as in surgical medicine, tactile feedback acquired from the handling and exploring of physical representations of intended products or biological structures has many clear advantages[50,51] compared to data obtainable from purely virtual or digital models. The use of physical prototypes in product development allows effective evaluation, verification and optimization of designs, communication between design team members, and running of marketing surveys. Moreover, physical prototypes are economical, as they can be fabricated at low cost through prototyping service bureaus and require no special skills with their use and handling. In comparison, the implementation of computer-enabled 3-dimensional visualization techniques entails significant amounts of investment due to the high cost of high-performance workstations and the need for specially trained personnel with good knowledge of the techniques. In medicine, the use of physical replicas of biological structures or internal organs and tissues are intuitive and can provide accurate assessments of the patient's condition, to conduct surgery planning and rehearsal, to provide intraoperative guidance, and to obtain better informed consent for surgery.

Rapid Prototyping (RP), also commonly referred to as Solid Freeform Fabrication (SFF), covers a group of new and advanced manufacturing technologies made possible by advances in computing techniques and manufacturing systems. RP technologies are capable of direct rapid production of highly complex 3-dimensional physical objects using input data derived from CAD systems, medical imaging modalities, digitizers, and other data makers.[52] Unlike conventional fabrication techniques such as machining, which involves the subtraction or removal of materials from a stock, casting, and molding to obtain the required 3-dimensional object, all RP processes are based on the same underlying concept of material incress manufacturing.[53] Using RP, the required objects are built following a layer-by-layer building method via the processing of solid sheet, liquid, or powder material stocks.

The emergence of RP technologies and the commercialization of the first RP system in 1988 revolutionized and created a huge impact in the domains of product design, research, and development.[54,55] RP technologies are described as "watershed events"[56] because of the tremendous amounts of time and cost savings achievable with their application. Savings achievable in the product development cycle amounting to a reduction of up to 60% in time and 70% in cost have been reported.[57] Although originally intended for prototyping, design evaluation, verification, and optimization, the functionality of RP end products has expanded tremendously, owing to extensive research in the area of RP application. The increase in the popularity and widespread adoption of RP techniques is evident by its presence in most major industries (i.e., aerospace, automobile, manufacturing, and medical) and sciences (i.e., anthropology,[58] paleontology,[59] and medical forensics[60]).

Among the early applications of RP in medicine is for the fabrication of 3-dimensional physical models or replicas of human bone structures and body organs.[61] Such models have been successfully used for education and implemented in healthcare to aid in surgery planning[62] and rehearsal,[63] and to assist in the production of custom implants[64] and prostheses.[65] Recent advances and refinements in RP techniques, the formulation of new materials, and the implementation of comprehensive systems have led to an extension in the utilization of RP technologies in a wide variety of medical applications. RP has recently been introduced and adopted as a new fabrication technique in the area of drug delivery devices[66] and tissue engineering. In tissue engineering, the high degrees of control and predictability of the RP fabrication process have been exploited for the production of tissue engineering scaffolds[67] for use in human organ reconstruction and repair.

8.3.2 Rapid Prototyping Systems

All commercialized RP systems generally share the same key components in their setup. They include a system computer, which analyzes the STL input file format transferred from a CAD system, for errors and mathematically sections the models to generate cross-sectional or slice data, and a production machine, which faithfully reproduces the object layer-by-layer according to the slice data. All RP systems can be classified into 3 main categories depending on the original physical form of the stock material utilized. The three categories are liquid-based systems (e.g., Stereolithography (SL), Solid Ground Curing (SGC)), powder-based systems (e.g., Selective Laser Sintering (SLS)), and solid-based systems (e.g., Laminated Object Manufacturing (LOM), Fused Deposition Modeling (FDM)).

Figure I.8.6 is a schematic illustrating the process chain common to all RP methods. In the process chain of RP, a 3-dimensional digital model of the desired object has to be first created using a CAD system. Alternative methods for obtaining the digital model include the use of medical imaging modalities, surface digitizers, and coordinate measuring machines (CMM). However, when employing such alternatives, a 3-dimensional model reconstruction process has to be carried out on the acquired data using specialized software interfaces. During the model creation stage, it is essential that the created or acquired 3-dimensional digital models are represented by unambiguous closed surfaces that define enclosed volumes. This is to ensure that the subsequently generated

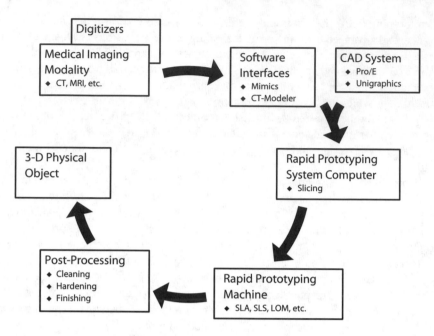

FIGURE I.8.6 Process chain for RP methods.

cross-sectional slice data are in the form of closed curves, which are necessary for the reproduction of solid replicas. The completed solid model must then be converted to a format dubbed the STL file format, which approximates the surfaces of the model using a large number of tiny triangles. The STL files are verified and checked for errors using customized software, such as MAGICS developed by Materialise, N.V., Belgium.[68] Although tedious and time consuming, errors in the STL model can be corrected manually using the software.

The prepared and error-free STL files are then transferred onto the RP system's computer via data storage media, e-mail, LAN (local-area network), or the Internet. The control software on board the RP system's computer analyzes the input file and creates the slice data, following a mathematical algorithm which automatically sections the STL model at specific intervals. The slice data are faithfully followed for the layer-by-layer reproduction of the object with the various RP processes. The fabrication stage will usually take a few hours, depending on the size of the object, and is fully automated for a vast majority of commercialized RP systems. For most RP processes, the completion of model fabrication is usually followed by post-processing stages whereby the "green" RP models are subjected to further downstream processes for cleaning, finishing, and hardening.

Since its introduction in the late 1980s, the number of RP system manufacturers has grown steadily. Currently, there are at least 23 RP system manufacturers worldwide,[69] making RP and its related processes an industry by itself. Among the more popular commercial RP methods include Stereolithography (SL), Selective Laser Sintering (SLS), Fused Deposition Modeling (FDM), Laminated Object Manufacturing (LOM), and Solid Ground Curing (SGC). The principles of operation of the various systems are briefly described in the following sections.

8.3.2.1 Stereolithography (SL)

Stereolithography (SL) is the pioneer, and to date remains as one of the most popular RP methods with the highest number of installed systems worldwide.[69] The SL method was patented by 3D Systems Inc.[70] in 1986 and commercialized in 1988. The method combines laser technology and photo-chemistry in its fabrication process. The material stocks used for SL are one component

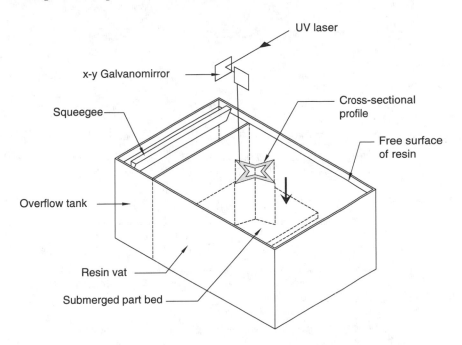

FIGURE I.8.7   The stereolithography (SL) method.

photo-sensitive liquid resins (i.e., epoxy- or urethane-based). Objects are fabricated by the selective solidification of polymeric material layer-by-layer using a UV laser beam following the paths generated from the slice data of each layer. Figure I.8.7 is a schematic illustrating the working principle of the SL method.

The SL system hardware setup consists of a vat of liquid resin, an elevator platform (i.e., part bed) on which the model is built, a sweeper to ensure the coating of thin layers of liquid resin, and a computer-controlled optical scanner to focus and direct the laser beam onto the free surface of the resin. To build an object, the elevator is lowered below the free surface of the resin by a distance equivalent to the thickness of one single layer. The sweeper traverses along the length of the vat to uniformly coat a thin layer of resin onto the elevator platform. The laser beam is then directed by the optical scanner to solidify a cross-sectional profile on the resin surface according to the slice data. Upon completion of laser scanning, the elevator platform is lowered through a distance sufficient for liquid resin to overflow and immerse the previously solidified profile. A new layer of resin is coated when the platform is raised such that the previously solidified layer is at a distance of one layer thickness below the free surface of the resin. The sweeping process is repeated to ensure the flatness of the newly re-coated layer of resin before laser scanning commences. Models fabricated using the SL process are not completely cured upon removal from the RP system. The "green" models have to be subjected to several post-curing processes, which include cleaning and draining of uncured resin, and subjecting the model to intense UV radiation to attain full curing.

8.3.2.2 Solid Ground Curing (SGC)

The Solid Ground Curing (SGC) method was developed by Cubital Ltd.[71] and commercialized in 1991. The SGC method is probably the most complex among all RP methods and involves the largest number of processing steps. A diagram illustrating the operations of a SGC method is given in Figure I.8.8.

The SGC method consists of two simultaneously operated cycles; the mask generation cycle and the model fabrication cycle. For mask generation, the cycle begins with the charging of a

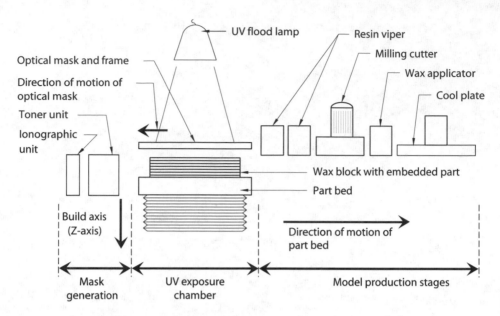

FIGURE I.8.8   The solid ground curing (SGC) method.

transparent mask plate using an ionographic unit. The resulting charged image is then developed with electrostatic toner. Regions on the mask plate defined by the cross-sectional profile of the slice data remain transparent. The prepared mask plate is then moved into a UV exposure chamber. A thin layer of photo-sensitive liquid resin is uniformly spread onto the part bed and exposed to a burst of UV light through the mask plate. Resin on the part bed not protected by the opaque regions on the mask plate will be solidified. A wiper connected to a vacuum is then used to remove all remaining liquid resin. The cavities left behind by the removal of the residual resin are filled with a thin coating of melted wax, which is subsequently cooled and solidified by a cool plate. All irregularities on the surface of the layer caused by the uneven deposition of liquid resin and wax are removed using a milling process. The milling stage ensures the exact thickness of each fabricated layer and prepares a flat surface for the deposition of the next layer. The image on the mask plate is erased and a new image corresponding to the next layer is generated before the cycle is repeated.

8.3.2.3   Selective Laser Sintering (SLS)

Selective Laser Sintering (SLS) was originally developed and patented by the University of Texas at Austin. Commercialization of SLS systems began in 1992 by DTM Corporation. The SLS method is a thermal process which employs a heat-generating CO_2 laser to coalesce finely powdered materials (e.g., plastic, wax, plastic-coated ceramics, or metal) to form solid masses making up the structure of the desired 3-dimensional object. The SLS process is illustrated by the schematic diagram presented in Figure I.8.9.

A thin, evenly distributed layer of heat-fusible powdered material is first deposited onto the part bed using a precision roller mechanism, which pushes the powdered materials from the feed cartridges located on either side of the part bed. Powder deposited on the part bed is raised to a temperature close to its melting point with the aid of pre-heaters. The laser is directed by an optical scanner to selectively trace the cross-sectional profile of the object on the surface of the powder. The temperature of the exposed powder is above its melting point, causing the individual powder grains to melt and fuse. At the end of the sintering stage, the part bed is lowered by one layer thickness and a new layer of powder is deposited. Laser scanning is repeated for the new layer, causing it

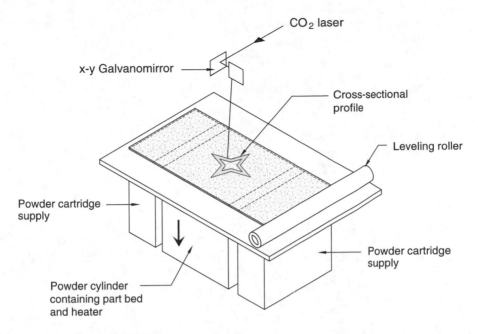

FIGURE I.8.9   The selective laser sintering (SLS) method.

to be fused to the previously sintered layer. The whole process repeats itself until the entire object is fabricated.

8.3.2.4   Fused Deposition Modeling (FDM)

The FDM method was developed by Advanced Ceramics Research (ACR), Tucson, AZ, patented and marketed by Stratasys Inc. Commercial sales of the FDM system began in 1991. The method uses the computer-controlled extrusion of thin polymeric or wax material roads to construct the cross-sectional profiles of the desired 3-dimensional object. A schematic drawing showing the operations of the FDM method is presented in Figure I.8.10.

The material stocks used are in the form of thin filament windings. The filament material is fed into a temperature-controlled liquifier head, whereby it is heated to a temperature close to its melting point. The softened material is extruded out of a nozzle tip and directed onto the building platform by the horizontal (x- and y-axis) translational movement of the liquifier head. The extruded material roads are arranged uniformly with a specified spacing to form the cross-sectional profile of the object. Supporting structures for overhanging regions are built using a different material extruded through a separate nozzle tip. Upon the completion of each layer, the nozzle tip is raised by a distance of one layer thickness before a new layer is drawn on top of the previously formed layer. The process is repeated until the entire object has been fabricated. The support structures are broken off to obtain the final object.

8.3.2.5   Laminated Object Manufacturing (LOM)

The LOM method was developed by Helisys Inc. and commercialized in 1991. The method uses the building up and lamination of cross-sectional profiles cut from adhesive-coated solid sheet materials (e.g., paper, plastics, polyester composites) in the construction of 3-dimensional objects. A schematic illustrating the LOM method is given in Figure I.8.11.

For each new layer, sheet material from the supply roll is advanced onto the building platform and caused to adhere to the stack by a heated roller, which activates the adhesive on the coated material. A CO_2 laser beam, guided by a moving optical head, is used to cut the cross-sectional

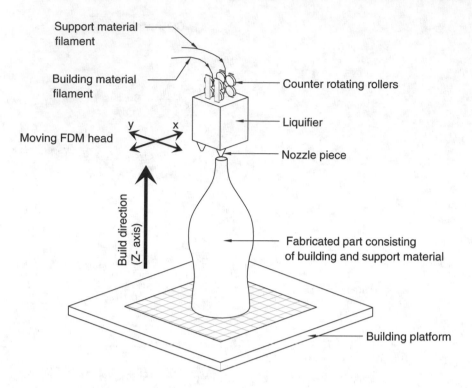

FIGURE I.8.10   The fused deposition modeling (FDM) method.

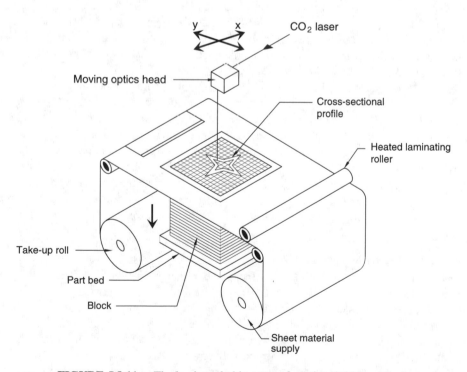

FIGURE I.8.11   The laminated object manufacturing (LOM) method.

profiles of the model. The power of the laser beam is adjusted such that the depth of cutting does not exceed the thickness of a single sheet of material. Upon completion of the cutting process, the building platform is lowered and a new section of the sheet material is positioned above the stack. The platform is raised and the new layer of material is again bonded onto the stack by the heated roller. The cutting process is repeated following the slice data generated for the new layer. The entire sequence is repeated until the fabrication of the model is complete. Unwanted regions on the material stack are cut using a hatch pattern to facilitate their removal after the fabrication process.

8.3.3 Applications of 3-Dimensional RP Fabricated Models for Surgery Planning

As mentioned, 3-dimensional reconstructed models derived from CT or MRI scans are valuable tools for diagnosis, surgery planning, and intervention. However, the usefulness of such models are limited as they are not readily available as guides for the surgeon in most operating theaters, nor are they portable enough to be used in clinical consultations or as a communication tool between the doctors, the patients, and their family members. Two-dimensional on-screen representations of real-life objects lack the tactile feedback and depth perception information associated with the handling and exploring of 3-dimensional physical models. The surgeons' understanding of the 2-dimensional images is not intuitive and interpretation skills have to be acquired through training and experience.

RP allows the fabrication of near-to-actual 3-dimensional physical prototype models of biological structures captured using either a CT or MRI scanner. However, to use the CT and MRI scanned data, a software interface is required to convert the data to the STL file format before being loaded into the RP system's computer. Data conversion can be carried out using specialized software such as Mimics, which performs 3-dimensional segmentation of CT scans and MRI scans via a selection of sophisticated editing tools. The processed data is then verified and corrected using a CT-Modeller (CTM) software system before being interfaced to any RP system. Both Mimics and CTM are developed and marketed by Materialise, N.V., Belgium. The benefits attainable with the application of RP fabricated physical models in medical sciences are immense. Among some of the benefits are as listed below:

- **Enhanced visualization and interpretation:** The availability of full-scale physical models is valuable in the design of surgical procedures and planning. Detailed visual and physical evaluation can be carried out using the RP models to enhance the surgeon's insight and interpretation on the complexity of the patient's condition before carrying out the surgical procedure.

- **Modeling of tumors and injuries:** In the case of tumor growth or injury (i.e., multiple fractures sustained in bone structures), full-scale RP replicas of the afflicted tissue, organ, or bone can be reproduced to provide realism in surgical planning and rehearsal. Vital areas that should be avoided during surgical intervention can be clearly marked onto the models to facilitate in trajectory planning of surgical instruments. The ability to visualize and evaluate the extent of damage caused by the spread of disease or injury can serve to guide a surgeon through a surgical procedure. Figure I.8.12 shows an RP model of the human skull fabricated using the SGC method with data derived from a CT scanner. The arrow in the photograph shows the location of tumor growth within the skull.

- **Surgical rehearsal and guides:** Full-scale RP models of biological structures can be used by surgeons in the rehearsal of complicated surgeries. Multiple rehearsals can be carried out on expandable RP models to optimize and simplify otherwise complicated procedures before invasive surgery is conducted. High levels of confidence can be attained from the

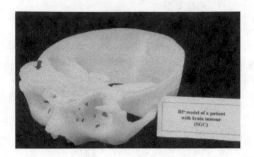

FIGURE I.8.12   RP model of the human skull showing location of tumor growth.

rehearsals, since the surgeons are better prepared and are aware of the consequences of adopting certain surgical routes and complications that may arise. The shortened intervention period resulting from simplified and optimized procedures reduces patient risk and morbidity. RP models are portable and can be brought into the operating theater to serve as templates or guides to surgeons working under difficult or obscured visual conditions. With the models, visualization of overlapping internal and hidden details is greatly enhanced.

- **Short turnaround time:** The time required for the fabrication of complex biological models with internal structures is much shorter using RP methods as compared to conventional fabrication techniques. Depending on the object's size, complexity, and the RP method chosen, the fabrication of most 3-dimensional physical replicas of biological structures can be completed in a matter of hours.
- **Messenger:** RP models can aid communication between surgical team members and medical practitioners. The models can also facilitate doctors in providing clear and detailed explanations, prognoses, and procedures to patients and their family members, and to obtain better informed surgical consent.

8.4   Conclusion

Surgical planning is a process instigated by increasing demands to maximize productivity of medical facilities and utilization of resources to provide high-quality, cost-effective treatments and patient care. The importance of surgical planning is evident in its implementation in complex surgical procedures whereby significant improvements are achieved in the quality and precision of applied treatment to eradicate patient trauma caused by the spread of disease or sustained injury. Accurate and precise preoperative assessments are important to pinpoint the site and extent of damage or abnormality in order to facilitate surgical repair or removal while avoiding damage to surrounding tissues. Besides promoting the healing of affected tissues, the use of surgery planning procedures can significantly reduce the probability of intra- and postsurgical complications, leading to surgical procedures that are safer with little risk imposed on the life of the patient.

In this chapter, the development and practice of surgery planning procedures based on biomedical imaging modalities, 3-dimensional computer-generated anatomical structures, VR, and modern CAM technology were reviewed. The benefits derivable from the utilization of such modern and state-of-the-art technologies were discussed. The progress of surgical planning techniques hinges on advancements and breakthroughs achieved in the areas of computer and information technology. With continued progress in computer and information technology, newer and more effective surgical planning techniques are expected to impact and change medical research and healthcare practice.

Acknowledgment

The authors gratefully acknowledge the generosity of the Department of Radiology, Singapore General Hospital for the kind contribution and permission for the use of images presented in this chapter.

References

1. Hidson, D., Computer-Aided Design and Bio-Engineering: A Review of the Literature (U), *Tech. Note 88–31*, Defence Research Establishment Ottawa, 1988.
2. *Imaging Systems for Medical Diagnosis: Fundamentals and Technical Solutions; X-ray diagnostics, computed tomography, nuclear medicine diagnostics, magnetic resonance imaging, sonography, biomagnetic diagnostics*, Krestel, E., Ed., Siemens-Aktienges, Berlin, 1990.
3. *Medical Imaging Systems Techniques and Applications: Computational Techniques, Gordon & Breach Int. Ser. Engineer., Tech. Appl. Sci.*, Leondes, C.T., Ed., Gordon & Breach Science Publishers, vol. 6, Amsterdam, 1998.
4. Guy, C. and Ffytche, G., *An Introduction to the Principles of Medical Imaging*, Imperial College Press, London, 2000.
5. Glombitza, G. et al., Virtual planning of liver resections: Image processing, visualization and volumetric evaluation, *Int. J. Med. Inform.*, 53(2–3), 225, 1999.
6. Kikinis, R. et al., Computer-assisted interactive three-dimensional planning for neurosurgical procedures, *Neurosurgery*, 38(4), 640, 1996.
7. Vahl, C.F., Meinzer, H.P., and Hagl, S., Three-dimensional representation of cardiac morphology, *The Thorac. Cardiovasc. Surgeon*, 39(Suppl. III), 198, 1991.
8. Vannier, M.W., Gutierrez, F.R., and Laschinger, J.C., Three-dimensional magnetic resonance imaging, *Magnetic Resonance Imaging*, 2(2), 61, 1990.
9. *The Strait Times*, Singapore, website: http://straitstimes.asia1.com.sg/mnt/html/webspecial/siamese/siamese3. html#op, 2001.
10. Robb, R.A., *Three-Dimensional Biomedical Imaging – Principles and Practice*, VCH Publishers, New York, 1994.
11. Cormack, A.M., Representation of a function by its line integrals with some radiological applications, *J. Appl. Physics*, 34(9), 2722, 1963.
12. Cormack, A.M., Representation of a function by its line integrals with some radiological applications. II, *J. Appl. Physics*, 35(10), 2908, 1964.
13. Hounsfield, G.N., Computerized transverse axial scanning (tomography): Part 1: Description of system, *Br. J. Radiology*, 46, 1016, 1973.
14. Ambrose, J.A., Computerized transverse axial scanning (tomography): Part 2: Clinical application, *Br. J. Radiology*, 46, 1023, 1973.
15. Perry, B.J. and Bridges, C., Computerized transverse axial scanning (tomography): Part 3: dose considerations, *Br. J. Radiology*, 46, 1048, 1973.
16. Ambrose, J.A. and Hounsfield, G.N., Computerized transverse axial scanning (tomography), *Br. J. Radiology*, 46, 148, 1973.
17. Lauterbur, P.G., Image formation by induced local interactions: Examples employing nuclear magnetic resonance, *Nature*, 242, 190, 1973.
18. Kuhl, D.E. and Edwards, R.Q., Image separation radioisotope scanning, *Radiology*, 80, 653, 1963.
19. Robb, R.A., Visualization, in *Handbook of Medical Imaging: Processing and Analysis*, Bankman, I.N., Ed., Academic Press, San Diego, 2000, 655.

20. Kaufman, A., Visualization, in *Handbook of Computer Science and Engineering*, Tucker, A., Ed., CRC Press, Boca Raton, FL, 1996.

21. Solaiyappan, M., Visualization pathways in biomedicine, in *Handbook of Medical Imaging: Processing and Analysis*, Bankman, I.N., Ed., Academic Press, San Diego, 2000, Chapter 41.

22. Castleman, K.R., *Digital Image Processing,* Prentice Hall, Upper Saddle River, 1996.

23. Clarke, L.P. et al., MRI segmentation: Methods and applications, *Mag. Res. Imaging*, 13(3), 334, 1995.

24. Shareef, N., Wand, D.L., and Yagel, R., Segmentation of medical images using LEGION, *IEEE Trans. Medical Imaging*, 18(1), 74, 1999.

25. Sahoo, P.K. et al., A survey of thresholding techniques, *Computer Vision, Graphics, and Image Processing*, 41, 233, 1988.

26. Mitiche, A. and Aggarwal, J.K., Image segmentation by conventional and information-integrating techniques: A synopsis, *Image and Vision Computing*, 3(2), 50, 1985.

27. Rogowska, J., Overview and fundamentals of medical image segmentation, *Handbook of Medical Imaging: Processing and Analysis*, Bankman, I.N., Ed., Academic Press, San Diego, 2000, Chapter 5.

28. Ameil, M. and Delattre, J.F., Computerized reconstruction of an anatomical structure based on digitized sections, *Clin. Anat.*, 5, 261, 1984.

29. Hendren, P. and Mayor, P., A computer graphics approach to investigating the architecture of icosahedral viruses, *Comp. BioMed Res.*, 13, 581, 1980.

30. Keepel, E., Approximating complex surface by triangulation of contour lines, *IBM J. Res. Develop.*, 19(1), 2, 1975.

31. Christiansen, H.N. and Sederberg, T.W., Construction of complex contour line definition into polygonal element mosaics, *Computer Graphics*, 12(3), 1978.

32. Ganapathy, S. and Dennely, T.G., A new triangulation method for planar contours, *Computer Graphics*, 16(3), 1982.

33. Harris, L.D. and Camp, J.J., Display and analysis of tomographic volumetric images utilizing a vari-focal mirror, *Proc. SPIE*, 507, 38, 1984.

34. Serra, L. et al., The brain bench: Virtual tools for stereotactic frame neurosurgery, *Med. Imag. Anal.*, 1(4), 317, 1997.

35. Lorenson, W.E. and Cline, H.E., Marching cubes: A high-resolution 3-D surface reconstruction algorithm, *Computer Graphics*, 21(3), 163, 1987.

36. Gibson, S. et al., Volumetric object modeling for surgical simulation, *Med. Image Anal.*, 2(2), 121, 1998.

37. Kaufman, A.E., Cohen, D., and Yagel, R., Volume graphics, *IEEE Comp.*, 26(7), 51, 1993.

38. Leroy, M., Display of surfaces from volume data, *Comp. Graphics Appl.*, 8(5), 29, 1988.

39. Robb, R.A., Three-dimensional visualization in medicine and biology, in *Handbook of Medical Imaging: Processing and Analysis*, Bankman, I.N., Ed., Academic Press, San Diego, 2000, Chapter 42.

40. John, J.F. et al., Use of acetabular models in planning complex acetabular reconstructions, *J. Anthroplasty*, 10(5), 661, 1995.

41. Serra, L. et al., Multi-modal volume-based tumor neuro-surgery planning in a virtual workbench, *Proc. 1st Int. Conf. Med. Image Comp. Computer-Assisted Intervention*, MIT, Cambridge, 1007, 1998.

42. Zonneveld, F.W., 3D imaging and its derivatives in clinical research and practice, in *3D Imaging in Medicine*, 2nd ed., Udupa, J.K. and Herman, G.T., Eds., CRC Press, Boca Raton, FL, 2000.

43. Satawa, R.M., Medical virtual reality: The current status of the future, in *Medicine Meets Virtual Reality 4: Health Care in the Information Age*, IOS Press, Amsterdam, 1996, Chapter 12.

44. Rosen, J.M. et al., The evolution of virtual reality from surgical training to the development of a simulator for health care delivery, in *Medicine Meets Virtual Reality 4: Health Care in the Information Age*, IOS Press, Amsterdam, 1996, Chapter 11.

45. Satava, R.M., Medical virtual reality: The current status of the future, in *Medicine Meets Virtual Reality 4: Health Care in the Information Age*, IOS Press, Amsterdam, 100, 1996, Chapter 12.
46. Vining, D.J., Virtual endoscopy: Is it reality?, *Radiology*, 200(1), 30, 1996.
47. Jones, S.B. and Satava, R.M., Virtual endoscopy of the head and neck: Diagnosis using three-dimensional visualization and virtual representation, in *Medicine Meets Virtual Reality 4: Health Care in the Information Age*, IOS Press, Amsterdam, 152, 1996, Chapter 18.
48. Davis, C.P. et al., Human aorta: Preliminary results with virtual endoscopy based on three-dimensional MR imaging data sets, *Radiology*, 199, 37, 1996.
49. Scrange, M., The culture(s) of prototyping, *Design Manage. J.*, 4(1), 55, 1993.
50. Bajura, M., Fuchs, H., and Ohbuchi, R., Merging virtual objects with the real world: Seeing ultrasound imagery within the patient, *Computer Graphics*, 26(2), 203, 1992.
51. Caponetti, L. and Fanelli, A.M., Computer-aided simulation for bone surgery, *IEEE Comp. Graphics Appl.*, 13(6), 86, 1993.
52. Chua, C.K. and Leong, K.F., *Rapid Prototyping: Principles of Applications and Manufacturing*, John Wiley & Sons, Singapore, 1997.
53. Kruth, J.P., Material Incress manufacturing by rapid prototyping techniques, *Ann. CIRP*, 40, 603, 1991.
54. Ashley, S., Prototyping with advanced tools, *Mech. Engineer.*, 116, 48, 1994.
55. Kochan, D. and Chua, C.K., State-of-the-art and future trends in advanced rapid prototyping and manufacturing, *Int. J. Inf. Tech.*, 1(2), 173, 1995.
56. Kochan, D., Solid freeform manufacturing—possibilities and restrictions, *Comp. Ind.*, 20, 133, 1992.
57. Styger, L., Rapid prototyping and tooling technologies, *Material World*, 1(12), 56, 1993.
58. Recheis, W. et al., Virtual reality and anthropology, *Eur. J. Radiology*, 31(2), 88, 1999.
59. Zollikofer, C.P.E. and Ponce de Leon, M.S., Tools for rapid prototyping in biosciences, *IEEE Comp. Graphics Appl.*, 15(6), 48, 1995.
60. Vanezi P. et al., Facial reconstruction using 3-D computer graphics, *Forensic Sci. Int.*, 108(2), 81, 2000.
61. *Anatomics Pty Ltd* website: http://www.qmi.asn.au/anatomics/, 2001.
62. Chua, C.K. et al., Rapid prototyping-assisted surgery planning, *Int. J. Adv. Manu. Techn.*, 14(9), 624, 1998.
63. D'Urso, P.S. et al., Stereolithographic biomodelling in cranio-maxillofacial surgery: A prospective trial, *J. Cranio-Maxillofacial Surgery*, 27(1), 30, 1999.
64. Porter, N.L., Pilliar, R.M., and Grynpas, M.D., Fabrication of porous calcium polyphosphate implants by solid freeform fabrication: A study of processing parameters and *in vitro* degradation characteristics, *J. Biomed. Mater. Res.*, 56(4), 504, 2001.
65. Rogers, W.E. et al., Fabrication of prosthetic sockets by selective laser sintering, *Proc. Solid Freeform Fabrication Symp.*, 158, 1991.
66. Leong, K.F. et al., Fabrication of porous polymeric matrix drug delivery devices using the selective laser sintering technique, *Proc. Inst. Mech. Engineers, Part H–J. Engineer. Med.*, 215(2), 191, 2001.
67. Hutmacher, D.W. et al., Mechanical properties and cell cultural response of polycaprolactone scaffolds designed and fabricated via fused deposition modeling, *J. Biomed. Mater. Res.*, 55(2), 203, 2001.
68. *Materialize NV* website: http:// www.materialise.com, 2001.
69. Wohlers, T., Wohlers Report: Rapid Prototyping and Tooling, State of the Industry, *Ann. Worldwide Progr. Rep.*, Wohlers Associates Inc., 2000.
70. *3D Systems Incorporate*, USA website: http://www.3dsystems.com, 2001.
71. *Cubital Ltd* website: http://www.cubital.com, 2001.

9

Intelligent Software Systems for CNC Machining

George-C. Vosniakos
National Technical University of Athens

Preface

Manufacturing is a very broad discipline encompassing a number of technical and managerial domains or sciences. One combination of criteria used to distinguish among sectors is manufactured product volume and variety. Discrete part or batch production is one such sector particularly important for the mechanical industry. Computer Numerical Control (CNC) is a key technology within that sector. This has enabled programmable automation and flexibility at the same time, translating into quite a wide range of volume and variety in production. Much of the automation and flexibility is achieved through software. Intelligent software has not really made its impact yet on a practical scale, but it has certainly come of age on a research scale. Intelligent software for CNC manufacturing cells is the focus of this chapter.s

The approach taken here in presenting achievements in intelligent CNC manufacturing differs from other attempts in that it is deliberately a wide-scope coverage of Artificial Intelligence (AI) applications in a vertically integrated CNC manufacturing activity chain starting with feature recognition and process planning, and ending with product quality control and machine scheduling.

0-8493-1121-7/03/$0.00+$1.50
© 2003 by CRC Press LLC

A number of activities comprise that chain and each of them is analyzed under the light of AI software support. Six non-AI hybrid paradigms are employed to demonstrate how such support can materialize.

This approach was deemed more interesting than detailed analysis of just one activity of the chain with a number of alternative AI paradigms and techniques. For each activity actually examined, a typical AI solution is presented based on the author's own work or a "digested" analysis of some other representative piece of work. This approach also proves how the same computation tools and techniques can be equally useful in modeling and solving problems of considerable diversity.

9.1   Introduction

In the manufacture of high variety and small batch size discrete mechanical parts, CNC is the key technology. The main reason is that it allows agility in responding to market needs. Agility is ensured at a first level by the programming capability of the machine in respect to part programs, programmable tools, and workpiece handling (tool magazines, pallet loading and unloading devices, workpiece handling manipulators, etc.), and is manifested in absolute minimization of nonproductive times (machine setup time, tool changing time, machine loading and unloading time, etc.). Equally important, agility is further ensured at a second level by soft-automating manufacturing system activities that support production on the CNC machines, including design activities that increase the level of automation of the machine itself. The main activities immediately supporting production on CNC machines are as follows (see also Figure I.9.1 for an IDEF0 schematic representation):

- Design of the manufacturing processes to be executed on the machine, resulting in executable part programs, a complex of activities formally known as process planning
- Preparation of the machine tools for execution of part programs pertaining to daily scheduling production and to in-process quality monitoring
- Advanced monitoring of the execution of part programs, pertaining to process monitoring, including tool condition monitoring and wider process state identification, machine tool error detection and possible compensation, and machine improvement through adaptive control

Conventionally, the above-mentioned activities are performed with the aid of software systems which, if nothing else, prefix the name of these activities by the well-known acronym CA, standing for "Computer-Aided." The conventional character is associated with a purely algorithmic approach to problem solving, irrespective of whether this is of numerical, geometric, or logical nature. Examples are: calculation of feed and speed in various turning operations, calculation of volumetric machine tool errors based on measurements on the machine tool, tool path derivation for pocketing based on Voronoi diagrams, heuristics-based machine loading, equation-based adaptive control to respect cutting force in turning.

Conventional approaches are satisfactory to the extent that they are complemented by human user intervention. However, they fail to solve problems characterized by:

- Combinatorial complexity, as in the selection of manufacturing operations, machines, tools, and parameters to perform part production.
- Non-numerical data, i.e., knowledge of logical nature, theoretically or empirically founded, as in the sequencing of manufacturing operations.
- Inherent imprecision, which is a feature of most manufacturing environments, as in many cases of adaptive machine tool control, of operations selection, of process monitoring, etc.

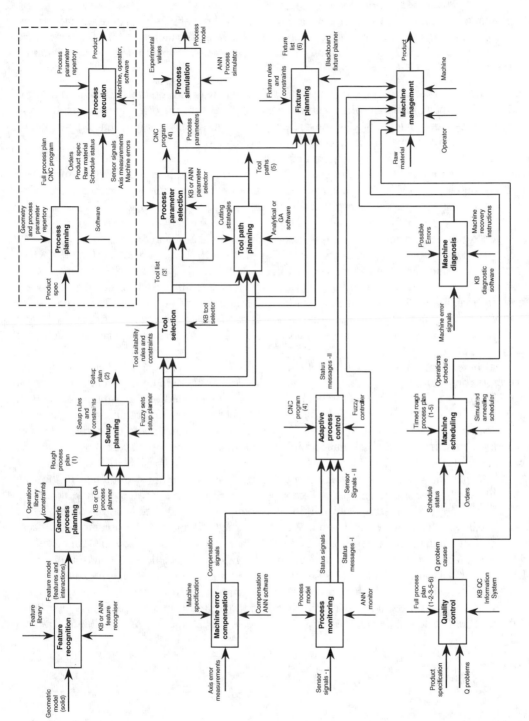

FIGURE I.9.1 IDEF0 diagram of the main activities in CNC manufacturing systems.

- Dynamic (changeable) situations in the time domain, as in machine loading (scheduling) or, more generally, whenever desired.
- Adaptation or adjustment is necessary in the face of ever-changing situations, as in adaptive process control, dynamic process planning, etc. Indeed, adaptability to largely unforeseen changes (equivalent to learning ability) is the very essence of human intelligence and a basic target of artificial intelligence.

Nonconventional software tackling the problems where conventional software performs inadequately is termed intelligent software. Intelligent software is the major component of intelligent manufacturing systems; intelligence is considered to be the next generation technical paradigm of manufacturing, agility being the conjunctive managerial paradigm.

In this chapter, a sub-class of intelligent manufacturing systems will be examined, namely, the one that is centered around CNC technology. CNC technology is built into many types of modern production equipment used in discrete part production; of considerable importance are turning centers, machining centers, and coordinate measuring machines (CMMs), to which most of the examples presented next will refer.

Intelligent software in manufacturing was first implemented at a research level. Although several research systems for robot task planning were developed in the 1970s, it was not until the 1980s that practical AI tools and applications emerged and gave rise to ever-increasing developments. Expert systems were perhaps the first intelligent applications that made an impact in the area of process planning for machining as well as in early Design for Manufacturing exercises. The 1980s can be characterized as the knowledge-based period, whereas the 1990s can be viewed as the neural and fuzzy periods. Similarly, but not as safely, the start of the 21st century can be characterized as the genetic or evolutionary period.

Artificial intelligence has dealt with a number of problems which recur in manufacturing, too. Major themes are planning, diagnosis, cost-based optimization, system configuration, and pattern recognition. All of these are pertinent to the main activities immediately supporting production on the CNC machines mentioned above. Central to these major themes are the fundamental research issues of search, choice making, knowledge representation, and learning with the derivative issues of reasoning about time, dealing with uncertainty, accommodating multiple sources of knowledge, and handling constraints.

Several AI paradigms have been developed along these lines. The most important of those in the light of CNC-based manufacturing are knowledge-based systems, artificial neural networks, genetic algorithms, fuzzy systems and multi-agent architectures and simulated annealing. Their technical basis will be reviewed in Section 9.2. In Sections 9.3 and 9.4 the domains of process planning and process implementation are examined; their nature is discussed and characteristic problems in sub-domains within each domain are analyzed. It is needless to note that for each of these problems a variety of intelligent solutions can be found in international literature. In this chapter, the purpose is not to provide an account of these approaches in a superficial literature-survey fashion, but to give some insight into the mechanics of specific intelligent solutions with respect to the problem's nature in a tutorial fashion. Section 9.5 presents concluding remarks and future research directions.

9.2 The Main AI Paradigms Applicable

9.2.1 Knowledge-Based Systems

Knowledge-based systems (KBS) replaced expert systems (ES), which had a somewhat narrower scope. They are computer programs embodying knowledge—as opposed to data—about a narrow domain structured in such a way as to solve a problem defined in this domain in a similar way to that

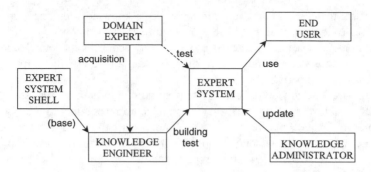

FIGURE I.9.2   Expert system factors.

of an expert human. A knowledge-based (expert) system comprises three basic parts (Figure I.9.2): the knowledge base, the inference engine, and the user interface.[1]

The knowledge base is probably the most important part of the system, because it contains the domain knowledge in suitable formalization, the most common paradigms being:

- Production rules. These contain a condition part and an action part: IF condition, THEN action. These parts may be simple or compound statements containing variables.
- Facts. These are simple unconditional statements about the domain at hand.
- Frames. These are descriptions of objects as collections of attributes. They can be used (referenced) by rules and facts. They can also use (reference) rules, but in that case these are called procedures or methods. Frames can be structured hierarchically, thereby implementing the notions of object orientation, most notably that of inheritance.
- Trees. These are used as rigid knowledge constructs providing rigid paths among pieces of knowledge. They can equally well accommodate objects, rules, and facts.

The inference engine offers the procedures according to which knowledge is combined to provide an answer to a given problem within the domain. The two basic paradigms are forward chaining and backward chaining. Forward chaining (data-driven approach) starts with what is known about the domain and induces new knowledge in the hope that, at some point, this will provide the answer to the problem posed. Backward chaining does exactly the opposite: starting from an assumed answer to the problem, it tries to work its way back through the knowledge base in order to arrive at a piece of new knowledge that is already contained in the knowledge base, thereby proving its initial assumption. Navigation through the knowledge base to combine pieces of knowledge has to be done in a structured manner, in view of the usually large amount of knowledge available. This is the search strategy, e.g., breadth-first or depth-first, referring to the possible new pieces of knowledge (or nodes in the search space) that can be immediately deduced from the current state of knowledge (search space state). The inference engine is coded as a conventional program, but it is often supplemented by so-called meta-rules, which reason about domain rules in a more abstract way.

The user interface in knowledge-based systems is much more important than in conventional software systems. This is because some pieces of knowledge, usually facts, may come from the user through structured dialogs supported by the user interface, but most importantly, because any solution offered by the system has to be explained in detail, again through the user interface.

The inference engine and the user interface, along with tools helping in knowledge elicitation and coding, are packaged together in development environments called "shells." These are mature enough to enable fast development of knowledge-based systems from scratch. The most important phase in this process is that of knowledge acquisition and elicitation.

There are several techniques that can be followed by a knowledge engineer, who is the person interfacing with the domain experts in order to acquire relevant knowledge: interviews, sample problem solving, on-site observation, etc. Basic issues after acquiring the knowledge are to structure it and to ensure its consistency, and to perform final validation, usually on past problems or in collaboration with other experts.

An interesting type of knowledge-based system is the one implementing case-based reasoning. This relies on coded knowledge in much smaller chunks, corresponding to individual cases within the problem domain. On presentation of a new problem or case the most relevant "archived" case is retrieved, and the respective knowledge is applied to modify it to suit the new case.[2]

A further interesting paradigm is a multi-expert architecture termed "blackboard." This is based on a number of knowledge bases collaborating to solve the problem. A common memory space (implementing the blackboard) performs communication of intermediate results to the knowledge bases concerned, updating of individual knowledge base results, and modification of them until a "unanimous" decision is reached by all knowledge bases.[3]

9.2.2   Artificial Neural Networks

An artificial neural network (ANN) is a massively interconnected network of a large number of simple processing elements called neurons. ANNs have been developed as generalizations of mathematical models of human cognition. Their operation is based on four basic assumptions:

- Information processing occurs at many simple processing elements (neurons).
- Signals are passed between neurons over connection links. The pattern of connections between the neurons constitutes the network architecture.
- Each connection link has an associated weight typically multiplying the signal transmitted. The method of determining the weights is called the training or learning algorithm.
- Each neuron applies a usually nonlinear activation function to its net input to determine its output signal, e.g., identity, binary step, binary sigmoid, and hyperbolic tangent function.

Typically, a neuron sends its internal state (activation) as a signal to several other neurons. The basic ANN architecture includes an input layer of neurons that receive the binary or continuous valued input signals, a number of hidden (intermediate) layers that are highly interconnected, and an output layer with one or more neurons (Figure I.9.3). A bias element is considered to be an input neuron with an output equal to unity, playing the role of an activation threshold.

Neural network classifications are not unique. In general, there are as many neural network types and variations as there are researchers in this area. Multi-layer perceptrons (MLP) and self organizing maps (SOM) are two widely used types of neural network.[4]

In the most typical neural network type, training is accomplished by presenting a sequence of training vectors, or patterns, each with an associated target output vector. The weights are then adjusted according to a learning algorithm, so that the network "learns" to produce the required output (dictated by the target vector) to its output layer. This is known as supervised training, and the most widely employed paradigm is the MLP.

According to the back-propagation learning algorithm, every time a training input is presented to the network, the algorithm compares the desired output and the actual output of the network and calculates the error. Information about this error is propagated backward to the hidden neurons and the weights are adjusted accordingly. Network training corresponds to finding the set of weight values that minimizes the total error on the whole of the examples presented. Minimization of the error function is most simply achieved by a lowering of gradient according to the weight vector (gradient back-propagation). After many repetitive presentations (epochs) of the training set, the output of the network may converge to the desired target outputs.

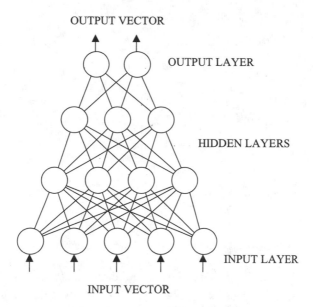

OUTPUT VECTOR

OUTPUT LAYER

HIDDEN LAYERS

INPUT LAYER

INPUT VECTOR

FIGURE I.9.3 Typical ANN architecture.

In unsupervised learning networks, only input patterns are supplied and the network organizes itself internally so that each hidden processing element responds to a different set of input patterns. These sets of patterns typically represent clusters or distinct features.

The typical SOM network has two layers. The input layer is fully connected to a two-dimensional so-called Kohonen layer, in which each neuron measures the Euclidian distance of its weights to the incoming input values.

In the operation mode, the Kohonen processing element with the minimum distance is termed winner and has a value of 1, while the other Kohonen elements have a value of 0. In the training mode, the Kohonen element with the smallest distance adjusts its weights to become closer to the input vector. The neighbors of the winner also adjust their weights to be closer to the same input data. The latter adjustment is crucial because it keeps the order of the input space. Kohonen networks exhibit a typically competitive learning character.

New training techniques and high-speed digital computers enable the construction, training, and implementation of very sophisticated neural networks. Currently large-scale networks can be built in hardware (VLSI chips) and their performance is extraordinary. Finally, it has to be noted that a neural network model can be developed on commercially available software platforms, providing easy means to specify topology, selection of activation function, and a variety of possible training algorithms. However, development of one's own software to simulate neural networks is not that troublesome, and can be the only option indeed, if new training algorithms need to be tried or more direct control on parameters of existing algorithms is desired.

9.2.3 Fuzzy Systems

Fuzzy systems make use of fuzzy logic for mapping the universe of discourse to a value in the [0, 1] interval and of fuzzy sets, which contain elements with only a partial degree of membership.[5]

Fuzzy systems are rule based, but unlike ordinary knowledge-based systems, they can handle situations not covered explicitly in the knowledge base and never fail completely to reach conclusions, but only gradually, similar to human decision makers.

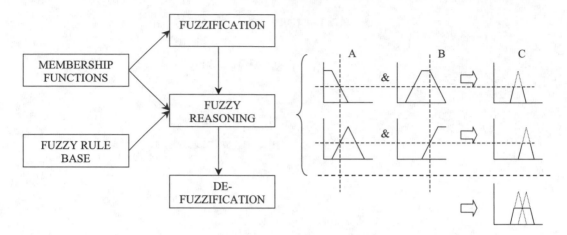

FIGURE I.9.4   Fuzzy systems functioning.

Fuzzy logic reflects the qualitative and inexact nature of human reasoning. The precise value of a variable is replaced by a linguistic description, the meaning of which is quantified by a membership number in the [0, 1] interval. Mapping between the universe of discourse (the variable value space) and the interval [0, 1] is done via membership functions. A common membership function is the triangular one, but other choices exist too, e.g., rectangular, trapezoidal, gaussian, sigmoidal.

Fuzzy inference is the process of mapping from a given input to an output, and involves application of fuzzy rules, fuzzy operators, and aggregation methods. Fuzzy rules form the main processing part of a fuzzy system. They have an antecedent and a consequent part. Fuzzy rule sets have several antecedents combined through fuzzy operators. The output of each rule is a fuzzy set and the output of a collection of fuzzy rules should be a single number resulting from combination of several fuzzy sets into one. This is done through aggregation operators such as "max," "probabilistic or," and "summation" of the respective fuzzy sets (Figure I.9.4).

The resulting set is defuzzified, i.e., converted to a single number. Several methods are possible in order to achieve this, reflecting the geometric representation of a number in terms of the membership function curves, e.g., center of area, bisector, largest of maximum, middle of maximum.

Fuzzy systems are conveniently programmed in conventional languages.

9.2.4   Genetic Algorithms

Genetic algorithms are first and foremost an optimization technique borrowing mechanisms from natural (Darwinian theory) evolution. The fundamental idea of genetic algorithms is that a population of structures, each of which represents a candidate solution to the problem, is set up, and inter-breeding produces new members of the population that cause its renewal toward better structures. Breeding continues indefinitely and is based on genetic operators. These act on single structures or pairs of structures and result in offspring inheriting some of the features of the ancestor(s). The fitter a structure, the more offspring it generates. Fitness of structures is evaluated by a function, which is essentially a measure of the quality of the evaluated structure as a solution to the problem.[6]

A structure in genetic algorithm terminology is called a chromosome, which is implemented as a string of fixed length. Each element of the string is called a gene and represents either a variable involved in the problem or an aspect (part) of it. Each gene may have a set of possible values (alleles) corresponding to various attributes. The fitness function, in essence, calculates a "cost" for combinations of genes and of alleles.

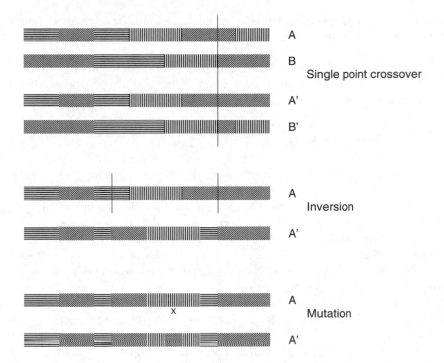

FIGURE I.9.5   Classic genetic operators.

Although a number of genetic operators have been defined in research literature, those are mostly based on the classic ones; namely, inversion, mutation, and crossover (Figure I.9.5). Inversion is applied to a single chromosome and reverses the order of the genes contained between two randomly selected points. Mutation is applied to a single chromosome, too, and changes the allele of a randomly selected single gene to another possible value. Crossover is applied to two 'mating' chromosomes. A crossover point is randomly selected for both of them and the second parts of each chromosome are swapped (simple crossover). This has also been subject of many variations, e.g., partially matched or linear order crossover.

A genetic algorithm specifies how genetic operators are applied to a population to generate offspring, how these replace members of the population, and when the process stops. The first step is to devise how a possible solution can be coded as a chromosome. The next step is to define an initial population. The third step is reproduction, i.e., copying chromosomes according to their fitness in order to "breed" them. The fourth step is application of genetic operators to the breeding population, e.g., crossover can be applied first with fairly high probability, then one of the new chromosomes can be chosen randomly and inversion can be applied to it with fairly low probability. The resulting chromosome can undergo mutation with low probability, perhaps decreasing with time, and the new chromosome can be taken as the offspring. Fitter offspring replace their ancestors.

Genetic algorithms are powerful when the problem tackled is combinatorial, and most effective when small changes result in very nonlinear behavior in the solution space. This is because they search simultaneously tiny distant parts of the solution space. They do not rely on heuristics, and due to the "subtlety" of their working, they do not get easily stuck in local optima. They are also suited to parallel computing.

In designing genetic algorithms (or genetic machinery) conflicting decisions may have to be balanced between. For instance, free combination of gene blocks should be allowed, but crossover

Intelligent Systems: Technology and Applications

must not kill many useful blocks potentially leading to good offspring. Also, mutation should not suppress inheritance of useful gene blocks, yet mutation rate (probability) should help avoid premature uniformity of the population, etc.

Finally, genetic algorithms are more practical to program from scratch, given that in most cases substantial modifications have to be implemented concerning genetic operators and the algorithmic procedures themselves.

9.2.5 Simulated Annealing

This is a method drawing from annealing of metals, which leads the material to a state corresponding to a global minimum of its internal stress energy by gradual cooling. By analogy, simulated annealing searches for the global minimum of an objective function in a space of solutions.[7] The fundamental idea is the acceptance of solutions with increased value of objective function (uphill moves) with small probability, which decreases as the search evolves. In this way, local optima are escaped from and the region of feasible solutions is further explored. The probability of acceptance is controlled by a parameter called temperature, which is gradually reduced according to a cooling schedule; e.g., temperature is held steady for a number of iterations and then it is multiplied by a constant (reduction) factor (geometric rule) (Figure I.9.6).

The initial temperature and the reduction factor can be pre-defined. The number of iterations performed at the same temperature has to be determined according to a criterion, e.g., expressing it as a function of the maximum size of the neighborhood. A neighborhood is a fundamental concept in local optimization methods. It is the set of all solutions that can be obtained from a given solution by performing one transition (legal move from one solution state to another). The neighborhood structure is the set of all pairs of antecedent-descendant solutions.

The stopping criterion has to be determined, too. For instance, it may be controlled by a counter which is initialized to zero each time the best solution found so far is updated, and is incremented by one each time the percentage of accepted moves in all iterations performed in the same temperature does not exceed a given threshold value.

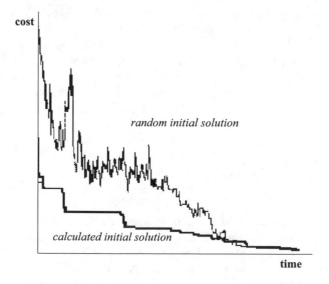

FIGURE I.9.6 Simulated annealing convergence characteristic curves.

A simulated annealing algorithm, apart from the decisions mentioned above, which are generic, has to specify the objective function, the neighborhood structure, and the initial solution to start with. These are choices related to the nature of the problem.

Simulated annealing is used for optimization of discrete variable problems and is invariably coded in custom-made software, given the relative simplicity of the algorithms.

9.2.6 Hybrid Paradigms

These are regarded as the most promising future development in AI because, in applications where they fit, they exploit the strengths of the individual paradigms which they combine. There are two main categories of hybrid AI systems, namely the "delimited cooperating" systems and the "combined paradigm" systems. Perhaps, only the second type deserve the characterization of hybrid.

An example of delimited cooperating system is a genetic algorithm-neural network system, where the neural model is used in order to improve a particular aspect of the genetic algorithm function, e.g., initial population generation or definition of an adaptive genetic operator.

An example of a combined paradigm is a connectionist expert system (or knowledge-based neural network), where the data attributes or variables are assigned to input elements of the neural network, target concepts are assigned to the output elements and intermediate concepts to hidden elements; the rule strength is represented by the respective weight of the premise-conclusion synapse.

Much research is expected into hybrid systems and their potential applications in manufacturing; however, for the time being, the lion's share of those applications is taken by single paradigm systems.

9.3 The Process Planning Domain

Process planning is the systematic specification and sequencing of processes by which a product is to be manufactured economically within a given manufacturing system. A process plan usually contains the route of operations and specification of each operation in the route, i.e., type of operation, operation parameters (feed, speed, etc.), machine, tool(s), jigs/fixtures, standard time, setup time, and additional instructions (for setup, in-process inspection, etc.).

Variant computer-aided process planning (CAPP) is based on the notion of group technology part families. All parts in a part family have similar shape and/or process plans (similar type, sequence, and number of operations). When a new process plan needs to be produced, the standard process plan of the appropriate part family is retrieved and modified to tailor the plan to the particular part.

Generative CAPP relies on manufacturing knowledge captured in the system and not on standard process plans. Thus, process plans are "generated" from scratch. Such knowledge requirements are usually product type-specific and company-specific, resulting in dedicated CAPP systems, e.g., for sheet metal parts, forged parts, machined parts, turned parts, etc.

The tasks involved in process planning can be distinguished into two basic categories: those strongly associated with part geometry and those weakly associated with it. Association implies a need for geometric reasoning, therefore a third task category arises; namely, that of automating precisely this geometric reasoning.

Intelligent process planning can be tackled at different levels of detail. At a relatively coarse level of detail, one may be interested to formulate a generic process plan, i.e., one that is not bound to specific machines and tools, but only to machine and tool types. Generic process planning encompasses the tasks of operation selection, operation sequencing, and setup planning, all at a generic level. Tool-type and machine-type selection can be thought of as secondary decisions. In reality, the complexity that results from possible "nesting" of these tasks in decision loops as well as advice from human process planners point toward a hierarchical approach.

Detailed process planning can take over from generic process planning or it can be performed directly for a sufficiently restricted family of parts, for which a skeletal process plan already exists, which is a "planning by analogy" paradigm. Decisions refer to setup, including fixture selection, tool path derivation, tool selection, and process parameter selection.

9.3.1   Generic Process Planning

The nature of generic process planning is threefold: first, it is experience intensive, second, it is open to multiple solutions (alternatives), and third, it is time dependent, i.e., it contains (manufacturing) steps which are arranged in time domain.

The first characteristic—experience base—directly points to a knowledge-based solution. Knowledge bases can be put together to allow derivation of feasible plans, or even good plans, according to criteria defined by human experts.

The second characteristic immediately poses the question as to how plan quality can be assessed, whether an optimum plan can be arrived at, and, of course, with which procedures and tools. This issue can be answered by AI optimization techniques, notably genetic algorithms.

The third characteristic is key to the previous two issues; namely, each part state is arrived at after a (manufacturing) operation is applied on the previous state and can lead to a number of alternative states associated with respective operations. Part states are mostly visualized as intermediate geometric shapes, and their time association is an expression of the constraints binding them. This is an issue of constraint modeling and fuzzy systems.

9.3.1.1   The Feasible and Open Process Plan: A Knowledge-Based Application

9.3.1.1.1   *Search Space Traversal Using Heuristic Pruning*

A process plan can be defined as a series of steps leading from an initial stage to a final stage (goal). There is a multitude of alternative steps and a multitude of levels, making the planning space very large. Each step can be represented as a node in a tree, and the possible next steps are its successor branches (Figure I.9.7). Each branch can be evaluated as to how promising a route it offers toward the goal—the final solution. Then, only the most promising branch can be examined for further expansion. If failure is encountered through one route, backtracking can "undo" the latest committed choice and an alternative route is taken. Alternatives can thus be examined in a breadth-first or a depth-first fashion with respect to the structure of the planning tree. Theoretically, forced backtracking can yield alternative plans.

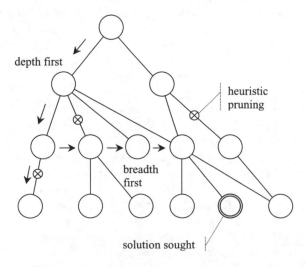

FIGURE I.9.7   State space traversal using heuristics.

Obviously, the hard part of the exercise is to attach weight values or certainty factors to the branches leaving each node. These have to be justifiable according to some criterion, but this criterion is hardly objectively justifiable itself.

In the CNC turning domain, where domain knowledge is very much empirical, each branch can be associated with a rule defining the conditions under which the rule and the respective operation are feasible and the result, which is the execution of an operation (shape modification or other) leading to a new state (node). Conditions can be proved either through evidence available in the working memory of the system or interactively by asking the user, thereby occasionally enriching the knowledge of the system.[8]

9.3.1.1.2 *Planning with Constraints Paradigm*

This paradigm is old, but well worth reviewing as, perhaps, the most classic one in operations sequencing. It works with terminal features, each of which corresponds to one or two manufacturing operations (preliminary—roughing and final—finishing). These operations initially constitute a flat plan, i.e., an unordered list. The goal of the planning process is to order this list, including additions of new intermediate operations and consolidations of operations as required by manufacturing knowledge. Ordering, additions, and consolidations are considered constraints. These are specified in rules of the type: conditions ⇒ pieces of advice. Conditions are matched against the working memory of the system, i.e., presence of features, machines, etc., and when a match is found, the relevant piece of advice can be applied. This piece of advice essentially imposes a constraint to the current state of the process plan.

Each piece of advice carries a weight factor between one and ten, representing the importance attached to its satisfaction. The final solution (transformed initial plan) might be reached by straightforward application of pieces of advice (constraints), although this is highly unlikely. In reality a piece of advice may contradict the current plan, thereby causing a conflict. The constraint imposed by a piece of advice is propagated through the whole current plan in order to check for more conflicts.

A conflict is resolved by detecting which piece of advice of those applied in a sequence should be removed—in fact, this is the one with the lowest weight. Removal means backtracking on the plan, i.e., removal of the constraints applied as a consequence of application of this piece of advice as well as the ones that followed it. Removal of a piece of advice also means that, in order to prevent looping, the opposite piece of advice has to be input with weight equal to the highest weight of its former successors. The best solution corresponds to a minimum total weight of the rejected pieces of advice.

Pieces of advice are applied until they are exhausted without causing conflicts; when they are exhausted there is still the possibility that pending rules exist providing pieces of advice which are not satisfied by the current plan. In that case, new pieces of advice are tried that make the pending rule(s) inactive.

This kind of reasoning has been implemented in LISP,[9] but it could also be implemented in Prolog, making use of its backtracking feature as well as its in-built database management system. The system clearly contains two parts: the planner (i.e., the rule manipulation engine) and the rule base itself. A sample rule in LISP specifying the sequence of drilling two co-axial connected holes looks like this:

(is-a &x hole) (is-a &y hole) (open-into &x &y) (not open-into &y &x)

⇒ (9 (before (roughing-cut &y) (roughing-cut &x)))

9.3.1.1.3 *Meta-Knowledge Paradigm*

This paradigm is heavily based on knowledge, in order to ensure that process plans derived are feasible. Their "optimality" is controlled via so-called "meta-knowledge," i.e., typically rules that direct the overall flow of the process planning tasks. The point made by this method is that "optimization" of plans is a logic-controlled task and it should be possible to imitate the corresponding flow of thinking

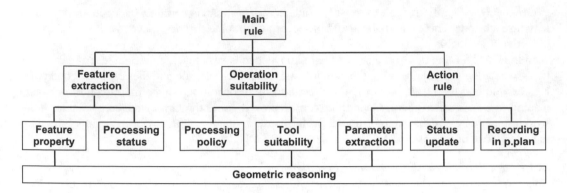

FIGURE I.9.8   Hierarchy of rule types in generic process planning.

of the human process planner. Therefore, the concept of weight factors and probability factors that are used by other paradigms to prioritize decisions and arrive at optimum plans are considered to be artificial constructs.[10]

Considering the machining domain, rules are categorized into several types (Figure I.9.8). Main rules are responsible for operation selection. Start and completion of all operations within a set-up are checked by separate rules. Another main rule removes the current tool from the working memory if no more work for it can be found. The body of a main rule is structured as follows:

Condition part: Feature extraction, tool suitability, feature properties, processing status, operation suitability

Action part: Operation execution

Feature extraction rules either select a feature directly or under certain conditions, e.g., in milling, a pocket being part of a shallow compound when the feature above has been roughed. This rule makes reference to another rule, defining how it can be found whether or not the feature above has been roughed, i.e., a current processing status rule. The rule also makes reference to a property of connected features called "critical depth"; a compound feature is considered to be deep if its minimum cross dimension is smaller than $\frac{1}{8}$ of its depth. Depth and cross dimension are primitive characteristics of the feature. Feature extraction rules also make sure that the feature selected is accessible in the current setup, thereby performing setup planning through feature accessibility directions.

Tool suitability is a check enforcing use of the tool already present in the tool holder.

Operation suitability rules are necessary but not sufficient conditions for a pertinent operation to be performed; e.g., boring is allowed if high diameter accuracy with nonstandard diameter is required (one case). These rules usually make reference to further rules defining processing policy of the company, e.g., the definition of a nonstandard diameter.

Action rules perform the following sub-tasks: obtaining operation parameters such as slot machining length, updating the feature processing status (e.g., "pocket has been roughed"), recording an operation in the process planning sequence or selecting the next set-up as the access direction of the majority of the remaining features.

There are also auxiliary geometric reasoning rules. Note that main rules are forward chained, i.e., start with available data and look for a conclusion, whereas other rules are backward chained, i.e., try to prove a given conclusion.

The process planning procedure examines the alternative operations one by one. If an operation can be executed, the corresponding shape feature is marked so that it is not reconsidered in the next round. If no more features are available for processing, the procedure stops. Constraints referring to inability or to necessity to execute two operations sequentially due to feature interactions are taken into account through rules, too. The difficulty in this example is that changes induced by the executed operations are effected as "flags," whereas the ideal would be that the geometric model of the part was updated.

A more suitable domain for this paradigm would be one that did not involve shape changes of the workpiece, e.g., inspection planning on a CMM. Rules in that case (for generic and specific process planning tasks) should select shape features for inspection, based on the tolerances allocated, determine accessibility of each inspection feature, select datums for each measurement, select tool (probe), and issue measurement cycle instruction. Probe path planning is normally outside the scope of such a system.[11]

9.3.1.1.4 Incremental Diagnosis Paradigm

Diagnostic type of process planning consists in letting design define the steps of the process plan and having a rule base in the background to check compatibility of the design actions with manufacturing practices. This is based on features which, are not recognized *a posteriori*, but used as design building blocks. Key to the implementation of such an approach is the definition of a library containing exactly the type of features that a process planner perceives in a part. For the approach to be open to extensions and independent of specific manufacturing "cultures," these features have to be user definable (UDFs).

The process begins at a random state (n) (Figure I.9.9), which could be a simple blank or a casting, if starting a new project from scratch, or a half-made product, respectively. The user selects from the UDF library the desired geometry with which to increment the part. After specifying the UDF dimensions and the position of the new feature, a new ($n + 1$) state results.

In this case though, together with the UDF, its rules apply, which prevent state ($n + 1$) if this is not possible to manufacture, diagnosing the problem and issuing a warning and advice on remedies.

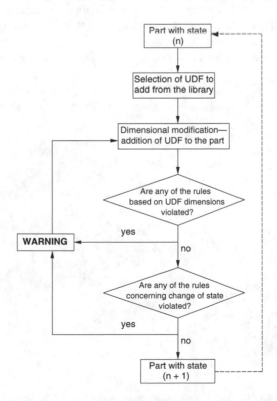

FIGURE I.9.9 Process planning by incremental diagnosis principle.

Rules fall generally into two categories: first, those influencing only the UDF dimensions, and second, those that govern the transition to state $(n + 1)$, taking into account the parameters of previously inserted features.

Rules have to be categorized to ensure fast response and hence practicality of the system, and also to further develop intelligence in a structured and extensible way.[12]

9.3.1.2   The Feasible Inherited Process Plan: A Case-Based Application

This approach is close to classification or clustering, because it exploits similarity of the new artifact with those in the "case base." A case-based process planning system consists of typically four elements. The first is the retriever, whose task is to compute a similarity index between the new planning case and the stored cases. The second element is the modifier, whose task is to adapt the plan corresponding to the most similar case to the new one. The third one is the simulator, whose task is to check feasibility of the new plan, and the fourth one is the repairer, whose task is to provide feasible solutions when the simulator gives notification of unfeasibility.

This approach resembles the variant process planning mode. However, variant process planning is based on rigidly defined standard process plans for part families (Figure I.9.10), represented as group technology composite parts, and therefore it suffers various drawbacks, e.g., the plans need to encompass all possible features to enable editing by removal, and it is hard to accommodate variations that are due to feature interactions.

The key issue in case-based process planning is the definition of a sufficiently comprehensive similarity matrix. This should be based on geometry as well as on other features, such as tolerances, surface finish, hardness. Shape features can be defined and any part can be described in terms of these. However, not only can features be part matched but also their interactions. The latter in some cases (e.g., in rotational parts), are closely associated with feature sequence (adjacency) and relative size, so this information is sufficient for complete representations. In other cases, feature interactions have to be explicitly defined as such, which is not a trivial task. Similarity index can be defined in terms of number of features that are identical in the two parts being compared, plus number of modifiable features (including material properties), and is usually normalized within a certain range, e.g., zero to one.

A parametric type of standard process plan is much more flexible and can accommodate variations of presence, dimension, and interaction of features. This needs to be based on a language, such as

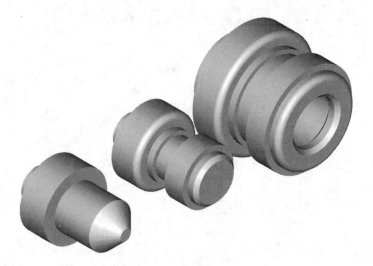

FIGURE I.9.10   Similar parts planned according to case-based reasoning.

ALPS,[13] which provides conditional and Boolean structures as well as grouping mechanisms applied to operations that make up a standard process plan.

Modification of features and plans is definitely possible in the interactive mode. Otherwise, automatic editing has to be based on automatic comparison of the features of the particular part and those of the archetype. The rules according to which modification is performed could be coded into an expert system. However, the complexity added in this way undermines the simplicity of the approach.

However, even after modification, it is not completely certain that the new process plan will work. The simulator can then be used to identify failures and send them to the repairer for further processing. The task of simulation and repair might be seen as superfluous to the modifier. If the latter is sophisticated enough (in terms of embedded knowledge), then simulation and repair might not be necessary. However, in general, new cases arising after modification or after repair are stored in the case base, thereby implementing a learning process.

Optimization or near optimization of the plans is ensured by the fact that the stored cases are already considered to be optimized and any discrepancy of the new plans are taken care of by the modifier or repairer.[14]

9.3.1.3   The Optimum Process Plan: A Genetic Algorithm Application

Process plan optimization is the ultimate target of process planning. In complex situations there may be a large number of feasible plans that satisfy all applicable constraints. Among those plans, one must be determined as the optimum plan according to one criterion or to multiple criteria. Enumerating all possible plans, i.e., developing the whole planning space explicitly in all but the simplest cases, is too large a task and certainly not elegant. Heuristic pruning of the search space (see 9.3.1.1.1) is a possibility, but powerful heuristics are always associated with the particular planning domain, and there is no guarantee that enough of those will be found for new domains.

Implicit, as opposed to explicit, generation of process plans and in-built evaluation with an aim to optimization is the approach offered by genetic algorithms. Whereas the genetic algorithms to be used may be just the classic ones (crossover, inversion, and mutation), perhaps with some variation depending on the particular domain, the objective function is a key issue and is associated with the particular model of the planning problem adopted.[15]

An example is a cost function adding up processing cost and setup cost for each feature. Processing cost for a feature is normally independent of the order in which this is processed, while setup cost depends on what feature was processed before and includes cost of tool change, cost of setup (orientation) change, etc.

Depending on the parameters that are taken into account in the optimization, representation of a process plan consists of consecutive n-tuples of the type $V_i\langle v_1, v_2, \ldots, v_n\rangle$ with $i = 1, \ldots, m$, where m is the number of features whose manufacture is planned and n is the number of significant variables, e.g., in the machining domain features, machines, tools, and setups are the four variables of interest. Sets of n-tuples that cannot be separated by crossover appear between delimiters, e.g., "$\$$."

A key issue is anteriority constraints between tuples. These may be taken into account either in the cost function with artificial (large) costs involved when such a constraint is violated, or at a preprocessing stage when each plan in the population examined is made up of individual nonbreakable sequences of features (corresponding to operations) that absorb any anteriority constraints. The former solution risks too many rejections to the point where the whole process of interbreeding is undermined. The latter solution has to be complemented by rules ensuring that changes introduced by mutation, inversion, and crossover operations are allowed with respect to tuple ordering and when they are not, then suitable changes in the tuple are applied to rectify the situation, if possible.

The sequence of features in the plan belongs to the genotype of an individual (i.e., the inherited traits), as opposed to the phenotype (i.e., all inherited plus acquired but not inheritable characteristics).

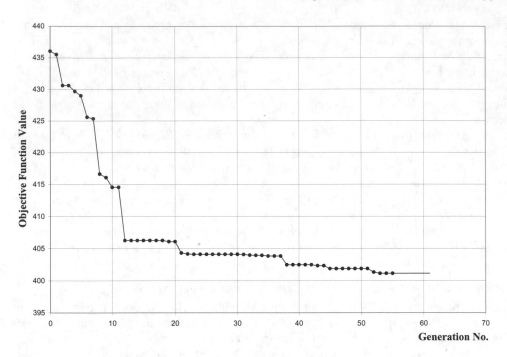

FIGURE I.9.11   Typical evolution history of genetic optimization.

Tools and setups can be seen as belonging to the phenotype, as more than one value of those may be alternatively allocated to each feature.[15]

Filling up any free spaces in the offspring chromosome in a random, but admissible fashion with unused tuples is a possibility; if there are still noncomplete offspring chromosomes, they are rejected.

A successful interbreeding session converges to a least cost after a number of generations, usually corresponding to more than one possible plan (Figure I.9.11). Optimization time depends on the size of the problem in a quadratic manner. It seems that a good compromise between which variables belong to the genotype and which to the phenotype of the individual is important for the quality of the solution.

9.3.2   Detailed Process Planning with Geometric Reasoning

Generic process planning does not look in detail into particular problems requiring further geometric reasoning, due to their complexity and the resulting inability to be considered at a generic planning level. A most characteristic task whose results might influence back generic process planning is setup and fixture planning mainly for machining center operations. Another characteristic task whose results build on generic process planning is tool path derivation for milling and inspection applications.

9.3.2.1   Setup Planning: A Fuzzy Sets Approach

Setup planning for CNC turning is inherently conducted in generic process planning at an acceptable level of detail. However, for processing on CNC machining centers, the picture is different. A setup is defined by an orientation vector of the part and a set of features to be processed. Definition of setups in this case is influenced by a multitude of factors which are often imprecisely interrelated, e.g., accessibility of the features to be machined, tolerances relating those features to others, and to reference features of the part, without special mention to fixture element design (Figure I.9.12). This is a sufficiently simplified picture.

FIGURE I.9.12 A set of setups for machining a prismatic part.

The objectives of setup planning are that the number of setups should be minimum, every setup should be easy to implement in fixturing terms, and the setup plan should ensure maximum dimensional accuracy of the part.

The fuzzy sets approach is an optimization approach which bases decision-making solely on the relations of features. Feature A may be dependent on feature B through a relation R with a dependency grade of strength x, which is a normalized value in the range zero to one. Relations may express geometry (e.g., entering-to-ending face), size tolerance (between features with the same or with different approach directions), datum (e.g., in orientation tolerances), fixturing, heuristic (expressing good practice rules of thumb), and some inferred statement (according to transitional logic).

Instrumental to the approach is the definition of a number of sets: FA as the set of all features available on the workpiece and FM as the set of features required and manufacturable.[16] FM features are related to FA features, and FM features may be related to other FM features. A fuzzy matrix contains all dependency grades for FM to FM dependencies, a mapping of FM × FM to [0, 1]. Similarly, there also exists an FM × FA mapping to [0, 1]. Zero values mean that the features can be made in any order. FM1 contains FM features related to other FM features through some relation, and FM2 contains FM features related to an existing set of FA features, a subset of FA marked as FA+. Set MP represents features from the FM1 set, which are most preferred to be machined next. MR represents features from the FM2 set, which are most closely related to the FA+ set.

The MP set is obtained through rules reasoning to obtain a dependency grade matrix from which columns corresponding to FM features with no relations are removed. The membership grade of ith feature FM1 in the MP set is defined as the ratio of the sum of i-column values to the sum of the i-row values plus the i-column values. The row values in the matrix represent the dependency of the respective feature on the rest of the features to be machined, whereas the column values represent the dependency of the rest of the features on the respective feature. An analogous procedure is conducted for the MR set.

The intersection of MP and MR sets gives the MPR set, which contains FM features that are most preferred and most closely related to an existing set of features FA+. The problem to be solved is to determine a best setup from the MPR set. If SU is noted as the set of setups containing features which are most preferred and most closely related to an existing set of features FA+, and *Si* is noted as the membership grade of setup *i*, then *Si* will be the sum of the membership grades of all features of setup *i* in the MPR set. The next setup is selected from set SU as the one with highest membership grade. When two setups have the same membership grade, the one with more features is selected. First, a vector (denoted by M($_-$)) containing the membership grades of the features in the MPR set in the order in which they appear in the MPR set is obtained. An approach direction matrix [AD] of these features is constructed and a multiplication is performed to find membership of a set of possible setups: M(SU) = M(MPR)[AD]. The setup with the highest value will be selected.[16]

The above procedure is quite mechanistic. The relations between features are taken into account, but more elaborate feasibility checks as to fixture planning cannot be performed. It is a strict setup planning optimization approach.

9.3.2.2   Fixture Planning: A Blackboard Approach

A fixturing plan is defined by a set of setups, and for each of them a set of positioning, clamping, and supporting features as well as the corresponding accompanying tools. In addition to setup planning influencing factors, positioning, locating, and supporting principles in 3D coordinate space, availability of appropriate fixturing tools, etc. come into play. In the general case, setup and fixture planning are essentially the same task, which is, unfortunately, closely related to the operations selection and sequencing task.

A single knowledge-based approach might make use of strategic rules and main body rules. The former emphasize one of the setup decision objectives at the expense of the others; e.g., for large batch sizes minimization of number of setups is most important, for small batch sizes this switches to ease of setting up, and for critically capable machines dimensional accuracy attainable by the setup is predominant. Main body rules might perform synthesis (e.g., for machines with an index table, a minimum number of setups is attained by starting with a setup where features approach directions are normal to the table axis), impose constraints (e.g., every feature belonging to a setup has to have a location face in that setup), or allow heuristic discrimination (e.g., location faces have to have an area larger than 10% of the average face area of machinable features in a setup). Geometric reasoning rules might require elaborate calculations, e.g., a setup is stable when the vertical projection of the workpiece center of gravity passes through the convex hull of its support face. Although rules in a single knowledge-base paradigm may be associated with various domains, their extent is not so great as to justify a "separate expert" character. In the opposite case a blackboard approach is in hand.

A blackboard system might make use of multiple separate knowledge bases and tools, each following a different paradigm, but all working together collaboratively to solve the setting-up problem by contributing their own piece of advice. Even procedural decision tools such as CAD or CAE packages may be connected, provided that a communication mechanism is available, typically based on the object orientation paradigm.

One knowledge base might provide design heuristics such as the 3-2-1 principle (three support points on one plane, two location points on a second orthogonal plane, and one location point on the third orthogonal plane), the maximum distance between fixturing elements principle, the principle of cutting forces absorption by supporting and locating elements, etc. (Figure I.9.13).

Another knowledge base may tackle the stability problem based on a kinematic model of the workpiece–fixture system, referring to three separate situations of resting, clamping, and processing (machining). Clamping force distribution (respecting the equilibrium conditions) for processing may be conducted using linear programming.

Yet another knowledge base might tackle feature accessibility by checking interference of tool paths and fixturing elements using a solid modeling engine.

Deformation due to clamping forces at contact regions, as well as globally, might be checked by another agent running on a finite element package.

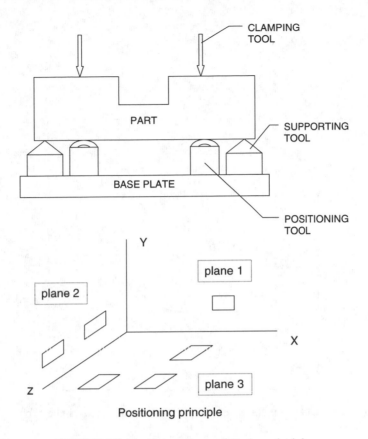

FIGURE I.9.13 Prismatic part fixturing principles.

A blackboard approach is thus "rich." The blackboard structure acts as a posting board of individual parts of the solution. A special processor activates the knowledge base that is necessary to resolve arising discrepancies based on the status of the blackboard, and modifies the current blackboard data according to the new execution results (also filtering out unacceptable solution parts). Blackboard reasoning stages pertain to modification and initial configuration after checks for interference, stability, and deformation. When all "agents" agree on solution feasibility or unfeasibility, the solution procedure stops.[17]

9.3.2.3 Tool Path Planning: A Nested Genetic Algorithm Approach

The derivation of a tool path is a task closely connected to tool selection. In milling tasks, especially pocket milling in the presence of internal protrusions, which in milling jargon are termed "islands," there are several alternative strategies that can be adopted (e.g., in flat pocket milling these are zigzag machining, offsetting, whereas in sculptured pocket roughing these may be constant z-height, 3D offsetting, etc.).

Calculation of tool path is strictly based on algorithmic mathematical methods, most notably Voronoi diagrams, convex hull methods, and other computational geometry algorithms. However, optimization-related tasks within toolpath derivation are open to AI approaches. A characteristic application concerns cutting pass design.

According to the perceived need of freeform surface roughing in industry, the criteria established for cutting pass design may be the least machining time and the least remaining material in the form of "scallops." An important factor is the correct tool choice, according to its geometric characteristics

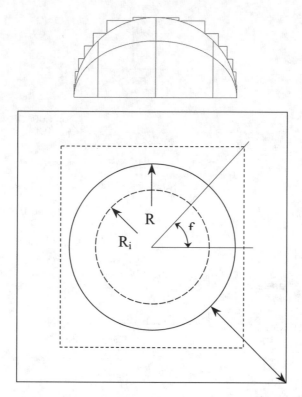

FIGURE I.9.14   Parameters of tool path optimization of hemisphere roughing.

(diameter, depth of cut, engagement, etc.) and the applied feed rates. The required solution is the optimal tool combination for the roughing of the given surface and pass (scallop) height. Hemisphere machining from a prismatic block is considered as an illustrative example.[18]

The objective function is $O = \frac{\lambda_1 \cdot T + \lambda_2 \cdot E}{\lambda_1 + \lambda_2}$, where λ_1 and λ_2 are weighing coefficients, T is a function summarizing the cutting time plus tool changing time, and E is a function representative of the total remaining volume to be removed by finishing. T sums up the length of tool path segments along X, Y, and Z axes and is clearly a function of tool diameter and allowable cutting edge length. E is derived as a sum of remaining volume of each scallop by subtracting the volume of a spherical slice from the volume of the corresponding cylindrical slice; due to axis symmetry, volumes can be simplified into areas.

An optimization process is designed centered around objective function O. Optimization is based on calculation of cutting time using angle f as variable, calculation of the remaining volume using number of scallops N as variable, and overall scallop strategy using the number of scallops and scallop height distribution (Figure I.9.14).

As an example, for a hemisphere with $R = 50$ mm and an original block $130 \times 150 \times 50$ mm^3 and $\lambda_2 = 0$, the scallop height distribution is 5.3999, 4.0370, 4.0534, 4.2331, 5.2504, 5.7116, 6.0862 mm for seven scallops and three tools are used from a pool of 21 available.

9.3.3   Feature Recognition

Invariably, geometric reasoning in planning tasks is carried out in terms of features, i.e., shape regions of specific significance in manufacturing method terms. An intelligent CAPP system should not rely on description of these regions by the user, but it should be able to recognize them from computer part models. The two schools of thought on this issue support either feature-based design or feature recognition. The former obliges the designer to model the workpiece as a collection, or even a sequence, of manufacturing features. The latter extracts such features according to pre-defined templates from a low level CAD representation, usually a *B*-rep solid model.

Manufacturing processes can be associated with shape-forming operations, e.g., volume removal, bending. This makes feature-based design very attractive for modeling "manufacturable" shapes. A library of such features for modeling the workpiece on a suitable feature-based CAD system streamlines the design to such a point that it resembles very closely a manufacturing process simulation starting with the blank and ending with the desired final shape. Such a paradigm is less straightforward when the shape obtained can be determined strictly through elaborate methods, such as finite element simulation.

It would seem at first that in such a context feature recognition does not provide any substantial advantage over feature-based design. However, there are two paradigms which necessarily involve feature recognition. First, the set of features available to the designer for modeling may not be representative of the actual operations possible to execute in a specific machine shop. Therefore, a feature transformation has to take place according to specific rules.

Second, there is always a need to discover specific types of relationships among features, e.g., proximity, axis parallelism, overlap. By construction, such relationships cannot be built into the feature definitions as such; they have to be discovered in retrospect, their discovery being a form of feature recognition. In the same context, the combined presence of two features may alter the nature of either one or both of them. This is often called the "feature interaction" problem.

Feature recognition requires extensive geometric reasoning, which draws on low-level routines such as Boolean operations, e.g., for comparing volumes, and ray casting, e.g., for determining feature access directions.

When feature recognition is performed in the planning realm, it is often desirable that this is done incrementally, i.e., the shape (blank or workpiece) is updated by actually simulating the shape-forming operation (subtraction, etc.) resulting in a new shape on which the next feature is to be recognized.

Part representation in terms of features consists of a tree whose leaves are instanced feature templates and whose nodes are operators acting on the features (usually Booleans).

Feature recognition can be knowledge based (using predicate logic) and learning based (using neural networks) or pattern based (using graphs or grammars). The first two will be examined in more detail next.

9.3.3.1 Knowledge-Based Feature Recognition: Rule-Predicate Logic Applications

Feature definition is considered the most crucial step in feature recognition. The most straightforward approach for defining features is the use of faces and their connections along edges characterized as either convex or concave, depending on whether the corresponding convex angle contains material or not. This is the so-called attributed adjacency graph (AAG) representation, where faces form nodes, edges form branches between nodes, and convexity information is entered as a binary label on branches. A complete part expressed in that form can be traversed and sub-graphs matched against the AAG structure defining particular feature types. Matching can be based on rules. However, the information contained in the AAG representation may prove to be poor for some applications. A partial remedy was the introduction of an augmented AAG, where additional information is included in the form of node and branch labeling, labels referring to face geometry and to possible tangency along edges.

A more open approach extends the notion of augmented AAGs further. Blocks (face sets) are used instead of just faces. "Key blocks" are defined for every feature, i.e., characteristic collections of faces used to immediately differentiate between various features, and there are also "normal blocks." A block is a set of entities grouped together according to some criterion. Relationships exist

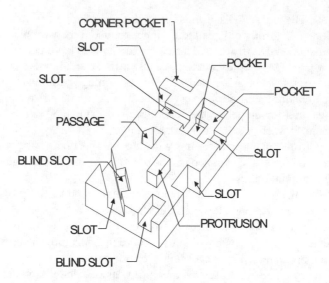

FIGURE I.9.15   Recognized features.

among members of the block as well as between entities of the block and entities outside it, either belonging to other feature blocks or being outside the feature altogether. Therefore, a hierarchical relationship system is used to define feature templates with the following levels: feature, block, solid modeling entity.[19]

Standard blocks can be defined and then assembled at will to create a variety of feature templates. Examples of building blocks relating to prismatic components are defined (in Prolog) as follows:

Floor: planar horizontal profile with concave edges and an upward outward vector
Open floor: floor with both convex and concave edges
Wall: a planar vertical profile
Wall loop: a recursive loop pattern of walls adjacent to each other

Predicates describing spatial relationships of entities within these blocks are, e.g., next_to, parallel_profile, etc. There are also parameter extraction predicates used to calculate the minimum corner radius of a planar profile, the minimum and maximum height of a wall, etc.

Feature definition is followed by the actual recognition phase, which is typically an attempt to match the actual component database with the feature templates. A subtle point in this process is that after a certain entity has been allocated within a key block to a (recognized) feature instance, it is reserved and cannot be allocated to any other feature instance subsequently.

Examples of simple features defined and recognized with the above building blocks are as follows (Figure I.9.15):

Corner pocket: an open floor nesting no other profiles, its concave edges being adjacent to wall
    blocks.
Slot: an open floor block nesting no profiles with its concave edges being adjacent to wall blocks
    which are parallel planes. Wall height (minimum and maximum) and slot width are characteristic
    parameters.
Pocket: a floor block nesting no other profiles and being adjacent to a wall loop. Characteristic
    parameters are minimum corner radius of the floor and minimum and maximum height of the
    wall loop.

An example of a feature recognition rule is as follows: If the feature has a floor block and the floor block nests no profiles and the floor block is adjacent to a wall loop, then the feature is pocket.

Addition of new feature templates and their subsequent matching with the component geometry are straightforward and efficient.

Note that features do not always have a concrete shape, but they may possess a generic (repeating) property, e.g., a stepped hole with any number of steps. Such features may be defined recursively using predicate logic in Prolog.

9.3.3.2 Machine Learning-Based Feature Recognition: ANN Application

Feature recognition based on ANN has the potential of allowing custom definition of features. This is feasible only when the representation of shape is appropriate, i.e., a transformation from 3D primitives to patterns containing the information required.

Based on AAGs, an adjacency matrix (AM) can be built expressing the edge relationships between faces of a feature object. The lower triangular region of the matrix corresponds to concave edges and the upper region to convex ones. If there is a connection between the ith and the jth face of the object (where $i \neq j$), either the element AM[i, j] or the element AM[j, i] will be made equal to 1, according to the nature of the edge. The faces of the AAG are assigned a mark, according to a specially designed marking scheme,[20] and the face with the highest mark is placed first in the sequence. The next place is taken by the face, which is connected to the previous one and has the highest mark among the remaining faces.

It can be observed that a particular pattern is created for each feature class. For example, pockets have the first column of the AM filled with "1" elements on the two sides of the first diagonal, and the second element of the last row is always equal to 1. Passages have elements on both sides of the diagonal and the first element of the last row is equal to 1. In order to resolve recognition ambiguity, in some cases the number of external faces have to be taken into account, too.

Based on the adjacency matrix, a "representation vector" of 20 bits is constructed, in order to circumvent the scale factor problem when features may contain either a few or a few hundred faces. Each of the first twelve bits of the representation vector is filled in after answering positively or negatively corresponding questions referring to "standardized" regions of the adjacency matrix defined appropriately; e.g., "first diagonal," "side diagonal," "side diagonal couples," "first normal,"

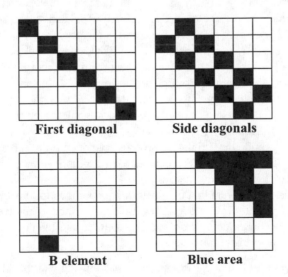

First diagonal   **Side diagonals**

B element   **Blue area**

FIGURE I.9.16   Adjacency matrix regions standardized.

FIGURE I.9.17   ANN training examples (pockets and passages).

"A element," "B element" of the matrix, "Blue area," "Red area" (Figure I.9.16). Elements 13 to 20 deal with the external faces of the particular AAG (up to 256 external faces).

A backpropagation feedforward ANN with three layers of 20, 10, and 8 neurons each was trained to classify patterns in eight feature classes: pockets, slots, passages, protrusions, corner pockets, holes, and steps for a training set of 120 examples (Figure I.9.17). The network uses the delta rule and hyperbolic tangent activation function and converges with a mapping error (RMS) of the order of 2%.[20]

9.3.4   Detailed Process Planning without Geometric Reasoning

9.3.4.1   Knowledge-Based Tool Selection

This task is performed after generic process planning has been completed and, therefore, each operation in it is associated with a type of cutting tool. Intelligent tool selection is mainly a knowledge-based application. A number of parameters are associated with each cutting tool, starting with tool holders and adapters, and extending to insert shape and geometry as well as insert material.

Because of the large number of tool instances in a CNC manufacturing environment and the relative standardization of material and dimension aspects of tools according to ISO, prerequisite to realistic tool selection is a comprehensive tooling database.[21] A feasible possibility of representation is object frames. Cutting tool data, for example, can be modeled with the following slots: a-kind-of, stamp, work-element, geom.dimensions, cutting_angles, possible_operations, possible_materials, relation_of_tool_wear, TMS_code, TMS_address.

Note that several slots provide direct interface to possible operations and materials, as well as to tool life determination, including generalized cost data. In addition, connection to tool management system entries is provided through relevant keys.

It may not be immediately apparent that a conventional approach to tool selection is inferior to a knowledge-based one. However, the number and complexity of factors involved in decision making reinforces the latter case. The guidelines implemented in an intelligent tool selection system for turning operations result in giving preference to:

The minimum number of tools that can completely machine the workpiece
The strongest insert and corresponding tool holder

The most rigid tool shank (holder)

The largest nose radius allowed by the machining conditions and surface finish

The most economical insert in terms of cost per cutting edge

Right-hand over left-hand tools and those over neutral tools depending on the machine

In the tool selection process several matches have to be made first in order to identify suitable tool holders and insert types. These refer to workpiece material, feature type, nature of operation (finishing, roughing, semi-finishing), etc. Part shape is indirectly taken into account in the form of constraints to maximum insert angle (e.g., for recesses in turning), to minimum length of cutting edge, to minimum useful tool length (e.g., in end-milling), etc. The maximum number of positions in the tool magazine is also influential and may point toward consolidating several tools into one. Several tools may fulfill given criteria for each operation, so a final decision may be made either at this stage or at a later stage, when cutting conditions are selected in a global optimization exercise.

Sample tool selection rules for turning are as follows :

IF feature is internal_bore AND overhang ratio is < 4 THEN select tool_type solid_boring_bar

IF feature is external roughing AND clamping device is chuck AND stability is fair THEN select insert type not (round_insert).

IF feature is through_hole AND diameter_area is between x and y AND machine is okuma_LB10 THEN select tool_holder-type K_holder AND select tool_type M8.

9.3.4.2 Process Parameter Selection

Selection of cutting conditions is conventionally based on theoretical models of cutting forces, surface finish, and dynamic system performance. These models incorporate a large number of parameters such as tool geometry, workpiece and tool material, entry and exit conditions, width and depth of cut, feed, speed, presence of cutting fluid, stiffness of the machine tool. In general, a number of constraints are used in order to reflect the real situation, i.e., machine tool power, chip breaking mode, allowable workpiece-tool deflection. Such constraints are usually traced back to feed, speed, and depth of cut restrictions. So are dimensional tolerances and surface integrity, although these are also moderated to a large or small extent by factors that are generally hard to predict, such as tool tip condition, tool wear, chip area variations due to NC program practicalities. Fully accounting for these factors leads to construction of systems for process modeling/simulation rather than for just parameter selection.

Given such constraints, it is tempting to regard optimization of machining conditions as a multi-variate/multi-objective optimization problem typically solved with linear programming and other operations research tools. However, the difficulties in formulating the cost function and the constraints render this approach relatively impractical for large-scale exercises. Heuristic optimization procedures have been used instead.

Knowledge-based approaches are by definition helpful in "pruning" the search space before mathematical approaches take over. Machine learning approaches are a realistic alternative provided that there is enough data covering the target domain.

9.3.4.2.1 Inductive Learning from Machinability Databases

Given a large volume of data referring to machinability, as for instance in the well-known infos databases, there is a need for formalization. Such a task can be tackled by a learning approach, e.g., formation of decision trees following an automatic algorithm. A decision tree will have nodes corresponding to attributes, arcs corresponding to values or ranges of values of these attributes, and leaves corresponding to predicted classes. Trees can be built in recursive steps, an attribute being selected and the corresponding sub-tree being built each time. The selection of an attribute should be based on some criterion, e.g., informativity of an attribute, which is a measure of the information gained by partitioning a set of examples E according to the values of the attribute A: informativity

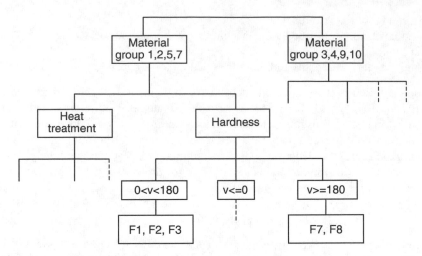

FIGURE I.9.18   Learning-from-examples tree for attribute "feed."

$(A, E) = \text{entropy}(E) - \sum [\text{probability of attribute value } v \times \text{entropy (subset of } E \text{ where } A = v)]$. The summation refers to all possible values v of attribute A. Entropy of a set of examples E can be defined as $\sum (p_c \times \log_2 \times p_c)$, where p_c is the probability of identifying an example as belonging to a class c in E, and the summation extends over all classes. Tree-pruning mechanisms for dealing with noisy (erroneous) data refer to stopping, if the number of examples in the current node or the majority class or the informativity of the best attribute fall below a threshold value.[22]

The ordering of attributes from the tree root toward the tree leaves reflects the relative importance of attributes for the predicted class in the leaf.

An example is the class feed with attributes such as material type, hardness, heat treatment, yield strength, which is expected to give as a result a tree similar to the one shown in Figure I.9.18. In essence, each attribute value space is partitioned into groups that take part as such in the decision-making process.

In the tree induction process some attributes might be consolidated from the original ones of the database. Also, parts of the domain space may not be classifiable, due to lack of examples.

9.3.4.2.2 *ANN Generalization of Catalogue Data*

Neural networks based on tool catalogue data can be constructed with an aim to determine initially acceptable combinations of feed, speed, and depth of cut for a certain specification of machining environment, i.e., tool and work material.

Various ANN models can be created, each taking different parameters into consideration. The training data are gathered from catalogues of tool manufacturers.[23]

For face milling one model includes three parameters: the material's hardness (expressed in HB), the axial rake angle (degrees), and the cutting edge angle (degrees). In total, 50 sets of data were used. Another model includes the same three parameters combined with the insert material (carbide or coated carbide), the presence of Ti (0 or 1), and the presence of TiCN (0 or 1), for a total of six parameters. In total, 80 sets of data were used. Yet another model includes four parameters: the material's hardness (expressed in HB), the axial rake angle (degrees), the cutting edge angle (degrees), and the depth of cut (mm). The outputs are feed (mm/tooth) and cutting speed (m/min). In total 234 sets of data were used. The best performing architecture was found to be $4 \times 7 \times 2$, i.e., with one hidden layer with 7 neurons (Figure I.9.19).

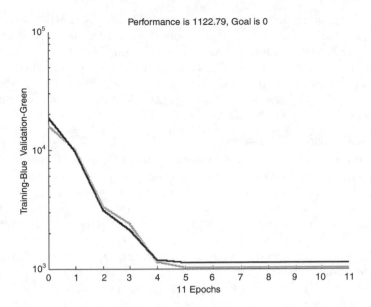

FIGURE I.9.19   Evolution of training and validation errors in the third ANN model.

It is a common problem for a neural network to overfit the training data and consequently make bad predictions when presented with new input data. For this reason, early stopping was implemented during the training of the ANNs. Training stopped after 22 epochs.

This technique requires that the training data be divided into three subsets: the training, the validation, and the checking subset. The first is used to train the ANN, while the third is used to check the network's performance after the training. The second subset's error during training is used as a stopping criterion. At first, this error decreases as well as the training error, but as the ANN begins to overfit the training data it increases. If this increase is present for a certain number of epochs, then the training stops and the weight values that occurred for the minimum validation error are returned. Approximately 70% of the available data were used for the training subset, 20% for the validation subset, and the remaining 10% for the checking data.

As the number of available data increases, so does the ability of the ANNs to generalize without large variations. The second ANN, which took into consideration six parameters, was not much more accurate than the third ANN that included four parameters. This is an indication of the qualitative nature of the ANNs. It is better to examine parameters that surely affect to a great extent the phenomenon under investigation than to embrace many more of lesser importance. Note that considerable experimentation is necessary to determine the final network topology as well as training parameters.

9.3.4.2.3   Genetic Algorithm Optimization of Numerical Process Parameters

Numerical parameters such as feed and cutting speed can be optimized following mathematical models, which is a classic theme in machining optimization, or using genetic algorithms.[24] There are several issues in the latter.

One issue is the type of chromosome used for representing the members of the population. The nature of the parameters under optimization (numerical) imply that the chromosome should have a binary form. The fact that there are at least two parameters for optimization means that the chromosome should represent them in combination. Scaled values for feed and speed should, therefore, be represented in tandem into one binary word or chromosome.

Another issue is the fitness evaluation function. According to classic mathematical optimization, the minimum production cost and the minimum production time formulations are equally common. Production cost function includes terms for material cost, setting up cost, machining cost (time), and tool cost (including tool changing time). This function can be used to calculate the unit production cost and serve as basis for promoting good individuals in the new population. This strategy presupposes that depth of cut is known or that a set of possible depths of cut is examined in order to choose the best solution that considers depth of cut, too.

9.3.4.3 Process Simulation Using ANNs

When process parameter values are to be determined, handbook data and manufacturers' catalogues are a possible source of information, but they always provide a conservative estimate. The actual values to be determined depend on the actual workpiece-machine-tool system and the safest way to determine them is to perform experiments. However, the factors influencing the parameters under investigation may be numerous and interdependent. Therefore, a full factorial experiment is impractical to conduct. This problem can be circumvented by the design of experiments method introduced by Taguchi, where specific levels of each parameter are examined, thereby reducing the experiment space substantially.

Furthermore, based on a limited set of experimental measurements, it is desirable to have the possibility to predict the effect of process parameters on process output indices, i.e., to perform generalization of measurements into models. This may be achieved through neural networks.

An example is prediction of surface finish in face milling operations. From analytical models constructed by various researchers, it is concluded that the parameters that may have a large or small influence on the surface finish attained are the depth of cut, the feed rate per cutter tooth, the cutting speed, the engagement of the cutting tool (ratio of cutting width to the cutting tool's diameter), the cutting tool wear, the use of cutting fluid and, possibly, the three components of the cutting force.

A relatively inexpensive experiment with 27 combinations of factor levels determined according to an L27 modified orthogonal array provides the experimental data necessary. The next question relates to which of the nine parameters examined should be used as inputs to the neural network model. The simplest solution would be to incorporate all of them, but the performance of that network might not be as good as some other network's with fewer factors included.

Again, in order to compare different ANN models, several factor combinations can be tried, according to an L32 orthogonal array. Thirty-two feed-forward back-propagation networks with three different numbers of hidden layer neurons were constructed and trained.

The Levenberg–Marquardt algorithm selected for training the ANNs is a variation of the classic back-propagation algorithm that, unlike other variations that use heuristics, relies on numerical optimization techniques to minimize and accelerate the required calculations, resulting in much faster training. Furthermore, in order to improve generalization, "early stopping" was implemented, based on the increase of the error of the validation subset of data.

The performance of each ANN was measured with the mean squared error (MSE) of the testing subset. The outcome of the training greatly depends on the initialization of the weights, done randomly according to the Nguyen–Widrow technique; therefore, repeated training and averaging techniques must be employed.

The best performing model proved to be one with three neurons in the hidden layer and the following inputs: the feed rate per tooth, the F_x component of the cutting force, the depth of cut, the engagement of the cutting tool, and the use of cutting fluid. The network's average mean squared error was equal to 1.86%.

It should be noted that this kind of model can be based, instead of experimental data, on simulated data, too. A simulation package can be used to generate parameter values and resulting values, with an aim to train ANNs. The distinct advantage of an ANN model over a traditional simulation model is speed of execution; compare, for instance, several hours of running a simulation on a finite element package with possible what-if repetitions of the same order of magnitude to a fraction of a second of execution time on a trained neural network. Even if training time is taken into account, the comparison is still overwhelmingly in favor of ANN models.

9.3.5 Integration in the Process Planning Domain

Essentially, many of the activities constituting process planning depend on one or more of the others; e.g., machining operations need to know to some extent the fixturing arrangements in order to avoid collisions, and fixturing operations need to know the machining forces in order to keep the part on the table. One way to circumvent this type of chicken-and-egg situation is to assume some sequence for considering selection of these operation types and check for mutual constraint satisfaction afterward. This philosophy is ideally implemented via a backtracking engine; the Prolog engine has been used for that purpose. If, however, a deadlock occurs, user interaction is provided to resolve problems. User-based coordination of the planning tasks is still most important, due to the complexity of the planning tasks and sub-tasks.

Another important issue in process planning integration relates to feature recognition. It is easy to recognize a feature for use in some task within process planning. However, for complete plan generation, all features are necessary to be recognized. This is easier to do incrementally, i.e., one feature at a time, each time an operation needs a feature. However, the result of the operation should update the feature model. For 3D features this means that a solid modeler has to be connected to the reasoning engine. The alternative is to provide complete representations of parts in terms of features, but this, in essence, predefines the process plan, because different feature combinations correspond to different operations and/or sequences. Therefore, the latter may be termed the "optimum feature coverage problem" and is equivalent to optimum process selection and sequencing at a generic planning level.

Finally, the notion of nonlinear plans, i.e., plans with alternative or even conjunctive (work distribution) branches, points to an important research direction offering flexibility that is definitely needed in a real workshop environment. This is actually a smart implementation of the link between process planning and machine loading or short-term scheduling. It also proves that strict optimization without taking into account a dynamically changing resource scenario is practically worthless.

9.4 The Process Implementation Domain

Process implementation, as opposed to process design, refers to all actions required to set the manufacturing process running. Intelligence in this domain pertains to the machine itself, i.e., its original design features or those added later as enhancements, notably controller features, as well as to the process execution in terms of procedures ensuring the proper execution within a manufacturing system. The limits of such procedures in our context are considered quite narrow, i.e., they are confined to scheduling and quality control with respect to a single CNC machine (see Figure I.9.1).

9.4.1 Machine Design Sub-Domain

9.4.1.1 Process Monitoring: Neural Network Applications

A hardware-software system to monitor process parameters is essential for automating manufacturing processes. The hardware part includes sensors and signal processing boards. The software system includes the decision-making engine relating the measured parameter values to the process state. The process state might be characterized as normal or abnormal. Abnormality needs to be further classified into discrete classes representing a particular problem class, e.g., vibration (chatter), chip jamming, tool breakage. A particular type of problem is the quantification of the abnormal state, e.g., estimation of the amount of tool flank wear.

Sensors used refer, of course, to the input parameters measured. They may be direct or indirect. Direct sensors measure the parameter under consideration directly, e.g., tool wear. In most cases this interferes with the process and is therefore not practical. As a solution, the parameter under consideration is measured indirectly by measuring other parameters somehow related to it. These include force (through piezoelectric rings or plates), torque (through strain gages), power (through motor current), acoustic emission (with special sensors, mainly for tool breakage), displacement (with capacitive or laser sensors), vibration (with accelerometers), or even shape (through vision sensors), etc. Each measurement has its own advantages and disadvantages depending on the process monitored, the range of values expected, the speed of phenomena, etc.

A process monitoring software system theoretically contains three parts: first, the feature extraction part in which "features," i.e., characteristic parameters of measured signals, are computed or estimated; second, the feature selection part in which the most important features influencing the monitored parameter are selected; and third, the monitored value computation part in which a model relating features and monitored parameter is—usually implicitly—built and executed to yield the expected result.

In conventional systems the first two parts are predetermined, i.e., for well-defined processes and operation domains, usually one signal is measured and simple processing is performed. For instance, feed force in turning is monitored and correlated with flank wear. The modeling and decision-making part is also usually simple: essentially, limits are defined outside of which the process is considered unacceptable. These limits may be a single value or a value profile, fixed or moving with time, or even dynamically changing according to specific detectable conditions. Such systems are in use in industry with sufficient acceptability, but in narrow application domains, i.e., for discrete events such as tool breakage, and not for gradual process variations.

A necessity in designing universal monitoring systems is intelligence. An intelligent monitoring system is primarily able to "decide" the state of a process, according to intelligent interpretation of usually a number of signals coming from more than one sensor and in the presence of signal noise. The system should, secondarily, be able to decide which of those features should be retained as basis for such intelligent interpretation. This type of intelligence could be termed "sub-symbolic" and is related to pattern recognition.[25] Symbolic type intelligence using knowledge would be impossible to work in real time; it is less amenable to quantification, and is inferior in terms of effective learning ability.[26]

Feature extraction can be done in the time or the frequency domain. In the time domain, some preliminary processing includes smoothing, using the moving average method or the true RMS method, calculation of the signal envelope, removal of the mean, or any trend, etc. The features extracted may be the mean, standard deviation, RMS value, skew, kurtosis, mean square value, or zero crossing rate. They are of statistical nature. Special features can also be considered in the time domain, e.g., the shape of the force locus during one cutter revolution (Figure I.9.20).

Feature extraction can also be done in the frequency domain. Preliminary processing is conducted using the Fast Fourier Transform in order to calculate the frequency spectrum, power spectrum, autocorrelation function, or cepstrum function. The features extracted may be the maximum of magnitudes or the RMS value for several frequency bands, the frequency corresponding to the maximum observed magnitude, etc.

Feature selection is essentially a procedure used to assess the relative importance of features extracted from signals. This is commonly based on scatter matrix (between and within classes of features) implementing the sequential forward search method.[26] This approach is independent of the downstream computation of the monitored parameter value(s). A feature selection approach, which is integrated to downstream computation is the use of neural nets. All features extracted are used as inputs, the net is trained, and the relative importance of features is assessed by considering the absolute values of the weights on outgoing synapses. Alternatively, sensitivity analysis is

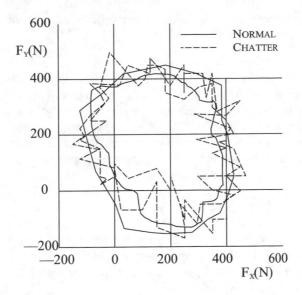

FIGURE I.9.20   Typical force locus patterns for normal and chatter cutting.

performed to determine the influence of inputs to the output. The least contributing inputs (features) are eliminated.

Despite the above tools that do streamline system design for process monitoring, there still remains open the issue of which indirect parameters to monitor in order to obtain credible modeled estimates of the direct parameter(s). In addition, process parameters (e.g., feed) can be related to indirect parameters (e.g., cutting force) in a similar way, as the latter are related to direct parameters (e.g., tool wear). Therefore, these should be added to the extracted features as inputs.

As a typical application of multi-layer perceptrons, three process states can be distinguished in milling: stable (normal), tool life end, and unstable (chatter).[27] The features used to decide which of the three states the process is in are extracted from force and vibration measurements. Force measurements are effected by mounting the workpiece on a 3-axis piezoelectric dynamometer and vibration measurements are taken using a triaxial accelerometer on the workpiece mounting. In order to train the multi-layer perceptron, 120 experimental samples were taken, 40 in each state class, with three spindle speed values, three feed rate values, and four plus two depth of cut values (for tool wear and chatter, respectively). Chatter is detected by growth and periodicity of vibration amplitude and tool life end is defined by average flank width equal to 0.5 mm as measured on a tool microscope.[27]

Adaptive autoregressive time series vectors of the sixth order, i.e., six-element vectors, were constructed at the feature extraction stage. This means that each value is expressed as a linear function of the six previous values. Analysis of the sensitivity of AR coefficients show that two different pairs of them can be used as discriminatory features for tool life end and for chatter, respectively. In addition, total power in spectral representation of force in the x-direction for tool life end is larger than that for the normal state. Similarly, in the power spectral representation of acceleration, contribution of chatter frequency grows with depth of cut and so does total power. Since six-element autoregressive force and power vectors may show some sensitivity to cutting conditions, feed and speed are considered, too, thereby forming 14 inputs to the neural network. Ten patterns from each of the three state classes were used as training sets. Target values for the corresponding output nodes at training were 0.99, and the remaining ones were 0.01. Out of three predominant architectures (14-10-3, 14-12-6-3, and 14-20-10-5), it was found by trials that the 14-12-6-3 network performed best with classification failure rates below 2%. The final decision—classification—is obtained by a "voting" machine, which considers the three outputs in the continuous range [0, 1]. When applying force and acceleration signals separately as inputs to the ANN, the error rises to 10–20%, proving that sensor fusion is superior to single sensor systems.[27]

9.4.1.2   Adaptive Process Control: A Fuzzy Sets Approach

CNC manufacturing processes are programmed off-line. Cutting parameters are set to specific values and they effectively cannot be changed at program execution time. The process, however, contains many sources of variability and there is no comprehensive global analytical model according to which the parameters could be set and controlled. Therefore, according to this deterministic control model, predefined parameters are, as a necessity, conservative and the process is conducted inefficiently.

Adaptive control provides on-line adjustment of operating parameters in order to either maximize or minimize a certain performance index (e.g., metal removal rate) under various constraints, or to regulate one or more "derived" parameters (e.g., cutting force) under the assumption that optimum conditions are found on the constraint boundary.

The adaptive control model requires a tight coupling with a monitoring scheme (see previous section) and most of the principles governing monitoring are applicable in control domain, too. The control model maps the input to the output space, where input is the required performance parameters of the process and output is the control parameters that are continuously adjustable. This mapping is the inverse of the usual predictive process model.

A formulation of a control model mapping is possible with neural networks.[26] At a symbolic level, knowledge-based control is also possible, emphasizing the empirical nature of the process model. However, both these approaches may be slow for fast-running processes where decision-making steps have to be faster than CNC block execution steps. A faster approach is implemented via fuzzy sets.

As an example, consider a grinding process, where surface finish is governed by feed rate.[28] Relative error E and change of relative error C in surface finish (in two consecutive measurements) are the two inputs to, and change of feed rate F is the output from, the fuzzy control system (Figure I.9.21).

Eleven levels are defined for error and change of error of surface finish[28] mapped to a −25 to 25% error and −10 to 10% change of error. Similarly, eleven levels are defined for change of feed mapped to −5 ΔF to 5 ΔF where ΔF is the feed adjustment step defined as a fraction of the absolute feed range. The membership functions are μ_E, μ_C and μ_F. For convenience, the discrete universes E, C, and F are defined as {−5, −4, −3, −2, −1, 0, 1, 2, 3, 4, 5} and the linguistic values of the corresponding variables are *HB* (high big), *VB* (very big), *B* (big), *NB* (near big), *LB* (little big), *Z* (zero), *LS* (little small), *NS* (near small), *S* (small), *VS* (very small), *HS* (high small). The membership vectors are defined in a table where each of the eleven linguistic values has membership value 1 corresponding to one of the eleven variable levels, 0.6 to its immediate neighbors on either side and 0.2 to the ones next (if applicable).

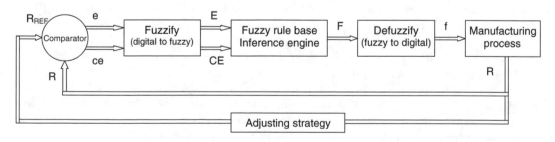

FIGURE I.9.21   Fuzzy controller architecture.

The linguistic rules of a fuzzy controller are a base of rules of the form: if E is Ei, then if C is Ci, then F is Fi, where Ei, Ci, and Fi are the values of the fuzzy variables E, C, and F in the ith rule. 31 rules are determined. For instance:

$$\text{if } E = HB \text{ then if } C = HS \text{ then } F = Z$$

The control rules adopted reflect the operator experience, the particular features of the process; they are initially subjective and undergo correction through experimentation.

The output from the fuzzy controller needs defuzzification. The mean of the maxima of membership values can be used to this end. Fuzzy reasoning on the linguistic control rules can be performed according to the max-min method as follows.

For simplicity, assume that n rules are used of the form:

$$\text{if } E = Ei \text{ then if } C = Ci \text{ then } F = Fi$$

The firing strength of rule i will be determined by the minimum of $\mu_{Ei}(e)$ and $\mu_{Ci}(c)$ where e and c denote sensor readings. The membership function of the combined control action F is determined by the maximum of $\mu_{Fi}(f)$ for all i.

9.4.1.3 Machine Diagnosis

Diagnostic tasks are by nature ideal as expert system applications. This is especially true for large systems. Small systems or sub-systems are perhaps as conveniently treated diagnostically by neural networks in a similar way to process monitoring (see Section 9.4.1.1).

The knowledge-based approach makes use of two types of reasoning: shallow and deep. Shallow reasoning, which is also called fault-tree type reasoning, is based on specific expertise and it is highly domain-specific and also fast but at the same time rigid; new rules are hard to accommodate, if consistency is sought. Deep reasoning, which is also called causal reasoning, requires no detailed expertise but just first principles and a model of the system to be diagnosed. Deep reasoning may be slow but is also generative, i.e., it can diagnose new faults; therefore, it is harder to develop. In practice, a combination of shallow and deep reasoning seems effective, typically applying first shallow reasoning for relatively small systems or deep reasoning for relatively large systems.

In the CNC machine tool domain there are several sub-systems cooperating with each other. There are usually three diagnostic levels recognized: diagnosis by alarm number and manufacturers' manual, diagnosis through simple tests known as troubleshooting, and diagnosis by direct reference to design specification (electrical and ladder logic diagrams). Of those three levels, the third one and partly the second one can be the subject of soft automation.

A possible strategy to that end is application of shallow reasoning first, encompassing experience of most common faults and their causes. If this fails, deep reasoning may be triggered, starting from a suitable point of the fault tree (Figure I.9.22).

Knowledge for deep reasoning is acquired from machine operation and maintenance manuals, as well as technical literature. By contrast, knowledge for shallow reasoning is acquired from observation of troubleshooting and repair procedures, as well as from interviews with the repair personnel. The expert system constructed is expected to provide fault classification/identification instructions, cause diagnosis, and repair instructions. These should be based on a highly interactive environment.

As an example, consider a CNC turning center[29] consisting of the machine and the controller systems. The machine system consists of several modules (sub-systems): spindle, turret, high-pressure coolant, manual slides movement, etc. The controller system also consists of modules: power unit, jog operation, servo system, manual pulse generator, operation mode, etc. Each of these modules has sub-modules; e.g., operation mode sub-modules are edit manual, MDI. There are interdependencies between modules, and therefore module diagnostic models are linked.

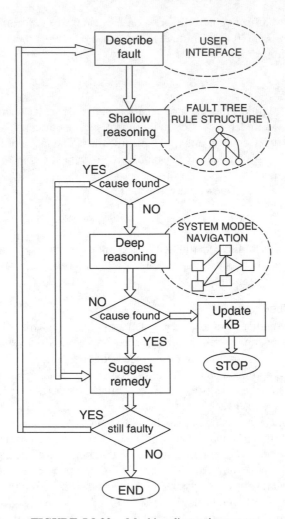

FIGURE I.9.22   Machine diagnosis strategy.

Shallow reasoning diagnosis is implemented by forward chaining, but rules are structured (ordered) so that search time is kept low. Several nodes in this structure of rules are pointed at according to heuristic rules acting as navigators depending on the context of the fault identified, so that further reduction in search time is achieved. However, too many heuristics may lengthen the search, thereby being rendered redundant.[29]

Deep reasoning diagnosis in effect searches through interconnected components in order to find the faulty one. PLC logic provides the interconnections and interdependencies of components. The question as to where to start traversal of interconnected components may be addressed by strategies such as most likely to be a faulty component or a component in the middle of a chain.

An issue in knowledge-based diagnosis is the possibility to find multiple faults. One solution is to find one fault at a time and repeat execution for finding the next fault; otherwise things may get too complex.

Another issue is the level at which a fault is defined. Usually, substantial constructs are examined and their constituents, e.g., relays and switches, are considered not reachable at all.

9.4.1.4 Machine Error Compensation

Only ideal machine tools give 100% accurate products. There are errors associated even with CNC machines and they are broadly classified as quasi-static and dynamic. The first type of error may be geometric or thermal in nature. Geometric errors are due to manufacturing discrepancies (including assembly defects) and also due to wear (natural or accelerated by manufacturing defects). Thermal errors are associated with heat generated by motors, friction and actual operation of the machine, as well as environmental temperature variations. The second type of error is associated with axes and spindle motion, and manifests itself in the presence of inertial forces, cutting force variation, and other similar situations.

Compensation is a common technique applied by CNC machine tool manufacturers and users, by which error maps are constructed and stored in controller memory so that they can be used for correction of the intended motion signal to the respective axes whenever applicable. The difficulty in such an approach lies in determining the error map. The error for one axis is simpler to determine by measurement, compared to the composite error resulting from combined movements, or to the volumetric error, which is the result of individual errors in the machine tool workspace.

Due to the fact that the tool tip position is not possible to measure directly because of interference with the cutting process, indirect measurements are taken first at places related to the tool tip position, e.g., interferometric measurement of the tool slides position. Then these are related to the tool tip position by a model, e.g., an analytical geometric-kinematic model or a mechanics model including the effects of cutting forces.

In the absence of analytical models, empirical or neural network models are used. The latter, in particular, are widely employed due to the number of factors affecting the accuracy of a machining operation and the still-existing difficulty to quantify them analytically. In neural network models, these factors, as well as their effects on accuracy, are captured through a number of sensors. This is initially done off-line in order to train the network and obtain the model. Then, in normal operation, on-line sensor measurements are fed to the model, which runs on the machine controller, in order to derive the actual corresponding error and perform compensation (Figure I.9.23).

A simplified approach to overall compensation for a turning center which has just two axes, X and Z, makes reference to four types of models:[30]

- Geometric model, where a $(\Delta x, \Delta z)_g$ error vector results from linear displacement errors δxx and δzz and straightness errors δxz and δzx, where δ AB denotes linear error along axis A owing to movement along axis B.
- Thermal drift model, assuming linear effects only, where a $(\Delta x, \Delta y, \Delta z)_t$ error vector is associated with the thermal state of the machine.
- Spindle thermal drift model, capturing axial, radial, and angular (in x-z plane) deviations, i.e., $(\Delta x, \Delta z)_s$ and ε_Y, respectively.
- Dynamic error, capturing cutting force effects and inertial effects, i.e., giving two error vectors $(\Delta x, \Delta y, \Delta z)_f$ and $(\Delta x, \Delta y, \Delta z)_i$, respectively.

Time effects, i.e., the history of changes, which may be important, especially in thermal error manifestation, are too little understood to be included in such an approach.

A total error is synthesized from the above-mentioned individual errors by simply adding up all components along the same axis.

Sensor fusion gives promising results; e.g., error reduction of one or two orders of magnitude is not uncommon. However, in order to arrive at such impressive results there is an issue to be invariably addressed, namely, which sensor signals are most influential in approximating the total error. One selection technique may use orthogonal arrays to design various possible models (combinations of sensors). Performance of each neural network model is computed according to indices such as sum of squared error for the training or checking data set, variance of residual error, total variance, etc.

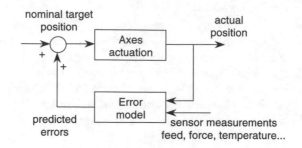

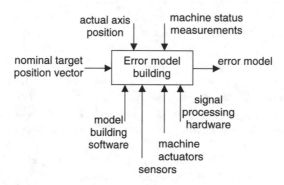

FIGURE I.9.23   Error compensation system architecture.

Also, the effect of each sensor (input) on each performance index can be determined by omitting the respective input. This, of course, implies that a maximal model has to be created and implemented with sensors in order to arrive at a minimal model gradually.

Another potentially troublesome issue is the form that the output is expected to have in a neural network model for machine tool error prediction. Instead of "analogue" error values, an accuracy class may be chosen as output, e.g., in multiples of 1μm, which is expected to improve performance of the model.

A problem belonging to the sensor fusion category is determination of the number and position of temperature probing points when constructing a thermal drift model. Time effects in this case cannot be ignored; in fact, a number of operation cycle patterns may have to be defined according to standards, where available, e.g., DIN 8602 which defines spindle duty cycles. Note that simulation using finite elements may help in interpolating on measured temperatures-error results (Figure I.9.24).

9.4.2   Process Execution Sub-Domain

9.4.2.1   Quality Control

Quality control is not considered here at the device level (e.g., automation of surface quality monitoring), as this is considered to be a topic exceeding the limits of CNC manufacturing. Instead, quality control is considered in the realm of quality information systems. A quality information system contains various sectors, some of which can underpin computer intelligence (Figure I.9.25).

Several areas can be intelligently supported by software. One is quality tool selection. Examples of quality tools are: graphs, process flow charts, Pareto diagrams, Ishikawa diagrams, histograms, and scatter diagrams, as well as the well-known control charts and cusum charts.[31] These are matched to prospective applications, e.g., to understand a quality problem some tools are more appropriate for determining the problem, others to visualize variation over time. Also, to analyze the causes, tool suitability ranges from determining the factors involved to selecting the most important ones and to

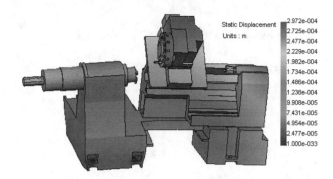

FIGURE I.9.24 Thermally induced machine tool deformation.

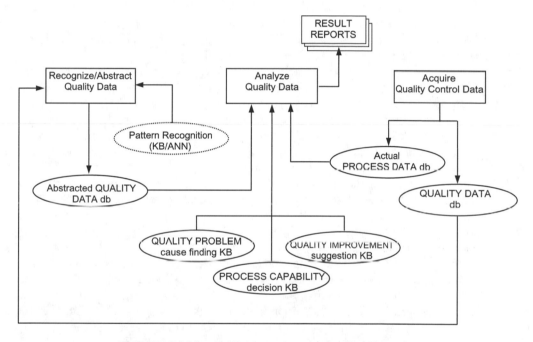

FIGURE I.9.25 Quality control system working principle.

looking at relationships among them. Finally, taking improvement measures also needs quality tools, e.g., for generating ideas, evaluating improvement proposals, making plans. All these decisions can be supported by a rule base.

Another area open to intelligent computing is specification of the quality control rig. The key areas of the configuration and examples of the factors to be investigated, which also determine the type of knowledge necessary, are process (parameters, machine alarms), sensors (suitability, level of technology, accuracy, investment level), data collection (volume of data in- and out-flow, type of down-line equipment, required data accuracy), system control (level of feedback and data archiving, method of corrective action at any of six levels, (i.e., "reflex" changes by the operator, adaptive NC for changing conditions, machine compensation, process planning changes, design changes, or production management level changes), SPC system (need, hardware, filters, sampling method), and system connections (cabling, LAN need).

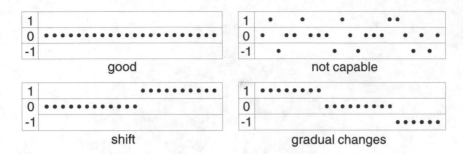

FIGURE I.9.26   Quality control attribute data pattern examples.

A general configuration for quality control rig layout for a machining center has been described.[31] Forty-nine objects were defined and 102 rules were associated with them, forming 25 rule sets. The knowledge base was structured at three levels: general configuration of the QC system, functional specifications, and element connection. A top level rule follows: IF tolerance is low or medium AND space is limited THEN sensor is capacitive.

The third area, which can benefit from intelligent software, is control chart pattern interpretation and remedy suggestion, if needed. Quality data can be classified into two types; i.e., attribute data arising from Boolean type inspection (go/no go) or the improved "Minus one-O-One" record (i.e., −1 for no go—smaller and 1 for no go—bigger) and variable data arising from parameter measurement.[31]

The patterns observed/recognized in quality tools are universal, i.e., the same patterns can stem from entirely different processes. For instance, in X-Bar R chart, patterns defined include cycles (periodicity), drifting or trends (7 consecutive rising or falling points), jumps, bias (7, 10 of 11, 12 of 14, etc. consecutive points on one side of the center line), plotting outside warning limits (2 out of 3, 3 or more out of 7 points plotted outside warning limits), plotting outside action limits (control limits), and mixture patterns. Patterns applicable to minus-one-O-one charts are shown in Figure I.9.26.

The above patterns can be recognized on the basis of rules coded in a pattern recognition knowledge base. Simple patterns are relatively straightforward to recognize. Mixture patterns consist of a number of overlapping simple patterns and are not as easy to recognize. Neural networks are considered appropriate for generic pattern recognition.

Continuous value (variable) chart recognition has been tackled with ANNs, but attribute charts have not been enjoying such attention. ANNs are used mainly to detect structural pattern changes, i.e., in process mean or variance, to identify nonrandomness, and to predict process means for future data groups. Input is usually standardized or divided in 7, 9, or 11 zones to avoid continuous/analog values. Signal-to-noise ratio is important as is inclusion of in-control points. An important issue is the width of the "observation window," i.e., the number of consecutive points that are considered in order to recognize a pattern. Sixteen is considered a sufficient number, but this is certainly a point of discussion. Performance of recognition is typically judged using "in-control ARL" (average run length), i.e., the average number of points that are recorded before a point triggers a false out-of-control warning when no special cause applies.

Recognition of generic patterns from quality tools can be considered as a clue toward investigating a certain sub-domain for more specific assignable causes. By contrast, specific pattern interpretation refers to a particular situation and quality characteristic, e.g., dimension, which suffers a particular problem. This is a straightforward diagnostic task (see Section 9.4.1.3). A sample rule leading toward this type of diagnosis follows: IF the pattern is "Jump" AND the data refers to "dimension" THEN look for "cutting tool breakage."

Having identified the assignable cause, corrective action can also be suggested, e.g.: IF cause is "measuring equipment calibration" AND frequency of appearance is $< 20\%$ THEN double frequency of measuring equipment calibration.

The action is usually straightforward when the causes identified are independent of each other, otherwise an action should be suggested by a rule summarizing the results of a Taguchi experiment design.

9.4.2.2 Machine Scheduling

Generic job shop scheduling is out of the scope of the process implementation domain as this is defined in Section 9.4. Single machine scheduling, however, is within scope and it is considered as an optimization problem. A simple formulation of this problem might be the sequencing of a number of jobs with certain due dates so as to minimize mean tardiness; i.e., the collective jobs' difference between completion time and due dates. Other optimization criteria may be cost, flow time, work in progress, or any other quality-of-schedule evaluators. There may also exist constraints, e.g., maximum processing time available in a day, maximum number of tools to process parts, or maximum number of pallets available to mount parts on. This problem is NP-hard with combinatorial explosion difficulty and has been tackled by a large number of operations researchers with dynamic and branch-and-bound programming approaches. To make for tractable solutions they applied heuristics.

AI techniques are promising, too, and often faster than classic operations research approaches, providing near-optimum solutions (within 1 to 2% of the optimum value). Two interesting techniques are genetic algorithms and simulated annealing.

In the genetic algorithm approach, representation of the individual may be in matrix form, with rows referring to parts index, and columns referring to dates index, where a_{ij} represents the number of pallets used in processing a part i on a date j before the due date. The fitness function may be the sum of squares of the deviations between the due dates and the actual processing dates.[32]

The crossover and mutation operators refer to columns and they may have to be re-defined in order to make the results respect the problem constraints. For instance, if a column exchange does not satisfy the constraint on the maximum number of available pallets per day, it is dropped and another exchange is performed randomly. Such redefinition of operators is strongly dependent on the nature of the applicable constraints. In the same sense, the initial population must be generated according to a specific procedure yielding feasible individuals.

The simulated annealing process works from an initial job sequence and generates a new sequence of jobs by interchanging two of them.[33] These may be picked at random or according to some rules to improve the solution convergence characteristics. The new sequence is accepted if it corresponds to a better value of the optimization function. Otherwise, it is accepted only with some probability, which decreases as the process evolves. In this way, the danger of getting trapped in a local minimum is lower. The acceptance probability may be exponential with exponent equal to $-\alpha \, \Delta t$, where Δt is the increase in the optimization function value and α is a control parameter, which is decreased with the number of iterations according to some law, e.g., linearly. A threshold for acceptance probability may also be imposed.

The initial sequence might be generated according to earliest due date, or to smallest processing time, etc. The stopping criterion refers to some kind of convergence, e.g., a large number of nonimproving iterations. Dependence of processing times on processing sequence can be taken into account by a suitable optimization function.

9.4.3 Integration in Process Implementation Domain

Intelligent machine tools are the integration test-bed of all hardware and software technologies in the domain of process implementation. If quality control and machine scheduling integration are counted within this domain, then "intelligent workstations" might be a more representative term.

Typically, an intelligent machine tool might contain several modules as follows (Figure I.9.27):

First, a tool monitoring module based on neural network and fuzzy identification for protecting the cutting tool and for providing relevant information to other modules.

Second, error compensation module(s) for mechanical actuators, such as stepping motors and feed motors.

Third, self-diagnosis and error recovery module.

Fourth, adaptive control module to optimize cutting conditions (primarily feed rate) on-line, e.g., on the basis of cutting forces, metal removal rate, tool deflection, tool wear.

Fifth, communication module linking the machine tool to the production system first at a cell level.

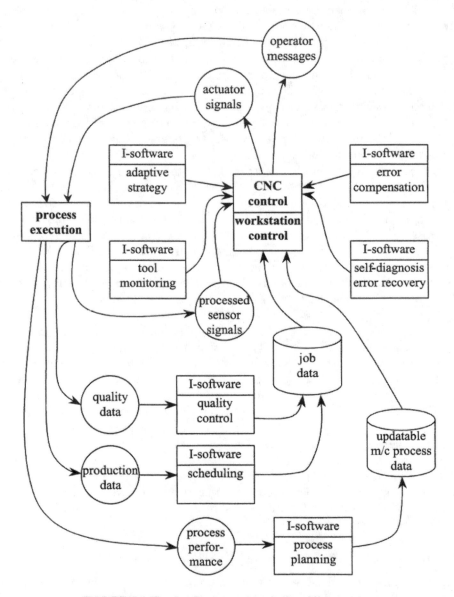

FIGURE I.9.27   Intelligent machine tool architecture.

Several of the above-mentioned modules are being increasingly found on high-end commercially available machines, perhaps not in full specification and implemented with conventional software philosophy, but certainly pointing to a promising near future that will certainly incorporate intelligent paradigms.

As for quality control activities, these are certainly planned by major manufacturers to be addressed in machine tools of the next generation that are being designed.[34] Quality control functionality is identified as implementation of SPC for monitoring, as well as for diagnosis of machine tool error causes. It is also extended to encapsulate planning in terms of inspection program generation for the machined regions. Scheduling has not been mentioned[34] possibly because it is left to a dedicated multiple-machine controller, i.e., at the DNC level.

9.5 Concluding Remarks

Intelligent software aids in CNC manufacturing systems as designated in Sections 9.3 and 9.4 are presented as independent systems, and they could certainly constitute independent modules in commercial developments in the near future. Some of these modules are foreseen to function in an unmanned environment, e.g., adaptive control, whereas others are still regarded as aids to the human operator, e.g., process planning.

This is the view taken in Yamazaki et al.,[34] too, where planning functionality is integrated into the new generation open controller but is termed "semi-interactive" in view of automation infeasibility for the whole range of geometry types of workpieces. In our view, turning shapes might form the exception and enable full integration of intelligent process planning, tool selection, and cutting condition selection into CNC control software. For analysis tasks the evolutionary build up of a database is advocated with tools, geometry features, cutting conditions, etc., which will be enriched every time a CNC program is executed successfully.[34] However, in our view, intelligence can certainly be integrated as described in Section 9.2 above.

A view toward encapsulation of knowledge in CNC controllers is implied in the description of the "evolving consultation" module,[34] which states that decision optimization might be sought based on information gathered from the process specialist automatically.

However, the most important implication toward possibilities of integrating intelligent modules of both planning and implementation (execution) nature in a single machining workstation controller is the trend of open controllers. An open controller should be transparent to the machine tool builder and user alike, transportable in the sense of module portability on remote computers, transplantable in the sense of upgrade compatibility, revivable in that outdated parts should be directly replaceable, user configurable in that custom modules can be added by the (experienced) user, and all in all evolving during the system life.[34]

Such a description of openness holds much promise for all ambitious developers who have been confined to prototypes and reinvention of the wheel (CNC controller) in order to implement and test just one module, for users who believe that the current state-of-the-art is way behind the state-of-the-art of other sectors/application domains of software, and for the manufacturing research community which needs a much more open and practical platform if its research results are to make any impact at all.

The matching of intelligent computing paradigms and application tasks is by no means one-to-one. In addition, there are many circumstances where conventional computing offers more practical solutions. However, with the advent of hybrid paradigms, e.g., neuro-fuzzy systems, new potential of AI for manufacturing applications has been discovered.[35]

Machine learning paradigms seem to have started putting more emphasis than before on self-learning compared to learning from examples. This paradigm, although somewhat less understood and relatively restricted in use, may add a new dimension in extensibility and openness of CNC intelligent systems where it becomes practical to employ.

References

1. Durkin, J., *Expert System Design and Development*, MacMillan, New York, 1994.
2. Leake, D.B., *Case-Based Reasoning: Experiences, Lessons and Future Directions*, MIT Press, Boston, 1996.
3. Engelmore, R. and Morgan, T. (Eds.), *Blackboard Systems*, Addison Wesley, New York, 1988.
4. Fausett, L., *Fundamentals of Neural Networks: Architectures, Algorithms and Applications*, Prentice Hall, Upper Saddle River, New Jersey, 1994.
5. Zimmermann, H.J., *Fuzzy Set Theory and Its Applications*, Kluwer, Boston, 1991.
6. Goldberg, D.E., *Genetic Algorithms in Search, Optimisation, and Machine Learning*, Addison-Wesley, New York, 1989.
7. Kirkpatrick, S., Gelatt, C., and Vecchi, M., Optimisation by simulated annealing, *Science*, 20/4598, 671, 1983.
8. Davies, B.J. and Darbyshire, I.L., The use of expert systems in process planning, *Ann. CIRP*, 33(1), 303, 1984.
9. Descotte, Y. and Latombe, J.-C., Making compromises among antagonist constraints in a planner, *Artif. Intell.*, 27, 183, 1985.
10. Vosniakos, G.-C. and Davies, B.J., Knowledge-based selection and sequencing of hole-making operations for prismatic parts, *Int. J. Adv. Manuf. Tech.* 8(1), 9, 1993.
11. ElMaraghy, H.A. and Gu, P.H., Expert system for inspection planning, *Ann. CIRP*, 36(1), 85, 1987.
12. Denzel, H. and Vosniakos, G.-C., A feature-based design system and its potential to unify CAD and CAM, *IFIP Trans. B: Comp. Appl. Tech.* B-10, 131, 1993.
13. Mamalis A.G. and Vosniakos, G., Intelligence in computer-aided process planning for machining, in *Proc. 3rd European Robotics, Intelligent Systems and Control Conf.*, Tzafestas, S.G., Ed., Springer Verlag, London, 13, 1999.
14. Yang, H., Lu, W.F., and Lin A.C., PROCASE: An intelligent case-based process planning system for machining of rotational parts, *J. Intell. Manuf.* 5, 411, 1994.
15. Horvath, M., Markus, A., and Vancza, J., Process planning with genetic algorithms on results of knowledge-based reasoning, *Int. J. Com. Integr. Manuf.* 9(2), 145, 1996.
16. Ong, S.K. and Nee, A.Y.C., Application of fuzzy set theory to setup planning, *Ann. CIRP*, 43(1), 137, 1994.
17. Roy U. and Liao J., Application of a blackboard framework to a cooperative fixture design system, *Comp. Ind.* 37, 67, 1998.
18. Kribenis, A. and Vosniakos, G.C., Optimization of multiple-tool CNC rough-machining of a hemisphere as a genetic algorithm paradigm application, *Int. J. Adv. Manuf. Tech.*, 2002 (to be published).
19. Vosniakos, G.-C. and Davies, B.J., A shape feature recognition framework and its application to holes in prismatic parts, *Int. J. Adv. Manuf. Tech.* 8(5), 345, 1993.
20. Nezis, C. and Vosniakos, G.-C., Recognising 2-1/2-D features using a neural network and heuristics, *Computer-Aided Design*, 29(7), 523, 1997.
21. Vosniakos, G.-C., An intelligent software system for the automatic generation of NC programs from wireframe models of 2-1/2-D mechanical parts, *J. Comp. Integr. Manufact. Syst.*, 11(1–2), 53, 1998.
22. Sluga, A. et al., Machine learning approach to machinability analysis, *Comp. Ind.* 37, 185, 1998.
23. Benardos, P., Vosniakos, G.-C., and Bozonas, G., Neural network models of technical literature for selection of cutting conditions in face milling, *Proc. ASME First Nat. Conf. Recent Adv. Mech. Engineer.* P. Drakatos, Ed., University of Patras, 2001, P168.

24. Teti, R. and Kumara S.R.T., Intelligent computing methods for manufacturing systems, *Ann. CIRP*, 46(2), 629, 1997.

25. Monostori, L., A step towards intelligent manufacturing: Modelling and monitoring of manufacturing processes through artificial neural networks, *Ann. CIRP*, 42(1), 485, 1993.

26. Monostori, L. et al., Machine learning approaches to manufacturing, *Ann. CIRP*, 45(2), 675, 1996.

27. Ko, T.J. and Cho, W., Cutting state monitoring in milling by a neural network, *Int. J. Machine Tools Manufact.* 34(5), 659, 1994.

28. Zhu, J.Y., Shumsheruddin, A.A., and Bollinger, J.G., Control of machine tools using the fuzzy control technique, *Ann. CIRP*, 31(1), 347, 1982.

29. Bohez, E.L.J. and Thieravarut, M., Expert system for diagnosing computer numerically controlled machines: A case-study, *Comp. Ind.* 32, 233, 1997.

30. El Ouafi, A., Guillot M., and Bedrouni, A., Accuracy enhancement of multi-axis CNC machines through on-line neurocompensation, *J. Intell. Manufact.* 11, 535, 2000.

31. Vosniakos G.-C. and Wang, J., A software system framework for planning and operation of quality control in discrete part manufacturing, *Comp. Integr. Manufact. Syst.* 10(1), 9, 1997.

32. Sakawa M., Kato K., and Mori, T., Flexible scheduling in a machining center through genetic algorithms, *Comp. Ind. Engineer.* 30(4), 931, 1996.

33. Ben-Daya, M. and Al-Fawzan, M., A simulated annealing approach for the one-machine mean tardiness scheduling problem, *Eur. J. Oper. Res.* 93, 61, 1996.

34. Yamazaki, K. et al., Autonomously proficient CNC controller for high-performance machine tools based on open architecture concept, *Ann. CIRP*, 41(1), 275, 1997.

35. Dini, G., Literature database on application of artificial intelligent methods in manufacturing engineering, *Ann. CIRP*, 46(1), 681, 1997.

10

Intelligent Systems Techniques and Applications in Product Forecasting

James Jiang
University of Central Florida

Gary Klein
University of Colorado

Roger A. Pick
University of Missouri

10.1 Introduction

Product forecasting is a critical function in the rapid turnover of goods and products through retail outlets. Accurate forecasts allow retail managers to dramatically increase inventory turnover and reduce the number of markdowns. Such ability is more responsive to the market, allowing a better match to consumer demand, fewer outages, and lower costs. The improvements spill over into the manufacturing side as well. Advance knowledge of demand allows more complete production planning, reducing costs, lowering inventory, and determining capacity requirements. Producers use forecasts as hedges against lost control of forward scheduling (Curry, 1993). Therefore, forecasts are beneficial for the consumer, the retailer, and the producer in the distribution chain.

The data, however, do not lend themselves to simple forecasting techniques (Jiang, 1995). Much retail sales data are collected via scanner technology. Scanner data provide volumes of data that make complex forecasts possible, but are resource intensive for the information technology required in support. The quantities of data lead to issues of storage and computation, especially for forecasting techniques that are time consuming or utilize intensive intermediate results. The data are also messy in terms of the properties exhibited. Seldom are sales data linear, as assumed by many forecasting techniques. Promotions, shortages, seasonality, and competitor actions all distort any linear properties that exist. In addition, multiple variables and parameters are also critical in the manipulation of sales that must be brought into consideration. Forecasts based solely on previous sales are not dependable; the manager must take price, promotions, advertising, and the competition into account when preparing a market plan. Complex relations between each set of variables may exist, and the lag time between implementation and impact on the sales is not known with any certainty.

In spite of these problems with the data, econometric approaches dominate product forecasting applications, permitting parsimonious modeling of competitive activity in a given product category and producing accurate short-term forecasts (Curry et al., 1995). This is a descriptive modeling approach—seeking to uncover marketing phenomena and represent them in mathematical form. One technique that has emerged as accurate and quite robust is vector autoregression (VAR). This technique has been compared favorably to its rivals such as multiple regression, Exponential Smoothing, and univariate and multivariate Box-Jenkins. More recently, VAR has been used to examine price-promotion effects on category demand (Nijs et al., 2001) and consumer spending effects on long-term marketing profitability (Dekimpe and Hanssens, 1999).

Neural networks (NN) are being considered as replacements for econometric methods by researchers in product forecasting and other applied fields. The attraction of NN technology is the ability to respond to recent changes to the data and an ability to see through noise better than traditional approaches (Hill, O'Connor, and Remus, 1996). Much of this ability comes from the learning characteristics of neural networks; thus, when the large amount of scanner data becomes available directly from electronic checkout systems, the NN can incorporate the new data into the forecast. Thus, NN technology offers a way to improve on high-volume forecasts.

Several applied fields have examined the performance of neural networks in time-series forecasting. Market researchers have found neural networks to be accurate and flexible because of the ability to handle missing data and category variables (Venugopal and Baets, 1994). Financial forecasts are made more commonly using NN technology because of the ability to handle nonlinearities in the data (Aiken and Bsat, 1999). Applications in finance have also attempted hybrid technology to use NN technology for forecasting as directed by rule-based expert systems (Tsaih, Yenshan, and Lai, 1998). Marketing applications include fashion forecasting, market share forecasting, retail assortment determination, and promotion strategy (Agrawal and Schorling, 1996; Belt, 1993; Dragstedt, 1991). The results have generally been encouraging for NN as a replacement for traditional econometric methods (Adya and Collopy, 1998; Balkin, 2000).

Still, even if NN technology provides the solution to forecasting in the product arena, it must be manipulable by the managers making the decisions regarding the products. The rigors of the method require a simplistic interface to those making the product decisions. Incorporation of the forecasting technology into a marketing decision support system will be critical to the success of any application (Abraham and Lodish, 1993; Little, 1979). Until then, forecasts will only be used for isolated problems, such as those focusing on a single variable of price or promotion event (Curry, 1993; Zenor, 1994). The incorporation of forecasting technology into a comprehensive decision support system (DSS) will allow broader application of the technology by a wider range of managers, down to the store level where many product decisions must be made (Jiang, 1995).

We address these issues in this chapter. First, we explore the data characteristics that inhabit the product forecasting arena, in particular the forecasting of brands at the retail outlet level. The examination describes the data and illustrates the problems that exist due to the properties of market fluctuations and frequent update. From there, we describe the technology commonly employed to produce forecasts for this problem, and outline the NN technology that could replace econometric approaches. We report on an investigation of the competing methods for three data sets. Last, a DSS framework is proposed for which a prototype was used to generate the results reported.

10.2   Characteristics of the Data

Data for product and brand forecasting usually come from single-source systems. Single-source systems are those databases containing measures of variables on single units of analysis, such as stores or households. Measures may include sales as well as information on prices and promotions for the unit over a lengthy period of time, or household information on purchases and television viewing. Two major suppliers of single-source data include Information Resources Inc. (IRI) and A.C. Nielsen. These two organizations have data sales approaching 1.5 billion every year.

The volume of the data is progressively expanding, with a 15,000% growth in the number of data items available over a 12-year period. Data are collected at various stages in a product's flow. Shipment data are collected on the flow from manufacturer to retailer. Consumer data are collected at the time of purchase as well as retailer information regarding price and promotions. This amount of data presents problems for effective use and challenges information-system designers and quantitative forecasters to develop methods to deal with the large quantities of data. Automated systems are necessary, both from the standpoint of the data and the complexity of the quantitative analysis techniques available today.

Quantity and dimensionality of the data do not present the only problems. In order for many forecasting techniques to apply, certain assumptions about the distribution and properties of the data must be met. Single-source data systems also present difficulties along these lines for decision makers. To investigate the potential difficulties, we use actual data sets for sales of soups in Denver, and national sales of potato chips and facial tissues for a national grocery chain. Four brands of soup are represented and five brands of chips and tissues. Generic and in-house brands were combined into a single brand. Time series variables collected include sales volume, price, and TV advertising time. Deterministic variables include seasonal indices, percent of sales made on special display, and percent of items sold on feature. Variables under the control of the decision maker include future pricing, future advertising, and future placement of items on display or feature. Data was collected over a two-year period.

Figure I.10.1 is a graph of one brand of soup (solid line) and one brand of tissue (dashed line). Visual inspection of the graph shows potential problems with unusual spikes and fluctuations in demand. A more formal analysis shows greater potential problems for econometric methods that require well-behaved data. We will look at some violations of assumptions for a method variation of VAR commonly applied to acquire brand forecasts, Bayesian Vector Autoregression (BVAR). BVAR

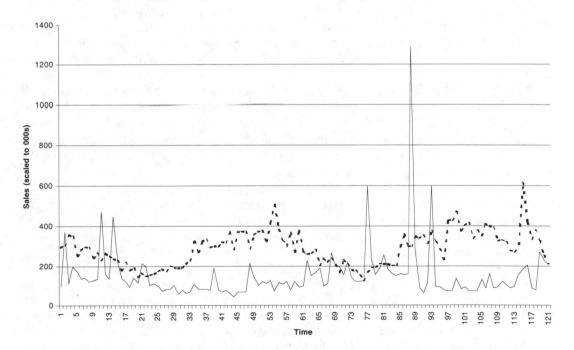

FIGURE I.10.1 Sales patterns.

TABLE I.10.1 Percentage of Specification Test Violations

	LAGS	ARCH (%)	Normality (%)	Linearity (%)	Heteroscedasticity (%)	Serial Correlation (%)
Chips	2	0	100	100	0	10
Chips	3	0	80	100	0	15
Chips	4	20	40	100	0	50
Tissue	2	0	80	80	40	50
Tissue	3	0	60	80	0	65
Tissue	4	30	100	80	0	25
Soup	2	25	75	25	50	50
Soup	3	12.5	50	75	0	50
Soup	4	0	25	100	0	0

assumes linearity, normality of error terms, homoscedasticity (constant residuals), no autocorrelation of error terms, and no autoregressive conditional heteroscedasticity. Although BVAR is a robust model, violation of these statistical assumptions may result in a biased estimation. The underlying methodology for BVAR is discussed in the next section.

Tests for each of the assumptions exist. Table I.10.1 shows results of a battery of tests done on the data used in this study. For each product category, a number of different time lags are tested. In the cells of the table, we present the percentage of failed tests, that is, the percentage of tests that indicate an assumption is violated for each category of data. The soup category has four different brands that could each fail the test, and the tissue and chip category each has five different brands. Each cell, thus represents multiple tests, at least one for each brand. As can be seen, the assumptions of the model are frequently violated according to the tests applied.

The test for autoregressive conditional heteroscedasticity is a Lagrange Multiplier test (Engle, 1982). First, one regresses the dependent variable (sales volume in our case) on the independent variables. Second, the squared residuals are regressed on an intercept and its set of lagged squared residuals. The sample size T times the centered R^2 from the squared residual regression is asymptotically distributed as a chi-square with degrees of freedom equal to the number of lagged squared residuals.

Normality is tested with the Jarque-Bera test for normality (JB). The normal distribution is symmetric and mesokurtic. The symmetry implies that the third moment of residuals, skewness, is zero. The mesokurtic value is the kurtosis of the normal distribution, which is three. Therefore, one might compare a distribution to the normal distribution by comparing this skewness to zero and its kurtosis to three. From this, Jarque and Bera (1982) derived a Wald statistic that is asymptotically distributed as chi-square with two degrees of freedom.

Ramsey (1969) proposed the regression specification error test (RESET) for linearity based on the simple notion that various powers of predicted values should add little explanatory power to a regression equation. One first regresses the dependent variable on the independent variables, and then predicts values using this regression. Second, one adds powers of these predicted values into the independent variable pool and runs this auxiliary regression. If the linear specification was correct, the coefficients of the additional variables would not be significantly different from zero.

Breusch and Pagan (1980) showed that a test statistic for heteroscedasticity can be computed as one half the regression (i.e., explained) sum of squares from a linear regression of $u_w^2/(u'u/n)$ on a constant, and the variables thought to affect the error variance. Here u_w is the OLS residuals, and the statistic is distributed asymptotically as a chi-square with degrees of freedom equal to the number of variables thought to affect the error variance (usually the number of independent variables). This test does not require prior knowledge of the functional form involved.

Two tests for autrocorrelation are applied. Breusch (1978) and Godfrey (1978) proposed a Lagrange multiplier test of autocorrelation. The test can be carried out simply by regressing the OLS residuals u_w on the independent variables and p lagged residuals, $u_{w-1}, \ldots, u_{w-p}$, and referring TR^2 to the critical value for the chi-square distribution with p degrees of freedom. The test is a joint test of the first p orders of autocorrelation, not just the first order. Ljung and Box (1979) proposed the Q-test for autocorrelation. This statistic is distributed as a chi-square and has been found to be reasonably powerful (Greene, 2000).

10.3 Basis of the Methods

We provide a cursory overview of the BVAR and neural network models. For more detail we recommend Curry et al. (1995) for a review of BVAR and Wasserman (1989) for an introduction to artificial neural networks.

10.3.1 BVAR

Vector autoregressions are dynamic models of time series. VAR has been proposed as an alternative to large simultaneous equations models for studying the relationship among important aggregate variables. However, the unrestricted VAR are not well suited for use in forecasting. Litterman (1980) developed a Bayesian procedure for estimating the VAR, which greatly improves forecasting performance. Formally, a vector autoregression may be written as:

$$\mathbf{x}_{w+1} = \sum_{j=0}^{l-1} \mathbf{A_j} x_{w-j} + \mathbf{D} \mathbf{m}_{w+1} + \varepsilon_{w+1} \qquad (\text{I}.10.1)$$

where

$\mathbf{x}_w$ is an $n \times 1$ state vector for time period w,

$\mathbf{A_j}$ is an $n \times n$ state transition matrix,

$n =$ number of endogenous time series variables in the system,

$l =$ number of lags,

$\mathbf{D}$ is an $n \times q$ coefficient matrix,

$\mathbf{m}_w$ is a $q \times 1$ vector of merchandising variables for time period w,

ε_w is an $n \times 1$ white noise vector for time period w, and

w is an index for time period (weeks).

The system consists of n endogenous variables and q exogenous variables. All the relations in the system are assumed to be linear. Forecasts adopting the above unrestricted VAR often suffer from over-parameterization of the models. The number of observations available may not be adequate for obtaining precise estimates of the large number of free coefficients in the VAR. This over-parameterization may cause significant out-of-sample forecast errors.

The Bayesian approach to this problem is to specify "fuzzy" restrictions on the coefficients. Litterman (1980) suggests that longer lag effects are more likely to be close to zero than shorter lags; however, the data should be permitted to override this suggestion if the evidence regarding a large lag is strong. Formally, this approach is implemented by placing a normal prior distribution with mean of zero and small standard deviations on the lags.

In VAR, we must examine not only lags of the dependent variable (e.g., sale volume), but also lags of the other endogenous variables (e.g., price). The size of lag coefficients on these endogenous variables, however, is often not so clear, and depends on the relative scales of the variables involved. The specification of the prior distribution on the lags of the endogenous variables also assumes normality. The default means of the prior distributions for all endogenous variables are zero, except the first lag of the dependent variable in each equation, which has a prior mean of one. Flat priors are put on the exogenous variables in each equation. These specifications lead to the BVAR model.

10.3.2 NN

Artificial neurons (nodes) form connections in neural networks. The simplest network is a group of nodes arranged into a layer. A multilayer NN is formed by simply cascading a group of single layers. Figure I.10.2 shows a three-layer neural network: an input layer, a hidden layer, and an output layer. The nodes at different layers are densely interconnected through directed links. The nodes at the input layer receive the values of the input variables (e.g., sales volume, deal price, or seasonal index) and propagate concurrently through the network, layer by layer. The hidden layer nodes are often characterized as feature detectors. The number of nodes in the hidden layer can be selected arbitrarily. However, too many nodes in the hidden layer may produce a system that merely memorizes the input data but lacks the ability to generalize. In most applications, the hidden layer includes about $\frac{2}{3}$ of the number of input nodes. The initial weights of the links are chosen randomly. Each weight represents the strength of a synaptic connection. Nodes in the hidden layer take multiple values, combine them into a single value, and then transform them via an explicit transfer function into an output value.

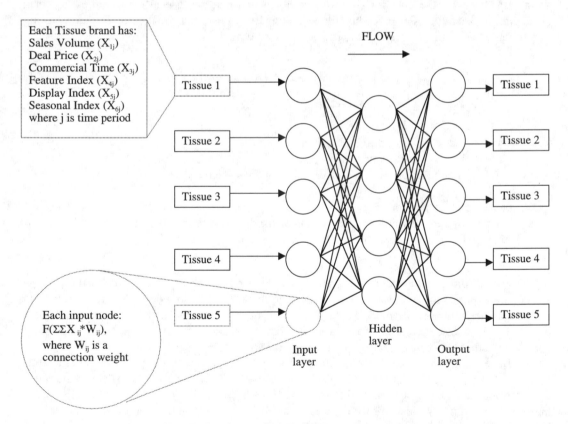

FIGURE I.10.2   A feedforward neural network.

In supervised learning, the network is trained by giving the input values as well as the target output values, and then fine-tuning the network weights until the trained output approximates the target output. Three training parameters are required in training an NN: learning rate, momentum, and training tolerance. A high learning rate (one that reacts very quickly to changes) will allow the network system to take radical weight changes and be unable to predict accurately. On the other hand, a low learning rate may cause the network system to take excessive time to learn. The momentum factor determines the proportion of the last weight change that is added into the new weight change. High momentum rates force a major change to the weights each time new input is entered, possibly preventing completion of the learning phase. Very low momentum rates prevent the model from making improvements in the face of new information. The tolerance factor specifies the precision in comparing network output with the target output. A small training tolerance results in less ability to generalize, since an extremely high degree of accuracy in the model is required.

10.4   Performance of the Methods

The most important criterion used in the forecasting ability of the model is its out-of-sample forecast accuracy. We summarize a model's forecasting ability using the Theil U statistic (Greene, 2000, p. 310), which scales a model's root mean squared error (RMSE) by the square of the actual observed value, and the MAPE statistic, which is simply the mean absolute percentage error. Using a 16-week forecast horizon, there are 16 one-step-ahead forecasts, 15 two-step-ahead forecasts, and in general, $k(17 - k)$-step-ahead forecasts ($k = 1, \ldots, 16$). In all, 136 forecasts are available for each sale volume. The Theil U statistic takes the following form:

$$\text{Min } U \left(\sum c_i \mid l \right) = \sum_{j=1}^{n} \sum_{f=1}^{F} u_{jf} \tag{I.10.2}$$

where

f = forecast steps ahead;
l = lag length;
u_{jf} = Theil U for time-series j with the following form:

$$u_{jf} = \frac{\sqrt{\frac{1}{F} \sum_{f=1}^{F} (\hat{S}_{jf} - S_{jf})^2}}{\sqrt{\frac{1}{F} \sum_{f=1}^{F} (S_{jf})^2}} \tag{I.10.3}$$

where $\hat{S}_j$ and S_j are forecast and actual values of sales volume for brand j and F is the number of forecast horizons. Sixteen observations were withheld from each data set and out-of-sample forecasting is used to initial the parameters of the methods.

TABLE I.10.2   Theil Sum Values for a 16-Week Forecast

	Soups		Tissues		Chips	
Lag Periods	NN	BVAR	NN	BVAR	NN	BVAR
2	9.53	9.78	23.75	26.97	22.79	18.57
3	10.17	10.07	24.16	26.95	23.14	17.60
4	9.76	9.55	23.50	26.70	29.25	17.24

TABLE I.10.3   MAPE Values for a 16-Week Forecast

	Soups		Tissues		Chips	
Lag Periods	NN	BVAR	NN	BVAR	NN	BVAR
2	7.64	8.62	19.28	21.98	20.13	18.15
3	7.19	8.52	20.99	21.58	20.55	16.43
4	7.26	7.98	21.02	21.66	21.04	15.84

Tables I.10.2 and I.10.3 summarize the forecast performance of the BVAR and NN models for the 16-week horizon. Only the best representative from each class of model are considered. The numbers in the cells represent the average value for the performance statistics over the categories and lag periods shown. It is evident that the NN and BVAR models perform similarly on the Theil-U and MAPE criteria. Additional testing indicated the models appear to be very stable in variations on parameters, which indicates a robust character important to complex forecasting models that require parameter specification by the analyst.

10.5   Development of a Decision Support System

Managers controlling the product decisions require an effective system to make their decisions using single-source data (Little, 1979). Today's decisions require market response reports that allow the decision maker to examine details about promotional effectiveness, price elasticity, and advertising response. The models used for providing the complex information are large and unwieldy, requiring full support of the software to assist managers not knowledgeable in the mathematical techniques required. Decision support systems (DSS) are designed to allow naive users access to more sophisticated models and larger qualities of data. We outline the architecture for a DSS that supports product decisions.

Figure I.10.3 shows a conceptual model of a DSS prototype developed to assist in marketing decisions. The DSS must interact with two types of users in this case, the marketing analyst responsible for building models that represent the data and the brand manager who manipulates the models to make stocking decisions. The market analyst will control the parameters of the models in order to arrive at the best fitting model. The brand managers utilize the models to analyze different pricing and promotional structures. The system itself utilizes four fundamental processes and two stored components. The stored components are the IRI database (or other single-source database) and the models being developed or manipulated by the user of the DSS.

The dialog manager must be able to handle the functions associated with building models and using the models to make decisions. The model fitting system works with a user to build models that best fit the data. The data management system is a standard database application package used to acquire the needed single-source data from the IRI database. The model application system applies specific models to data sets and decision variables. The output from the model application system is the output used by the brand manager to set process, advertising levels, and determine promotions.

The model fitting system is shown in more depth in Figures I.10.4 and I.10.5. As with most DSS, models are stored for comparison purposes or future access. This is done by a model storage system that preserves model instances in the model base. The model storage system is accessed by the three other components of the model fitting system shown in Figure I.10.4. Model generation takes the model specifications from the dialog manager, data from the database manager, and generates the

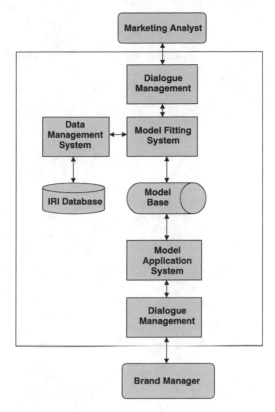

FIGURE I.10.3   Conceptual structure of a marketing decision support system.

correct input to the neural network software (for execution of the model). Once the neural network is completed, the results are sent to the dialog manager by the output generator.

The steps required of the user by the model fitting system are shown in Figure I.10.5. First, the user must select the time-series variables to be included in the model. The dialog manager uses a spreadsheet format to collect this and other requirements. The process of selecting variables is interactive, which allows for ready changes to the model. Once the variables have been chosen, the parameters required of the neural network system are entered. Here, it may be advantageous to use common default parameters. Once the variables are set, the parameters initialized, and the data collected, the model is run through the neural network repetitively to generate an optimal output. Performance measures for an out-of-sample forecast are generated. The final step involves the decision maker determining if the result is satisfactory. If so, the model fitting is completed, or else the cycle is repeated. The user determines when to stop, based on forecasting goodness of fit reports on the Theil-U and MAPE statistics like those shown in Tables I.10.2 and I.10.3.

The remainder of the functions are under the control of the product manager, who must manipulate the model by varying price and promotional variables until a satisfactory solution is found. In the model application system shown in Figure I.10.6, the models built by the analyst are applied to price and promotional variables. Input similar to that in Table I.10.4 is made through the dialog manager, and the appropriate model collected from the model base. These are used to generate a forecast using the optimal neural net model as previously determined. The model is executed and forecasts, similar to those shown in Tables I.10.5 and I.10.6, are derived. The specific models are saved for future reference and the dialog manager presents the results to the user. The decision maker can decide to employ the strategies tested or can elect to try another set of values for the variables under his/her control.

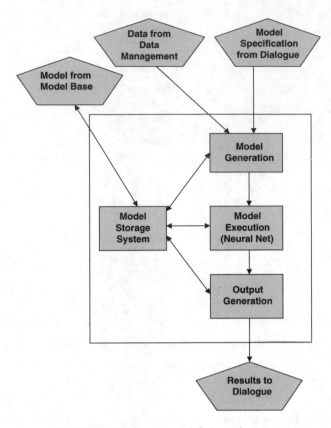

FIGURE I.10.4   A model-fitting subsystem.

TABLE I.10.4   Scenario Input

	Conditional Values								Baseline Values
	Week 1	Week 2	Week 3	Week 4	Week 5	Week 6	Week 7	Week 8	
Price 1	.80	.83	.79	.79	.69	.69	.69	.79	.69
Price 2	1.02	1.02	1.02	1.02	1.02	1.02	1.02	1.02	1.02
Price 3	1.12	1.15	1.19	1.19	1.19	1.19	1.19	1.19	1.12
Price 4	.99	.97	.97	.99	.89	.89	.89	.99	.99
Price 5	1.29	1.19	1.19	1.33	1.33	1.33	1.33	1.33	1.33
Commercial 1	5	5	10	10	5	5	5	5	5
Commercial 2	35	35	60	60	60	60	60	35	35
Commercial 3	50	60	50	50	50	45	60	60	50
Commercial 4	20	20	30	30	30	30	30	35	25
Commercial 5	30	50	50	20	20	20	20	20	40
Feature 1	ON	ON	ON	ON	ON	ON	ON	ON	ON
Feature 2	ON	ON	OFF	OFF	OFF	OFF	OFF	OFF	ON
Feature 3	OFF	OFF	OFF	OFF	OFF	OFF	OFF	OFF	OFF
Feature 4	OFF	OFF	ON	ON	ON	ON	OFF	OFF	OFF
Feature 5	ON	ON	ON	ON	ON	ON	ON	ON	ON

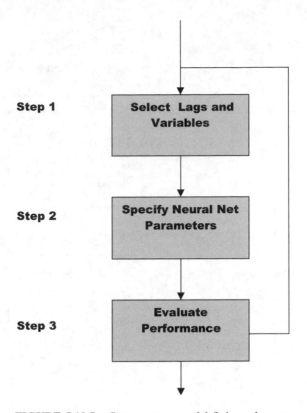

FIGURE I.10.5   Steps to use a model-fitting subsytem.

TABLE I.10.5   Single Brand Forecast – Tissue 1

	Baseline Forecast								
	Week 1	Week 2	Week 3	Week 4	Week 5	Week 6	Week 7	Week 8	Total
Volume	419.7	420.8	480.2	477.6	412	433.5	433.1	422.8	3499.70
Price	0.69	0.69	0.69	0.69	0.69	0.69	0.69	0.69	
Revenue	289.59	290.35	331.34	329.54	284.28	299.12	298.84	291.73	2414.79
Market Share	10.49	10.52	12.01	11.94	10.30	10.84	10.83	10.57	10.94
	Conditional Forecast								
	Week 1	Week 2	Week 3	Week 4	Week 5	Week 6	Week 7	Week 8	Total
Volume	380.1	366.1	379.3	390.2	440.8	430.2	433.1	368.6	3188.40
Price	0.8	0.83	0.79	0.79	0.69	0.69	0.69	0.79	
Revenue	304.08	303.86	299.65	308.26	304.15	296.84	298.84	291.19	2406.87
Market Share	9.50	9.15	9.48	9.76	11.02	10.76	10.83	9.22	9.96

10.6   Conclusion

Retail sales forecasts are inherently complex and error prone because of the large number of consumer and environmental factors determining the performance of marketing programs. These decisions are likely to become even more difficult and important for retail managers in the future. Because of

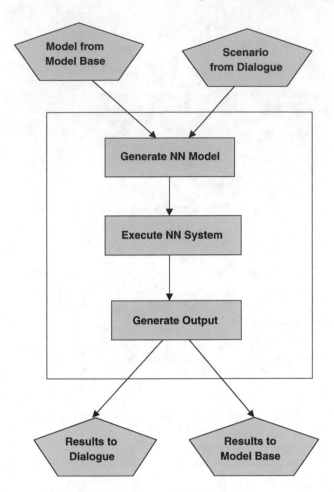

FIGURE I.10.6   A model application subsystem.

TABLE I.10.6   Sales Volume Summary Report

	Week 1	Week 2	Week 3	Week 4	Week 5	Week 6	Week 7	Week 8	Total
Tissue 1	380.1	366.1	379.3	390.2	440.8	430.2	433.1	368.6	3188.4
Tissue 2	432.6	490.5	475.5	489.2	417.3	420.4	432.9	476.8	3635.2
Tissue 3	1688.3	1498.2	1512	1449.8	1422.4	1417.4	1455.9	1634.2	12078.2
Tissue 4	742.7	799.2	802.3	830.2	869.4	832.2	834.9	792.7	6503.6
Tissue 5	758.3	878	823.9	852.6	852.1	901.8	864.2	729.7	6660.6

technological changes, consumers are getting information faster, so their knowledge, interests, and lifestyle are changing more quickly. The complexity of the marketplace is also increasing with respect to the volume of new product introductions, advertising, and promotional activities. Meanwhile, marketing managers need to forecast sales to make decisions about acquiring raw materials, inventory, and scheduling production. Retail forecasts are critical for efficient management of retail inventory while maintaining the flexibility to meet the demands at multiple stores.

Fortunately, information technology developments have provided product managers with more detailed information on consumers and markets. However, the data set is huge and not easy to use. Academics have developed sophisticated mathematical models as alternatives to help managers utilize these data more effectively. VAR-related applications have recently received a lot of attention from marketing scientists. VAR are used to study the dynamics of the marketing phenomenon. To avoid the over-parameterization problem in the use of unrestricted VAR, a Bayesian approach was proposed and proved to provide a more accurate forecast performance (Curry et al., 1995). Unfortunately, due to its mathematical complexity, much of this knowledge is still locked up in textbooks and journal articles. To utilize Bayesian VAR models, marketing forecasters would require understanding in the complex statistical relationship of the marketing variables involved in the system to specify prior distributions of lags. Under normal circumstances, a sales forecaster would have difficulty knowing the precise prior distribution for each variable involved in VAR.

However, an NN can be constructed in which the number of nodes at the input layer is equal to the number of variables involved in the system. The number of output layer nodes is equal to the number of dependent variables. The number of hidden layers and the nodes in the hidden layer can be selected arbitrarily. The network can be trained using a common back propagation model. Thus, there are few model manipulations that must be done by the manager. NNs are advantageous when marketing data structure (e.g., linear or nonlinear) is unclear. The sales forecasters don't have to make any assumptions while using NNs. Rather, NNs build their own models by learning the pattern from the data. As a result, they often handle nonlinear data better than traditional approaches. Furthermore, the use of NNs doesn't require managers to have any prior assumptions about underlying distributions or lags. As a result, better forecasting performance might be expected with NNs when the relationship between variables does not fit an assumed linear model.

BVAR and NN approaches on sales forecasts were compared for this report. It was found that the NN and BVAR models perform similarly on the Theil-U and MAPE criteria for several real data sets. Additional testing indicated both models appear to be very stable in variations on parameters, which indicates a robust character important to complex forecasting models that require parameter specification by the analyst. These results indicate that both BVAR and NN could be alternatives for marketing managers' forecasting needs. The NN approach has the advantage in simplicity. However, BVAR's coefficients are directly interpretable as response elasticities where NN connection weights are not directly interpretable.

The objective behind the development of decision support system in marketing is to provide a decision tool for marketing managers to make better use of the vast amount of available scanner data. Such systems also allow one to design more effective marketing programs without the burden of understanding the mathematical complexities of the models used in the systems. A prototype system was developed to allow marketers to utilize the more complex models. The system relies on database systems to manage the massive amounts of data, dialog systems to handle the communication with the user, and a model system that helps manipulate the model parameters and controllable variables. Models are stored for future reference and comparisons.

References

1. Abraham, M.M. and Lodish, L.M., An implemented system for improving promotion productivity using store scanner data, *Marketing Sci.*, 12, 248, 1993.
2. Adya, M. and Collopy, F., How effective are neural networks at forecasting and prediction? A review and evaluation, *J. Forecast.*, 17, 481, 1998.
3. Aiken, M. and Bsat, M., Forecasting market trends with neural networks, *Inf. Syst. Manage.*, 16, 42, 1999.

4. Agrawal, D. and Schorling, C., Marketing share forecasting: An empirical comparison of artificial neural networks and multinomial logit model, *J. Retail.*, 72, 383, 1996.

5. Balkin, S.D., Automatic neural network modeling for univariate time series, *Int. J. Forecast.*, 16, 497, 2000.

6. Belt, D., Neural networks: Practical retail applications, *Discount Merchandiser*, 9, 9, 1993.

7. Breusch, T., Testing for autocorrelation in dynamic linear models, *Austral. Econ. Pap.*, 17, 334, 1978.

8. Breusch, T.S. and Pagan, A.R., The Lagrange multiplier test and its application to model specification in econometrics, *Rev. Econ. Studies*, 47, 239, 1980.

9. Curry, D., *The New Marketing Research System*, New York, NY: John Wiley & Sons, 1993.

10. Curry, D. et al., BVAR as a category management tool: An illustration and comparison with alternative techniques, *J. Forecast.*, 14, 181, 1995.

11. Dekimpe, M.G. and Hanssens, D.M., Sustained spending and persistent response: A new look at long-term marketing profitability, *J. Market. Res.*, 36, 397, 1999.

12. Dragstedt, C., Shopping in the year 2000: Neural net technology is the brain of retail's future, *Discount Merchandiser Tech.*, 31, 7, 1991.

13. Engle, R.F., Autoregressive conditional heteroscedasticity with estimates of the variance of United Kingdom inflation, *Econometrica*, 50, 987, 1982.

14. Godfrey, L., Testing against general autoregressive and moving average error models when the regressors include lagged dependent variables, *Econometrica*, 46, 1293, 1978.

15. Greene, W.H., *Econometric Analysis*, 4th ed., Upper Saddle River, NJ: Prentice Hall, 2000.

16. Hill, T., O'Connor, M., and Remus, W., Neural network models for time series forecasts, *Management Sci.*, 42, 1082, 1996.

17. Jarque, C.M. and Bera, A.K., Model specification tests: A simultaneous approach. *J. Econ.*, 20, 59, 1982.

18. Jiang, J.J., Using scanner data: IS in the consumer goods industry, *Inf. Syst. Manage.*, 12, 61, 1995.

19. Litterman, R.S., A Bayesian procedure for forecasting with vector autoregressions. Working paper, Massachusetts Institute of Technology, Department of Economics, Cambridge, MA, 1980.

20. Little, J.D., Decision support systems for marketing managers, *J. Market.*, 43, 9, 1979.

21. Ljung, G.M. and Box, G.E.P., On a measure of lack of fit in time series models, *Biometrika*, 66, 265, 1979.

22. Nijs, V.R. et al., The category-demand effects of price promotions, *Marketing Sci.*, 20, 1, 2001.

23. Ramsey, J.B., Tests for specification error in classical linear least squares regression analysis, *J. R. Stat. Soc.*, B31, 250, 1998.

24. Tsaih, R., Hsu, Y., and Lai C.C., Forecasting S&P 500 stock index futures with a hybrid AI system, *Decision Support Systems*, 23, 161, 1998.

25. Venugopal, V. and Baets, W., Neural networks and statistical techniques in marketing research, *Marketing Intelligence & Planning*, 12, 30, 1994.

26. Wasserman, P.D., *Neural Computing: Theory and Practice*, New York, NY: Van Nostrand Reinhold, 1989.

27. Zenor, M.J., The profit benefits of category management, *J. Market. Res.*, 31, 202, 1994.

11

Neural Network Systems and Their Applications in Software Sensor Systems for Chemical and Biotechnological Processes

Svetla Vassileva
Bulgarian Academy of Sciences

Xue Z. Wang
The University of Leeds

Abstract. Modern process industries demand advanced control and optimization techniques for precision manufacturing of products and minimization of environmental and safety risks. The basis for success of advanced control and optimization is availability of on-line measurements of variables characterizing product quality and relating to environmental and safety performance. Despite the development in new sensor techniques, e.g., through adapting laboratory analytical techniques for use in industrial processes, the measurement is often not reliable on-line, or frequent enough to establish timely monitoring and adequate

0-8493-1121-7/03/$0.00+$1.50
© 2003 by CRC Press LLC

control and optimization. Software sensors, which use readily available measurement to estimate hard-to-measure variables, have received much attention in recent years. In this contribution, we intend to provide a comprehensive discussion on developing software sensors for chemical and biotechnological processes. The emphasis is on the necessary considerations and guidelines for developing neural networks (NN) and hybrid NN-knowledge-based software sensors as well as system architecture. Five practical implementations of software sensor development for industrial processes are presented, including refinery fluid catalytic cracking, toxicity prediction of aqueous effluents in specialized organic chemicals manufacturing, prediction of yeast population morphology and vitality, yeast growth and optimization in beer manufacturing, and prediction of sparkling wine composition.

11.1   Introduction

Modern chemical manufacturing and industrial biotechnology is subject to not only an increasingly competitive global market, but also to more and more stringent regulations on operational safety and environmental protection, which calls for precision control of product quality and variables relating to safety and the environment. This has led to a growing interest in techniques which can directly measure, predict, and control the physical, chemical, and hydrodynamic changes inside the process, in addition to the traditional variables of control such as temperature, pressure, flowrates, and levels of materials in vessels. Examples of the new variables include

- Product quality variables such as molecular weight and molecular weight distributions of polymers, viscosity and condensation temperature of oil products, and composition of trace impurities in food and pharmaceutical processes
- The size and size distributions as well as shapes of particulate products from emulsion polymerization and crystallization as well as precipitation of colloidal processes
- NOx, SOx, and particulate emissions from industrial processes and COD (chemical oxygen demand), BOD (biological oxygen demand), odor, and toxicity of aqueous effluents from specialized organic chemicals processes
- Rate of heat generation of exothermic reaction systems
- Morphophysiological characteristics of yeast populations and its vitality, control of the fermentation activity of the cultivated yeast populations, and its influence on final product composition and sensory rating in biotechnology.

A notable development in recent years has been the application of traditional laboratory analytical techniques to on-line monitoring of industrial processes, such as the use of near infrared spectroscopy (NIR), X-ray diffraction (XRD), ultrasonic spectroscopy (USS), and scattered light techniques. Another promising technique, initially adapted from the medical field, is process tomography, which can visualize through the wall of a vessel in real time the mixing and flow inside the process. Despite these developments, the measurement is often not reliable on-line, or frequent enough to establish timely monitoring and adequate control and optimization.

Software sensors (also known as soft sensors or inferential measurement) that use readily available real-time measurements to estimate the hard-to-measure variables provide an attractive technique when hardware sensors are not reliable on-line or frequently enough to establish timely monitoring and adequate control. In addition to being able to facilitate real-time monitoring and control, software sensors also bring another advantage that traditional hardware sensors can't offer, which is that using the software sensor model, it is able to carry out sensitivity studies to discover the most influential operational variables to the target control variable, to help develop the most effective control strategies.

Software sensors can either be developed from first principles or using data-driven modes. First-principle models require a deep understanding of the physical, chemical, and biological features and hydrodynamics of the process. Unfortunately, for many chemical and biotechnological processes, such fundamental models are not available due to lack of knowledge about these processes. First-principle models are also effort-intensive to develop. It is very common for a comprehensive mechanistic process model to involve several dozens to hundreds of differential and algebraic equations. Data-driven techniques can effectively make use of the data resources collected over the operational history of a process and continuously improve in performance as more data are made available. Various data-driven techniques have been studied for developing software sensors including the well-established statistical techniques, such as multiple linear regression (MLR), principal component regression (PCR), partial least squares (PLS), and nonlinear techniques, such as nonlinear principal component analysis, nonlinear autoregressive moving average with exogenous input (NARMAX) model, as well as artificial neural networks. Neural networks offer the distinctive ability to learn complex nonlinear relations between multiple inputs and outputs without requiring specific knowledge of the model structure. They have demonstrated surprisingly good performance in various applications.

The main purpose of this chapter is to describe approaches for developing neural network (NN) and hybrid NN knowledge-based software sensors for industrial processes, particularly chemical and biotechnological processes. The emphasis is on providing guidelines for working solutions and on practical experience in implementation. An attempt has been made to use representative industrial applications to illustrate the concepts. Most of them are drawn from the authors' own work and some from literature. They cover such processes as refinery reaction and distillation, wastewater treatment, control of the morphophysiological characteristics of yeast populations and its vitality, control of the fermentation activity of the cultivated yeast populations, and its influence on final product composition and sensory rating in biotechnology.

This chapter is organized as follows. In Section 11.2, we give a state-of-the-art survey of the current research on software sensors in chemical and biotechnological industries. In Section 11.3, by reference to two industrial applications, one concerned with a software sensor for a refinery reaction distillation process and one with toxicity of aqueous effluents produced from specialized organic chemicals plant, we illustrate the techniques for feature extraction from input variables, selection of training, and test data as well as the use of software sensor models to carry out sensitivity studies. Section 11.4 reviews software applications in biotechnological processes. Sections 11.5, 11.6, and 11.7 describe three practical applications of hybrid systems for software sensors for prediction of yeast population morphology and vitality, prediction and optimization of yeast growth in beer manufacturing, and prediction of sparkling wine composition and sensory rating. Conclusions and suggestions for future work are given in Section 11.8.

11.2   Neural Network-Based Software Sensors—An Overview

11.2.1   Feedforward Neural Networks (FFNN)

Simply speaking, an FFNN neural network is an algorithm or computer program that can learn to identify the complex nonlinear relationships between multiple inputs and outputs. The learning process has a number of characteristics. Firstly, FFNN does not need fundamental domain problem models and is easy to be set up and trained. This is different from conventional statistical methods that usually require the user to specify the functions over which the data is to be regressed. Secondly, data examples used for training are allowed to be imprecise or noisy. In addition, it mimics the human learning process: learning from examples through repeatedly updating the performance.

An FFNN neural network consists of a number of processing elements called neurons. These neurons are divided into layers. Figure I.11.1 shows a three-layered FFNN architecture including an input, a hidden, and an output layer. Typically, the input layer nodes correspond to input variables

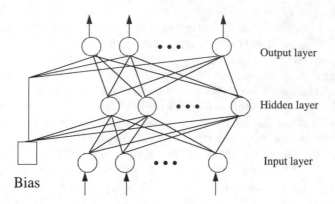

FIGURE I.11.1   A three-layer feedforward neural network.

and the output layer to output variables. Hidden neurons do not have physical meanings. Neurons between two adjacent layers are fully connected by branches. A transfer function (or activation function) describes each neuron in the hidden and output layer. Usually a sigmoidal function is used:

$$f(z) = \frac{1}{1 + e^{-aZ}} \qquad (I.11.1)$$

$f(z)$ transforms an input z to the neuron to the range of [0.0, 1.0] and the parameter a is used to change the shape of the sigmoidal function. Each connection branch is described by a weight representing the strength of connection between two linked nodes. The so-called learning or training process is the procedure to adjust the weights. A bias neuron that supplies an invariant output is connected to each neuron in the hidden and output layers. The bias provides a threshold force activation of the neuron, and is essential in order to classify network input patterns into various subspaces.

Given some arbitrary values for all the connection weights, for a specific data pattern, the FFNN makes use of the weights and input values to predict the outputs. The training is intended to gradually update the connection weights to minimize the mean square error E:

$$E = \sum_{m=1}^{M} \sum_{i=1}^{N} (t_i^{(m)} - y_i^{(m)})^2 \qquad (I.11.2)$$

where M is the number of training data patterns, N is the number of neurons in the output layer, $t_i^{(m)}$ the target value of the ith output neuron for the given mth data pattern, $y_i^{(m)}$ the prediction for the ith output neuron given the mth data pattern.

The process involves a forward path calculation to predict the outputs and backward path calculation to update the weights. For a neuron in the input layer, its output is equal to the input, so there is in fact no activation function for an input neuron. For a neuron in the hidden and output layers, it receives the values of the outputs of its front layer nodes and takes the weighted sum as its input. The weighted sum is then transformed by the activation function to give an output. The outputs of the output layer neurons are compared with the target values using Equation (I.11.2) to calculate an error. The error is used for backward updating of the weights. Early FFNNs use an algorithm called Delta rules to update the weights through backpropagating the errors. Later, more efficient optimization methods were studied. Leonard and Kramer[1] studied the use of a conjugate gradient approach in order to speed up convergence. Alternative algorithms have also been proposed by Brent,[2] Chen and Billings,[3] and Peel et al.[4] An attractive approach to the backpropagation learning algorithm is the quasi-Newton method.[5,6] FFNN has shown surprisingly good performance in solving many complex problems. Essentially, a three-layer perceptron with $N(2N + 1)$ nodes using continuously increasing nonlinearities can compute any continuous function of N variables.

It is generally accepted that only one hidden layer is necessary in a network using a sigmoid activation function, and no more than two are necessary if a step activation function is used. There are no generally accepted methods for determining the optimum number of hidden nodes in an FFNN with one hidden layer. The number of hidden neurons depends on the nonlinearity of the problem and error tolerance. Empirically, the number of neurons in the hidden layer is of the same order as the number of neurons in the input and output layers. The number of hidden neurons must be large enough to form a decision region that is as complex as required by a given problem; too few hidden neurons hinder the learning process and may not be able to achieve the required accuracy. However, the number of hidden neurons must not be so large that many weights required can not be reliably estimated from available training data patterns. An unnecessarily large hidden layer can lead to poor generality. A practical method is to start with a small number of neurons and gradually increase the number.

Chemical process models are multidimensional with peaks and valleys,[7] which can trap the gradient descent process before it reaches the system minimum. There are several methods of combating the problem of local minima.[8,9] The momentum factor α, which tends to keep the weight changes moving in the same direction, allows the algorithm to slip over small minima. Another approach is to start the learning again with a different set of initial weights if it is found that the network keeps oscillating around a set of weights due to lack of improvement in the error. Sometimes adjusting the shape of the activation function (e.g., through adjusting the constant a in Equation (1.11.1) can have an effect on the network susceptibility to local minima. Some new optimization approaches have been applied to multilayer neural networks, which prove to be able to address the local minima significantly, such as the simulated annealing[9] and genetic algorithms.[10]

Overfitting occurs when the network learns the classification of specific training points, but fails to capture the relative probability densities of the classes (Figure I.11.2).[11] This can be caused by two situations: (1) oversized network, e.g., due to inclusion of irrelevant inputs in the network structure or too many hidden layers or neurons; and (2) insufficient number of training data patterns. Overfitting in training a neural network deteriorates the generalization ability of the network. Generalization ability of a neural network is a very important issue because neural networks are trained using data instead of being obtained from first-principle models. Shao et al.[12] and Zhang et al.[13] developed a complicated procedure for calculating the confidence bounds of applying an FFNN model to predict a new data pattern. The most important procedure is variable selection and feature extraction from potential inputs of an FFNN model, which will be dealt with in detail in Section 11.3 by reference to two industrial case studies.

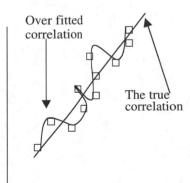

FIGURE I.11.2 Overfitting caused by over-parameterization.

11.2.2 FFNN Applications

FFNN is the most widely studied neural model for developing inferential models in chemical and biotechnological applications. Examples include:

- Refinery processes, for butane splitter towers and gasoline stabilizers,[14] solvent extraction processes,[15] predicting jet fuel endpoints,[16] product quality prediction in reforming processes,[17] prediction of condensation temperature of light diesel,[18] the light cycle gas oil 90% point,[19] and product distribution[21] of fluid catalytic cracking processes, crude oil distillation processes,[22,23,24] and fluid catalytic cracking processes for fault diagnosis and model predictive control [20,46,47]
- Polymerization processes, for predicting average molecular weight of polymers,[13,25] controlling molecular weight distribution of polymers,[26] and controlling the melt index of polyolefin[27,28]
- Fermentation processes, for real-time biomass concentration control,[29–32,113] prediction of the biomass and consumed limiting substrate and products,[33,34,35] on-line monitoring of limiting substrate and indirect measurement of the biomass,[36] estimation and control of antibody production using hybridoma cells,[37] adaptive pH control,[38] and long-range predictive control of the manipulated variables—temperature and dilution rate[39,40]—prediction of pressure of EtOH, pH, and inverse model-based prediction of manipulative variables—feed of molasses and NH_3[41]—as well as for control of pressure in a fermentor[42]

11.2.3 Recurrent Neural Networks

A feedforward neural network is static in nature and is not designed to deal with dynamics, e.g., time delays. Recurrent neural networks are designed to address dynamics.[43] Figure I.11.3(a) shows the architecture of a typical externally recurrent neural network (ERN). It indicates that the values of the output variables at the current time interval are predicted using the values of the input variables at the current time interval and the output values at the previous or delayed time intervals. The predicted values of the outputs will be used as the inputs to the network for predicting the future values of the

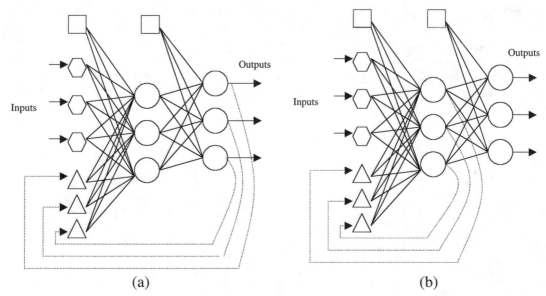

(a)   (b)

FIGURE I.11.3   Recurrent neural networks. (a) Externally recurrent network (ERN) and (b) Internally recurrent network (INN). ○ = Hidden and output nodes; □ = bias unit; ⬡ = input nodes; △ = context unit.

outputs. This process may cause prediction errors to propagate from one cycle to the next, resulting in poor long-range forecasting. However, there are techniques to improve this deficiency.[44]

Alternative network architecture is called internally recurrent network (IRN), in which the time-delayed feedback connections are to the hidden layer, instead of the output layer (Figure I.11.3(b)). Although the ERN and IRN can exhibit comparable modeling performance, they have different features that may make one more desirable than the other for a particular problem.[43] The IRN does not have any structure limit on the number of model states because the number of hidden nodes can be freely varied. The ERN, however, can only have the same number of states as model outputs because the outputs are the states. The IRN thus tends to be more flexible in modeling. The ERN has the advantage that the model states have a clear physical interpretation in that they are the variables of the process itself, whereas the states of the IRN are hypothetical and neither unique nor canonical. Initialization of ERN models is simple because the user can observe the current values of the process and use those values to initialize the states. IRN models are more difficult to initialize because the states lack physical meaning.

11.2.4   Radial Basis Function Neural Networks

Figure I.11.4 shows the structure of a radial basis neural network (RBFNN). Before an RBFNN is trained, it requires the data of input variables to be clustered, using a clustering method such as the *k*-means clustering approach. RBFNN is different from FFNN in several aspects. First, the inputs to the network are fed directly into the hidden nodes through connections with unity fixed weights. Second, the number of hidden nodes is determined by the number clusters obtained in clustering the input data. The transfer function of each hidden node is a symmetrical radial basis function (RBF), usually Gaussian or ellipsoidal.

$$G(\mathbf{I}) = \exp\left[-\frac{\|\mathbf{I} - \mathbf{c}\|^2}{\sigma^2} \right] \qquad (\text{I.11.3})$$

where $\mathbf{I}$ denotes the vector of inputs to the node, $\mathbf{c}$ is the vector of centers of all clusters in the input space, σ is a "span" parameter, and $\|.\|$ a vector norm. The weight sum of the outputs from the hidden nodes is fed to the output layer. Normally, the output of an output layer node simply equals its input. In other words, the transfer functions of the output layer nodes are linear.

Since the weight for each connection between an input node and a hidden node is fixed as 1, the training process is just to find the weights for the connections between the hidden and output layers. A linear regression function can be used to train the weights;[45] therefore, the time to train an RBFNN is much shorter than for an FFNN.

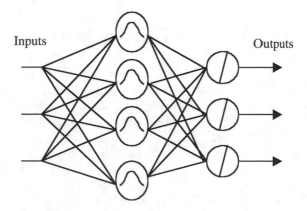

FIGURE I.11.4   Radial basis function networks.

Another advantage with RBFNN is that if a unique new input vector is encountered in testing, the network output will go to zero because of local properties of RBFNN in Equation I.11.3. This is a major advantage over FFNNs, since FFNN generally extrapolate for new inputs very poorly. While RBFNN suffers from this same problem, at least they are capable of detecting when the network is being asked to extrapolate.

Like FFNN, RBFNN is also static in nature. The recurrent neural networks will also have to be used to incorporate dynamics. Unfortunately, with RBFNN, the size of the network scales exponentially with the number of inputs to the network. Each of the hidden nodes represents a bump in the input space, and the number of bumps required to model a process over the entire space rises exponentially with the dimension of the input space. Even for steady-state models, RBFNN becomes impractical with more than four or five input variables.

RBFNNs have been applied to process fault diagnosis,[44] polymerization processes for model identification,[28] and fermentation processes.[29]

11.2.5 Hybrid Systems

Hybrid software sensor systems make use of information in heterogeneous (quantitative and quali- tative, numeric and linguistic) forms. Incorporating features of human-like perception, experience, and skill into neural network-based software sensor development is particularly important when the sensors have a "biological" nature, or are used for living systems monitoring and control. Due to the increasing volume of knowledge and degree of complexity and nonlinearity, it is necessary to effectively filter knowledge in order to extract only the relevant information, and maintain and update the available knowledge through learning dynamically, according to changes in system conditions. In this chapter, practical industrial examples will be used to illustrate the use of fuzzy and knowledge- based expert systems in incorporation with neural networks for development of software sensors in industrial biotechnology.

11.3 Software Sensor Development for Chemical Processes

In this section, we describe two examples of industrial applications of software sensors using FFNNs; one is concerned with a refinery reaction—distillation process—and the other with the prediction of toxicity of aqueous effluents produced from specialized organic chemicals plant. The purpose is to use these two industrial applications to illustrate the techniques used for feature extraction from input variables, selection of training and test data, and use of nonlinear FFNN models to extract cause-effect relationship knowledge.

11.3.1 Software Sensor Design for a Refinery Fluid Catalytic Cracking Process

Fluid catalytic cracking (FCC) is the most important and widely used refinery process for converting heavy oils into more valuable gasoline and lighter products. It represents on the order of 30% in product value of a typical refinery. It involves large, high throughput and complex equipment, which together with its economic significance makes it a very good candidate for advanced control. The FCC process is physically complex and proves difficult to both operate and control, not only because of operational constraints and process variable interactions dominated by the heat and momentum balances between the reactor and regenerator, but also because of the extreme high temperatures and pressures, and the hazards arising from potential fires and explosion arising from malfunctions. It is the nature of the system that important operating variables associated with product quality and process control performance cannot be measured directly or easily, despite the expenditure

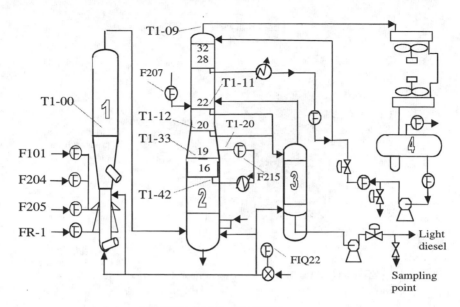

FIGURE I.11.5 The FCC process.

on special measurement techniques. Many of the important variables are normally determined by offline laboratory analysis or online product quality measurements, which suffer from excessive time delays and can only be done intermittently, thereby limiting their effectiveness as far as control is concerned.

Software sensors make it possible to compensate for relatively infrequent sampling and time delays by generating continuous real-time process information, which can be manually evaluated to give essentially continuous online assessment. Industrial applications of software sensors for the FCC process include product quality prediction,[18] prediction of catalyst circulation rate and heat of reaction,[20] and prediction of fluidization conditions as well as product distribution.[21] The FCC software sensors can be used for inputs for advanced control strategies as demonstrated by Yang et al.[20]

The relevant section of the process is shown in Figure I.11.5, where the oil–gas mixture leaving the reactor goes into the main fractionator to be separated into various products. The individual side draw products are further processed by downstream units before being sent to blending units. One of the products is light diesel, which is typically characterized by temperature of condensation. Previously, the temperature of condensation was monitored by offline laboratory analysis. The sampling interval is between 4 to 6 hours and it is not practical to sample it more frequently, since the procedure is time consuming. This deficiency of offline analysis is obvious, and the time delay is a cause of concern because control action is delayed. Moreover, laboratory analysis requires sophisticated equipment facilities and analytical technicians. Clearly, there is great scope for a software sensor to carry out such online monitoring.

Fourteen variables are used as inputs to the FFNN model as shown in Figure I.11.6. These variables are also summarized in Table I.11.1.

Initially, 146 data patterns (Dataset-1) from the refinery were used to develop the model (Model-1). Later, three different sets of data became available, having data patterns of 84 (Dataset-2), 48 (Dataset-3), and 25 (Dataset-4), respectively. Model-1 has therefore been modified three times with the versions being denoted by Model-2, Model-3, and Model-4. Interesting and different results for every model improvement are described below. In all cases, the prediction accuracy of the model is required to be within $\pm 2°C$.

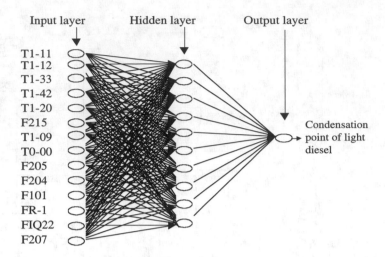

FIGURE I.11.6   The feedforward neural network structure.

TABLE I.11.1   The Fourteen Variables Used as Input to the FFNN Model

T1-11	- the temperature on tray 22 where the light diesel is withdrawn
T1-12	- the temperature on tray 20 where the light diesel is withdrawn
T1-33	- the temperature on tray 19
T1-42	- the temperature on tray 16, i.e., the initial temperature of the pumparound
T1-20	- the return temperature of the pumparound
F215	- the flowrate of the pumparound
T1-09	- column top temperature
T1-00	- reaction temperature
F205	- fresh feed flowrate to the reactor
F204	- flowrate of the recycle oil
F101	- steam flowrate
FR-1	- steam flowrate
FIQ22	- flowrate of the over-heated steam
F207	- flowrate of the rich-absorbent oil

11.3.1.1   Model Development (Model-1) Using Dataset-1

The data set (Dataset-1) used for model development comprises 146 patterns. The first step is to process the data using an automatic classification system named AutoClass. AutoClass is a tool developed by NASA[48] that can automatically divide multidimensional data into classes, so that data cases in one class are similar and are distinguished from cases in other classes. AutoClass predicts seven classes (numbered 0, 1, . . . , 6), as shown in Table I.11.2. For example, class 0 has 32 members (i.e., 32 data patterns). Test data are sampled from each class and they are indicated in bold and underlined in Table I.11.2. More data patterns are sampled from larger classes and fewer from smaller classes. Altogether, there are 30 data patterns used for testing and 116 for training.

The FFNN software sensor model (Model-1) is obtained when the training error reaches 0.424. With normalized [0, 1] training data, the error is calculated using $\frac{1}{2}\sum_{i=1}^{n}(y_i' - y_i)^2$, where y_i' is the prediction by the model for the ith training pattern and y_i the target value. There are three training patterns and two test patterns with absolute errors exceeding the required $\pm 2°C$. Therefore,

TABLE I.11.2   Classification of Dataset-1 and Test Data Selection for FFNN[a]

Class 0 Weight 32				Class 1 Weight 31				Class 2 Weight 29			
5	**6**	16	17	1	10	25	**26**	2	**3**	4	15
31	32	34	35	27	44	45	**46**	18	19	**64**	65
36	37–39	**40**	41–43	47–50	**51**	52	76	69	70	72	**73**
81	103	**104**	105	79	**80**	82	120	74	75	83	84
108	**109**	110	112	**121**	122–24	**125**	126	**85**	86–88	**89**	90
114	**115**	116	118	128	129	132	**133**	91	93	94	**95**
136	**137**	138	130	134	135			96	97	146	

Class 3 Weight 26				Class 4 Weight 11			Class 5 Weight 9			Class 6 Weight 8		
11	**12**	13	14									
53	54	**55**	56–59	7	9	28	**20**	21	30	8	22	**23**
60	61–63	66	67	**29**	33	111	101	102	106	24	107	113
68	71	77	78	117	127	**130**	140	**141**	142	119	144	
92	98	**99**	100	131	145							
143												

[a]Data in bold and underlined are test data; weight = number of members

the degree of confidence for training data is $100\% - (3/116)\% = 97.4\%$, and that for test data is $100\% - (2/30)\% = 93.3\%$. The refinery considers a degree of confidence of 90% acceptable.

11.3.1.2   Model-1 Improvement Using Dataset-2

The production strategy changes according to season. The accuracy of Model-1 originally developed was later found to be inadequate, and 84 more data patterns (Dataset-2) were provided in order to improve model performance. Initial application of Model-1 to the 84 new data patterns indicated that the degree of confidence is only $100\% - (63/84)\% = 25.0\%$.

The 84 new data patterns were combined with Dataset-1 and processed by AutoClass. It was found that the 230 data patterns $(84 + 146)$ were classified into 15 classes. Interestingly, all the 84 new data patterns in Dataset-2 were classified into seven new classes (classes 2, 5, 7, 8, 9, 13, 14), while all the 146 patterns in Dataset-1 are in eight classes (classes 0, 1, 3, 4, 6, 10, 11, 12). This is consistent with the poor predictions for the Dataset-2 using Model-1. A summary of the classification is given in Table I.11.3, where the data in italic are from Dataset-2. The data patterns in bold and underlined are chosen for testing and the rest for retraining, to generate Model-2. Altogether, 49 data patterns are chosen for testing, of which 19 are from Dataset-2 and 30 from Dataset-1. The degree of confidence for the training data is $100\% - (11/181)\% = 92.8\%$ and for test data $100\% - (3/49)\% = 93.9\%$.

11.3.1.3   Improvement Using Dataset-3

Later, 48 new data patterns (Dataset-3) were provided. Model-2 was applied to Dataset-3 and the degree of confidence was found to be $100\% - (27/48)\% = 43.8\%$. This is low, but better than the prediction to Dataset-2 using Model-1 (25.0%). The 48 new data patterns are then mixed with Dataset-1 and Dataset-2 and processed by AutoClass to give 16 classes. It is found that 25 of the 48 data patterns form a new class, but the rest are assigned to the classes combined with data from Dataset-1 and Dataset-2. The degree of confidence using Model-2 to predict the classes 1, 2, 3, 8, and 11 demonstrates that class 1 contains only new data patterns and has the lowest degree of confidence for prediction using Model-2. Of the 25 patterns in Dataset-1, 19 have deviations bigger than $\pm 2°C$. Classes with fewer new data patterns have a higher degree of confidence in the predictions. This result further demonstrates the advantage of using AutoClass for clustering data before training an

TABLE I.11.3   Classification Result of Dataset-1 plus Dataset- 2[a]

Class 0 Weight 38			Class 1 Weight 31			Class 2 Weight 26			Class 3 Weight 19		Class 4 Weight 17	
2	**3**	4	5	6	17						11	**12**
14	15	**18**	29	32	**35**	160	*161*	*175*	1	10	13	25
19	21	64	36–39	**40**	41–43	*177–179*	182	186	16	**44**	56	**57**
65	66–69	**70**	81	103	**104**	187	*192*	*193–195*	45–48	**49**	58–60	**61**
71–75	82–84	86	105	108	109	*196*	197	*209*	50–52	**53**	62	63
87	88–90	**91**	**110**	111	112	210	**211**	*212*	54	55	77	**78**
93	94	**95**	114	**115**	116	215	**216**	*217–219*	79	**80**	85	92
96–98	**99**	100	117	136	**137**	**220**	*221*		21	122	143	
146			138	139								

Class 5 Weight 15			Class 6 Weight 14			Class 7 Weight 12		Class 8 Weight 11		Class 9 Weight 10	
157	*176*	**180**	7	8	22	*162*	**163**	*158*	*159*		
181	*183*	**184**	**23**	24	31	*164*	*222*	*198*	**199**	*147*	*148*
185	*188*	**189**	33	**34**	76	*223*	**224**	*200–202*	**203**	*149*	*150–152*
190	*191*	*205*	107	113	**118**	*225–227*	**228**	*204*	*213*	**153**	*154–156*
206	*207*	*208*	119	144		*229*	*230*	*214*			

Class 10 Weight 10			Class 11 Weight 9			Class 12 Weight 8			Class 13 Weight 7		Class 14 Weight 3	
9	26	28	27	125	126	20	30	**101**	*166*	**167**	*165*	*173*
120	**123**	124	**128**	129	132	102	106	140	*168–170*	**171**	*174*	
127	130	**131**	133	**134**	135	**141**	142		*172*			
145												

[a]Italics refer to data patterns from Dataset-2; bold and underlined are used for tests.

FFNN model. It is also found that a confidence in a model for predicting new data is lower if more data are grouped into new classes. For example, all patterns in Dataset-2 are grouped into new classes and the confidence of predicting Dataset-2 using Model-1 is 25.0%; while some patterns in Dataset-3 are classified into old classes, with a confidence in predicting Dataset-3 using Model-2 of 43.8%.

The 48 new data patterns in Dataset-3 have been combined with Datasets-1 and -2 to develop Model-3. The sampling of test data patterns follows the same procedure as before. A total of 68 data patterns has been used for testing with the remaining 210 patterns for training. Model-3 has a degree of confidence of 92.6%($= 100\% - (5/68)\%$) for testing data and 93.8%($= 100\% - (13/210)\%$) for the training data.

11.3.1.4 Improvement Using Dataset-4

Dataset-4 was subsequently obtained from the refinery and has 25 new data patterns. The confidence of applying Model-3 to Dataset-4 is $100\% - (8/25)\% = 68\%$. Dataset-4 was then combined with previous data sets to give a database of 303 patterns. These are classified by AutoClass into 15 classes. seventy-seven data patterns have been selected as test data, and the rest for training to develop the FFNN Model-4. The model has a confidence of 92.5%($= 100\% - (17/226)\%$) for training data and 90.9%($= 100\% - (7/77)\%$) for test data. In order to make a comparison with the conventional sampling approach, 77 data patterns have been selected using random sampling as

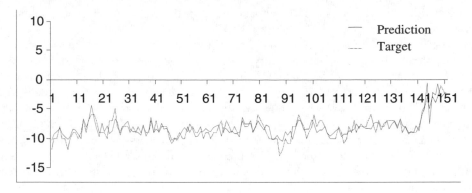

FIGURE I.11.7   Comparison between predictions using Model-4 and the target values for the first 151 data patterns.

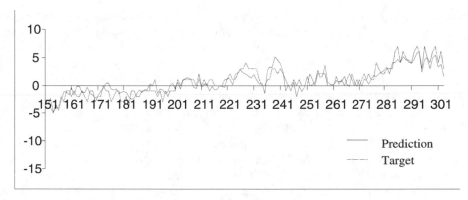

FIGURE I.11.8   Comparison between predictions using Model-4 and the target values for the last 152 data patterns.

test patterns. The training is terminated using the same criterion, i.e., a training error of $5.75\,e^{-1}$. On this basis, 11 test data patterns have errors exceeding $\pm 2^{\circ}C$. So the confidence for the test data is $100\% - (11/77)\% = 85.7\%$. This is lower than that of the test data used in this study, which is $90.9\%(100\% - (7/77)\%)$. The confidence of the training data for both approaches is the same, 92.5%.

Model-4 (i.e., all data sets) covers all the operational seasons and is being used very satisfactorily. It has proved to be robust and reliable over a wide range of operational conditions. Figures I.11.7 and I.11.8 show the differences between predictions using Model-4 and the target values of the 303 patterns. The plant has reduced the sampling frequency to twice a day compared with four to six times used previously, and the intention is to reduce it to once a day in the future.

11.3.1.5   Further Discussion on Selection of Training and Test Data and Model Generalization Ability

The above industrial application proved the importance of selecting training and test data, and the need to continuously and systematically improve the performance of an FFNN software sensor model. There are other important considerations, which have not been addressed in the above case study. One of the factors is that an FFNN model could be over-trained by data from the large clusters, but less trained by data from smaller clusters. This will cause model rigidity and will not have the required generalization ability. A useful approach to reduce this risk is to reduce the number of data patterns of large clusters.

In developing a system for pattern recognition for dynamic processes, Smaragdis[46] proved that using this approach, the generalization ability of the neural network model could be improved. In the same study, Smaragdis[46] argued that retraining a neural network to update its parameters could have an adverse effect. More specifically, the continuous change of the weights could lead to a phenomenon of "memory loss," which refers to the situation that the network model drifts away from the general model it was initially trained to approximate. He used the approach of classifying the mixed data and then reducing the data density in large clusters, and proved that it can help improve the problem of "memory loss." He classified the new measurements, which are dynamically received, and then only the data, which are classified separately from the rest of the data, need to be used in retraining.

11.3.2   Software Analyzer Design for Toxicity Prediction of Aqueous Effluents

Worldwide, more than 16.7 million distinct organic and inorganic chemicals are known, of which 70,000 are in commercial production.[49,50] Additionally, modern combinatorial chemistry techniques have led to the production of vast libraries of new chemicals at a very rapid rate. The specialized organic chemicals manufacturing sector in the U.K. estimates up to 10,000 novel chemicals for possible production; for most of them, toxicity data are not available. Toxicity has become one of the most important factors in production of specialized organic chemicals, because management of aqueous effluents is one of the most important business drivers, representing 20% of the production cost. Existing methods of toxicity testing used by the specialized organic chemicals manufacturers involve subcontracting work to environmental testing laboratories. The tests approved by the Environment Agency (EA) of the U.K. are trout, Daphnia (water flea), and green algae, each of which take several days to a week and cost on average £2–3K for a normal test and £10K for a full test. A complete carcinogenicity test can take even longer (3–5 years) and cost millions of pounds.[50,51] Apart from the time and cost involved, public opinion is an another factor that cannot be ignored. A promising technique that can minimize the use of animals is quantitative structure-activity relationships (QSARs), which try to map descriptors including physical, chemical, structural, and biological parameters characterizing a compound to the observed toxicity values. Several methods have been studied, including statistics-based regression, expert systems, and neural networks.

In the following, we present an example of applying FFNNs to predicting toxicity of aqueous effluents. The focus is on how to perform input dimension reduction and carry out a sensitivity study using the neural model. Although the toxicity measure Microtox is not an EA-approved measure, the principle is applicable to other toxicity measures.

11.3.2.1   The Effluents and Their Toxicity

The effluents are initially produced in a number of batch reaction units. After a series of separation and recovering units, they are mixed with aqueous effluents from other streams and go to an on-site treatment unit. The aqueous effluents are first extracted and then neutralized with a base. The neutralized wastewater is fed to holding tanks, where it is analyzed for compliance before discharge. Nineteen variables, as shown in Table I.11.4, are measured, including pH, COD (chemical oxygen demand), and individual chemical components, c1-c8 and x1-x8. Because of the confidentiality requirement, the names of c1-c8 are not given here; x1-x8 are known chemicals, and it is of interest to see if they make any contribution to its overall Microtox value. Some of the parameters, like pH, can be easily and accurately measured by conventional chemical analysis. However, some must be analyzed with special equipment, e.g., Microtox, an indicator of aqueous effluent toxicity. The results

TABLE I.11.4 Variables Measured

Number	Variable	Description	Number	Variable	Description
1	pH		11–18	x1–x8	x1–x8 are minor components
2	COD	Chemical Oxygen Demand			that are detected in the
3–10	c1–c8	Chemical species			chromatographic analysis
			19	Microtox	Toxicity measure

of analysis are recorded in a database system. The database used in this study comprises records of 180 days.

The measurement of Microtox requires the luminescent marine bacterium, as the test microorganism to observe the reduction of its emission light, instead of the observation of its death when it contacts with toxicant. The motivations for developing a software analyzer for Microtox toxicity prediction are, first, the current approach of Microtox analysis is expensive. Second, it requires experienced technicians. Third and probably more important, the current approach of measuring Microtox is not able to tell how to reduce the value of Microtox, because it does not indicate what factors are responsible for observed Microtox values. A software analyzer correlating Microtox value with other variables can provide the basis for developing methods to reduce the Microtox value.

11.3.2.2 The Software Analyzer

The strategy for developing the software analyzer can be illustrated using Figure I.11.9. The 18 inputs are pre-processed using principal component analysis (PCA) to remove the dependencies between the inputs. The principal components (PCs) are used as the inputs to a feedforward neural network (FFNN). The PCA processed data were clustered into six classes using AutoClass, and test data were selected from each cluster. In total, 150 data cases were used for training and 30 were selected for testing.[52] The comparisons of predictions and target values for both training and test data are shown in Figures I.11.10 and I.11.11. The agreement is considered as satisfactory by the company. However, there are nine patterns in the training data and three in the test data which have relative errors greater than 30%. There are various possibilities causing large errors. First of all, the measurement error of Microtox could reach as high as 50%. Second, there may be errors of measurement of other parameters or even human error in inputting data values. Therefore, it is important to have a mechanism to validate the results. This is done by plotting the error distribution for all the training and test data. It was found that the relative errors satisfy a Gaussian or normal distribution. From the distribution we can conclude that the large errors can be regarded as abnormal.

11.3.2.3 Variable Contribution Analysis

A neural network-based software analyzer has the potential advantage of identifying the compounds responsible for observed toxicity value, and helping develop strategies in process operation in order to reduce the toxicity value. There have been a few approaches proposed in literature for carrying out sensitivity studies to identify inputs which are irrelevant or not important so that simpler neural models can be developed.[53–66] However, they often give different results and none is widely recognized as more accurate than others. Sung[56] compared a number of approaches and concluded the best strategy is to combine the approaches with an expert's knowledge. However, the challenge is that in some areas of application, such knowledge is not available.

In this section, five approaches to contribution analysis are studied based on three sets of data. We have full knowledge for the first set of data about the relative importance of the inputs to the output, and some such knowledge for the second data set, but little knowledge for data set 3. In the following, various approaches are introduced and discussed:

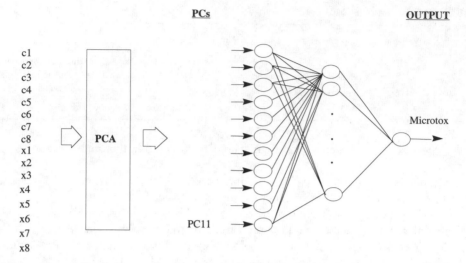

FIGURE I.11.9   The strategy for software analyzer development.

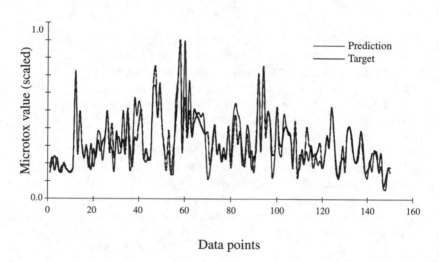

FIGURE I.11.10   Comparison of predictions and target values for the training data.

11.3.2.3.1   *The Methods*

Correlation analysis (CA). The correlation coefficient between two variables X and Y gives a measure of the linear relationship between them:

$$\rho_{xy} = \frac{\text{cov}\,(X, Y)}{\sqrt{\delta_x \delta_y}} \tag{I.11.4}$$

where ρ is the correlation coefficient which lies between -1 and 1, $\text{cov}(X, Y)$ is the covariance of variables X and Y, and δ standard deviation.

Sensitivity studies based on differential analysis (DA). In sensitivity analysis, each model variable is varied in turn by a small amount within the region of a best estimate or standard case. For each variable, the resulting relative change in the model output is divided by the relative variation in the variable to obtain sensitivity coefficients.[54–56,58] As a consequence, variables with a large sensitivity

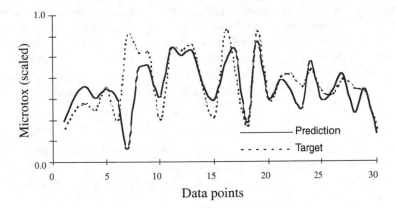

FIGURE I.11.11   Comparison of predictions and target values for the test data.

coefficient have a strong influence on the model output. For an extension of this method, we define the sensitivity of the ith input to the kth output (s_{ik}) as:

$$s_{ik} = f(x^1_{mean}, x^2_{mean}, \ldots, x^i_{max}, \ldots, x^I_{mean}) - f(x^1_{mean}, x^2_{mean}, \ldots, x^i_{min}, \ldots, x^I_{mean})$$
(I.11.5)

where x^{mean}_i is the mean, and x^{min}_i and x^{max}_i are the minimum and maximum of the ith input variable.

One input variable deletion (VD). Given a neural network model which has I inputs and just one output and has been trained using P data patterns, to study the contribution of the ith input to the output, we can delete the ith input from the network structure and examine the deviation caused in the output. Its relative importance can be measured using the following equation:

$$MSE_i = \frac{1}{P} \sum_{p=1}^{P} (COD_p - COD^i_P)^2$$
(I.11.6)

where P is the total number of training data patterns, MSE_i is the measure of the importance of the ith input, COD_p is the calculated output using the original network structure while, COD^i_p is the calculated output when the ith input is deleted from the network.

Fuzzy curves (FC). Lin and Cunningham[66] developed a fuzzy curve method for establishing the relationship between the input variables and an output variable. Suppose there are n inputs $x_i (i = 1, 2, \ldots, n)$ and one output y, and the number of training patterns is m. The method first determines the fuzzy membership function for x_i:

$$\phi_{ik}(x_i) = \exp\left(-\left(\frac{x_{ik} - x_i}{b}\right)^2\right), k = 1, 2, \ldots, m$$
(I.11.7)

A fuzzy curve c_i for each input variable x_i is then generated using the following equation:

$$c_i(x_i) = \frac{\sum_{k=1}^{m} \phi_{ik}(x_i) \cdot y_k}{\sum_{k=1}^{m} \phi_{ik}(x_i)}$$
(I.11.8)

We can then rank the importance of the input variables according to the range covered by their fuzzy curves. If the fuzzy curve for a given input is flat, then this input has little influence on the output data and it is not a significant input. If the range of a fuzzy curve is about the range of the output data, then it is the most important to the output variable.

Sensitivity analysis with PCA preprocessing. Drechsler[57] has raised concern that the input variables may be interrelated, but existing methods often assume that the inputs are independent of each other. A method to combat this is to pre-process the inputs using principal component analysis and then develop the neural network model using the PCs, which are linearly independent of each other. Sensitivity studies can then be applied for these PCs. The connection of a PC to the original variables can be found using the variable contribution diagram, which plots the loading coefficients of PCA analysis.

11.3.2.3.2   Dataset-1

Dataset-1 (Table I.11.5) was originally used by Lin and Cunningham[66] in developing the fuzzy curve approach, which was introduced above. The original data has three inputs, $X1$, $X2$, $X3$, and it was known that $X1$ and $X2$ are more important than $X3$ to Y. In this study, a fourth input $X4$ was added, which is equal to $X1 - X2$.

The results of applying various approaches to Dataset-1 are summarized in Table I.11.6. The numbers indicate the order of importance of an input variable to the output Y. For instance, the column with the heading of CA gives the result of correlation analysis, which indicates that the ranking of importance of inputs to the output variable Y is $X4$, $X1$, $X2$, and $X3$. If PCA processing of inputs was carried out before a sensitivity study approach was applied, then the numbers in the table indicate the order of importance of the PCs to the output Y. For instance, the column headed by DA_PCA indicates that the order of importance to Y is PC1, PC3, and PC2. Contribution plots can then be used to examine the links to the original variables. Table I.11.7 gives the loading coefficients that indicate the most important variables to PC1 are $X1$ and $X4$, and to PC2 is $X2$.

TABLE I.11.5   Dataset-1

No.	$X1$	$X2$	$X3$	$X4$	Y	No.	$X1$	$X2$	$X3$	$X4$	Y
1	14.9	2	9.3	12.9	3.8	11	6.9	2.3	10.3	4.6	3.3
2	16.6	1.7	5.8	14.9	3.7	12	7.4	1.9	9.4	5.5	2.8
3	21.3	2.1	9.1	19.2	3	13	11.3	2	8.4	9.3	3
4	24.3	2.9	7	21.4	4.7	14	17.6	2	8.2	15.6	2.7
5	26.6	1.7	4.8	24.9	4.1	15	19.5	2.4	6.5	17.1	3.3
6	23.2	1.3	4.7	21.9	3.1	16	21.6	2.5	5.1	19.1	4.9
7	22.2	2.5	4.5	19.7	3.1	17	26.5	1.9	6.3	24.6	4.8
8	18.1	2.5	5.9	15.6	2.5	18	26.4	2.2	8.1	24.2	4.4
9	13.7	2.8	7.9	10.9	3.3	19	23.1	3.8	4.9	19.3	2.2
10	7.7	2.6	9.3	5.1	2.3	20	21.1	2.5	5.1	18.6	1.8

TABLE I.11.6   Summary of the Results of Applying Various Methods to Dataset-1[a]

	Without PCA Processing					With PCA Processing		
	CA	DA	VD	FC		DA_PCA	VD_PCA	FC_PCA
$X1$	2	1	2	3	PC1	1($X1$)	1($X1$)	3
$X2$	3	2	4	2	PC2	3	2	1($X2$)
$X3$	4	4	1	4	PC3	2	3	2
$X4$	1	3	3	1				

[a]CA-correlation analysis, DA-sensitivity study based on differential analysis, VD-one variable deletion, FC-fuzzy curve. DA_PCA – principal component analysis plus DA.

TABLE I.11.7   Variable Loading Coefficients of Dataset-1

Variables	PC1	PC2	PC3	PC4
$X1$	**0.972891**	−0.02628	0.229765	−1.2 E − 08
$X2$	0.030296	**0.995129**	0.093815	1.01 E − 09
$X3$	−0.84403	−0.12057	0.522562	−2.5 E − 23
$X4$	0.96895	−0.10976	0.221561	1.21 E − 08

TABLE I.11.8   Summary of Contribution Analysis for the First Output for the Data of Chemical Mixture

	Without PCA Processing					With PCA Processing		
	CA	DA	VD	FC		DA_PCA	VD_PCA	FC_PCA
var1	5	5	2	5	PC1	1 (var5)	2	4
var2	2	4	1	4	PC2	5	5	1 (var1)
var3	1	1	4	1	PC3	2	1 (var3)	3
var4	3	2	3	3	PC4	3	3	5
var5	4	3	5	2	PC5	4	4	2

The result in Table I.11.4 can be summarized as follows. Without PCA processing, three approaches including correlation analysis (CA), sensitivity study based on differential analysis (DA), and fuzzy curve (FC) can all find the least important variable, $X3$. Though the order may be different, they also can identify the important variables $X1$, $X2$, and $X4$. However, one variable deletion (VD) wrongly identifies $X3$ as the most important variable.

It is also found that with PCA preprocessing, though the approaches can identify one of the important variables, they often fail to identify other important ones. For example, DA_PCA (sensitivity study based on differential analysis plus PCA) finds that the order of importance of PCs to Y is PC1, PC3, and PC2. It was found that $X1$ and $X4$ are the important ones to PC1, therefore it can be concluded that they are important to Y. However, it fails to identify $X2$ to be an influential variable to Y because $X2$ is the least influential to PC1, and though $X2$ is important to PC2, PC2 is less influential to Y than both PC1 and PC3. Similar observation can be made to the two data sets to be discussed next. Therefore, our discussion in the following focus on the methods without PCA preprocessing, though results with PCA preprocessing are also given in the summary tables.

11.3.2.3.3   Dataset-2

Dataset-2 is a set of data about a mixture of chemicals. The purpose of analysis is to find out the relative importance of five inputs (var1 to var5) to two input variables. Experts expect that the important variables to the outputs are the same and should be var3, var4, and var5. The result of applying the approaches to this set of data is summarized in Tables I.11.8 and I.11.9. It can be seen from these tables that correlation analysis (CA), sensitivity study based on differential analysis (DA), and fuzzy curve (FC) all identify var1 as the least influential variable to the two outputs. In Table I.11.8, DA and FC all identify var3, var4, and var5 as most influential (though the order is slightly different). CA can identify var3 and var4, but regards var2 as more important than var4. In Table I.11.9, both CA and FC can identify var3, var4, and var5 as the most important ones. DA correctly identifies var3 and var4, but regards var2 as more important than var5. The results indicate that CA, DA, and FC

TABLE I.11.9 Summary of Contribution analysis for the Second Output for the Data of Chemical Mixture

	Without PCA Processing					With PCA Processing		
	CA	DA	VD	FC		DA_PCA	VD_PCA	FC_PCA
var1	5	5	4	5	PC1	1 (var5)	2	1 (var5)
var2	4	3	1	4	PC2	4	4	5
var3	1	1	2	1	PC3	2	1 (var3)	2
var4	2	2	5	3	PC4	5	5	3
var5	3	4	3	2	PC5	3	3	4

TABLE I.11.10 Summary of Contribution Analysis for the Wastewater Data

	Without PCA Processing					With PCA Processing		
	CA	DA	VD	FC		DA_PCA	VD_PCA	FC_PCA
PH	9	8	9	10	PC1	3	4	1 (c7)
COD	11	6	1	8	PC2	5	1 (c3)	7
c1	7	7	4	6	PC3	8	3	8
c2	5	10	6	5	PC4	6	11	10
c3	8	3	3	11	PC5	2	8	4
c4	10	11	10	3	PC6	7	2	9
c5	2	1	2	2	PC7	11	9	11
c6	1	2	5	4	PC8	4	5	3
c7	3	4	7	1	PC9	1 (c5)	6	2
c8	6	9	11	9	PC10	9	10	5
x	4	5	8	7	PC11	10	7	6

Note: Variable x is the sum of x1~x8.

can give acceptable predictions regarding the contribution of inputs to outputs. However, from both Tables I.11.8 and I.11.9, it can be seen that, as for Dataset-1, one variable deletion (VD) approach gives incorrect results.

11.3.2.3.4 Dataset-3

Dataset-3 is the data collected from the organic manufacturing plant. The result of contribution analysis using various approaches is shown in Table I.11.10. Because it was proved in the above two data sets that the one variable deletion approach is the least reliable approach (this conclusion is consistent with that made by Sung[56]), we can ignore the result given by VD. Comparing the other three approaches, i.e., CA, DA, and FC, it can be seen that though they do not give identical results, they all regard c5, c6, and c7 as the most important variables to toxicity.

11.4   Neural Network-Based Software Sensors for Bioprocesses—The State of the Art

11.4.1   State of the Current Research

A notable trend in industrial microbiology is the increased demand on new technologies to facilitate monitoring and control of microbial growth and upstream and downstream processes.[29–42,69,71,73,74,79–82,84,85,89,94–96,97,102–106,111–113,115,116] Biotechnological processes are typical multivariable, nonlinear, and nonstationary dynamic systems in which the interconnections between the microbiological variables (usually slowly changing, indirectly controlled) and the physical and chemical (fast changing, directly controlled) variables are not well understood. The complex behavior of biotechnological processes is mainly due to the specific features of microbial populations that are self-organizing living systems, occasionally very sensitive to small changes in cultivation conditions, but at the same time capable of saving their vitality after a century-long existence in exceptionally unfavorable conditions.

For example, there exist microbial populations living in conditions of overpressure on the ocean bed or under high temperature in volcanoes. The most fascinating example is probably the long-time memory of yeast's microbial populations, found in the Egyptian mummies. The traditional methods of manufacturing bread, wine, and beer, known from the remote past, are considered up to the present as an art. As a result, the microbial populations' nonstationary behavior is often predicted on the basis of human skill and expert knowledge. The knowledge is often expressed in qualitative terms such as the use of "full-bodied" or "heavy" wine in describing the gradation of color, smell, or taste.

Conventional sensors used for measurement of physical (e.g., temperature, pressure, viscosity, dissolved oxygen content, other inlet and outlet gases, and redox potential) and chemical (e.g., pH and different ions) variables often work under high temperature, pressure, and acidity. The sensors can be contaminated by grease and other substrate components or damaged after a period of use. Since the requirement for sterile cultural medium is very strict, it is usually not allowed to replace sensors that are out of order during the operation of the cultivation process. In addition, the measurement of cultural medium content and product quality using composition analyzers in general still suffers excessive time delays, sample interval constraints, and lack of accuracy. In summary, there is a real industrial need to develop advanced techniques to facilitate an efficient monitoring and control of the microbial growth process. Known techniques can be broadly classified into two methods as described below.

The first method is based on the development and improvement of direct measurement sensors and methods for continuous determination of the physical and chemical process variables, for the organic and inorganic components of microbial cells, cultural medium, and metabolite products and ferments.[68,73,74,80–82,84,89,102,115]

The second approach is known as the indirect method, which can be further divided into the following three sub-groups:

- Indirect techniques based on calibration curves or calibration coefficient as well as nephelometric, viscosimetric, conductometric, and fluorimetric approaches.[68,73,96,115]
- Model-based indirect measurements, which use mass-balance expressions of the substrate and inlet gases transformation into biomass, metabolites, ferments, bioactive products, and outlet gases or relative physiological coefficients as well as microbial population key indicators, including respiratory quotient (RQ), dissolved oxygen content (DOC), oxygen uptake rate (OUR), carbon dioxide evolution rate (CER), ammonium uptake rate (AUR), redox potential rh or Eh, and hydrogen ion pH.[68,73,96,115]
- Observer-based methods, which implement state or parameter estimation approaches as well as extended Luenberger or Kalman observers and adaptive control techniques.[68,73,79–82,89]

The idea of biotechnological soft sensors and indirect model-based measurements was developed approximately in the middle of the 20th century.[68,73,96,116] Valuable results were obtained using classical modeling methods, based on differential equations, regression analysis, and mass-balance equations.[68,73,74,96,115,116] It was shown that by using model-based indirect measurements, certain biologically significant parameters could be extracted from the information of available sensors. Some researchers defined observers as a combination of models and measurements, which were applied to correct the state estimations provided by the kinetic model.[89,115]

The use of observers depends on the development of reliable mathematical models. For example, it was shown that extended Kalman filter leads to a better global estimation of state variables—biomass, substrate, and product—in comparison with the multivariable adaptive estimators, which could be safely used when the stress is laid upon biomass estimation.[105] An alternative estimation technique is the method of reduction, which produces a nonbiased estimate of a measured variable for both linear and nonlinear indirect measurements, but requires that all arguments be measured under a specific plan.[106]

Some key indicators of biotechnological variables are connected to the respiratory functions of certain microbial population. These respiratory functions are in turn related to the substrate consumption, the biomass, and the formation of intermediate metabolites or final bioproduct. Key indicators and respiratory quotient or dissolved oxygen content (DOC) indicate the changes in the cell metabolism.[68,95,115] The oxidative status of aerobic growth at low DOC indicates continuously measured redox potential (rH, Eh).[95,115] The CO_2 evolution rate is predicted using the CO_2 concentration in the outlet gas that is measured continuously using infrared CO_2 analyzer. Because CO_2 content influences the cultural medium acidity (pH), accurate control needs to be implemented for CER prediction.[95,115] Even so, the response of pH is usually not very fast; the pH relationship to the rates of substrate feeding can be used to control microbial growth and metabolism.[95,109,115] The metabolite formation rate is often closely related to the concentration of the fed substrate. This might be used as a control parameter, as well as ethanol formation in baker's yeast production where the glucose feed rate is controlled to maintain a constant ethanol concentration.[95,115]

The mean rate and standard error elimination from the indirect measurements is discussed in connection with the rate-of-change estimates obtaining several uncertainties[113] and fuzzy intervals implementation.[110]

The inherent complexity of biotechnological processes makes the measurement problem significantly different and more difficult than for conventional chemical processes. Because the measured variables are often interrelated, e.g., both the redox potential and oxygen dissolvability depend on the cultivation temperature, the measurement is inexact and uncertain. Another example of variable correlation is the measurement of oxygen, carbon dioxide, and other gaseous species in the exit gas, which frequently contains volatile substrates/products.[73,96]

The last 15 years have shown that the lack of accurate mathematical models limits the more realistic point of view and model accuracy considered in the sense of the complex characteristics of the mathematical expression of biotransformation processes. Scientific efforts have been focused on developing efficient methods for nonlinear system modeling and identification. Some successful applications by implementing inferential estimators[71] and adaptive control systems were compared and discussed.[79–82] These solutions are connected with determining the model structure and estimating or fitting the model parameters. Because of the complex features of biotechnological processes, model structure determination is often rarely available in practice. For this reason "black-box" models, model-based predictive control schemes and algorithms were tested.[67] Efficiency increased by introducing artificial intelligence-based methods and algorithms with computing properties such as "universal function approximation," parallel processing, learning capabilities, linguistic information, and expert knowledge processing comprised in fuzzy logic, neural networks, genetic algorithms, and its hybrid combinations. Such methods offer an attractive alternative to online software sensors and indirect measurement systems design.[69,85,94,97,102–104,110,112,116] The robustness of intelligent model-based control systems, implemented in softsensors and indirect measurement systems, ensures stability of the overall system in the presence of external disturbances and uncertain information.[38,69,85,94,97,99,102,103,104,111,115]

11.4.2 Design of Software Sensors for Biotechnological Applications

The purpose of intelligent software sensors is to use variables that can be readily measured in real time or obtained from record databases to predict the values of certain process variables that are difficult to measure directly in real time. These variables provide information about the following items:

- The cells
- The microbial growth
- The intercellular metabolites
- The final product composition and quality
- The economic parameters

Information about cell culture includes values for morphological and morphophysiological parameters. For example, morphological parameters of yeast cultures include the number or percentages of living, budding, weakened, or dead cells, and morphophysiological parameters are the budding energy, specific budding rate, relation between living and dead cells, or microbial culture vitality and respiratory characteristics.

Information about the microbial growth includes:

- Specific or integral growth rates
- Specific physiological coefficients, as well as specific oxygen consumption rate, specific consumption rates related to the carbohydrate, nitrogen, phosphorus, and other substrate components
- Yield coefficients
- Specific rates of product formation
- Specific rates of the cultural medium acidifying
- Mass transformation, e.g., in connection with the heat emission from exothermic fermentation processes
- Mass-balance characteristics (connected not only with microorganism functions, but also with process mass-exchange or hydrodynamic characteristics, or with the optimal cultural medium content)

Intercellular metabolites are traditionally not measured, but increasing interest on controlling metabolite transformations in the different stages of the biotechnological process requires their measurement. Therefore, software sensors can be developed for these variables. In the literature, most previous software sensors for this purpose are based on mathematical models. Neural networks can be an effective alternative technique, but have not been studied widely.

From the commercial point of view, there are two groups of the derived parameters that characterize the trade biotechnological product rating:

- Index of performance (quantitative measure), which presents the relation between obtained bioactive substances volume and substrate and/or power costs[68,73,84,96,115]
- Final product composition and quality (qualitative measure)—product purity, potency, stability, safety, and specific organoleptic particularities

Research efforts in predicting final product composition and quality are now focused on eliminating subjective factors and increasing objectivity. In the following, some results are given concerning sensory rating evaluations on the basis of biochemical analyses in wine and beer manufacturing. In order to reach a comprehensive evaluation of the final product quality, automation system could be equipped with the additional intelligent sensors for organoleptic qualities, as well as color, smell, and taste ratings.

11.4.3 The Techniques

Early efforts on developing software sensors for bioprocesses used knowledge-based systems. They typically involved the following steps: analyzing the causal relationships of input-output variables, knowledge acquisition normally from human experts, selection of an appropriate inference engine, and implementation, test, and validation. The main difficulty encountered was that the problem is not simply a mapping of multiple inputs and outputs; rather, the input variables (and the output variables) may also be intercorrelated. Knowledge-based expert systems using IF THEN descriptions were not always able to represent such complex relationships. The complexity of variable relationships also is a problem in knowledge acquisition, since even the most experienced human experts may find that they cannot describe complex variable relationships accurately.

A more effective strategy in developing soft sensors for biotechnological processes is the integration of various techniques including knowledge-based expert systems, fuzzy logic (FL), neural networks (NN), and neuro-fuzzy inference systems (NFIS).

11.4.3.1 Fuzzy Logic

The main advantages of fuzzy logic systems are that, unlike neural networks, they need only limited data to start, and are able to represent the behavior of multivariate nonlinear systems in an intuitive and accurate way.[100,110] Model-based expert advisers developed as fuzzy knowledge-based systems (FKBS)[76,88] are powerful in knowledge representation where each fuzzy rule is connected with linguistic terms representing process state variables.[90,100]

The principal structure of an FKBS is shown in Figure I.11.12. The main components of an FKBS are modules processing the input-output information (normalization, denormalization, fuzzification, and defuzzification), database (DB) and rule base (RB), and inference engine (IE), as well as user interfaces.

The knowledge base (KB) of an FKBS consists of a database (DB) and a rule base (RB). The DB provides the necessary information for the proper functioning of the fuzzification module, the RB, and the defuzzification module. The information includes fuzzy membership functions of process variables, physical domains, and their normalization-denormalization (scaling) factors.

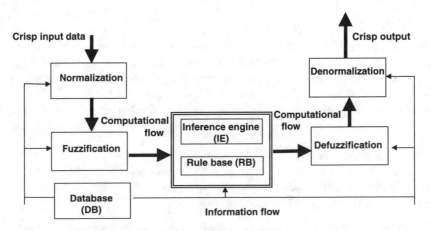

FIGURE I.11.12   General structure of FKBS.

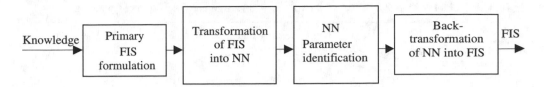

FIGURE I.11.13   Block scheme of neuro-fuzzy modeling approach.

The RB represents in a structural way the knowledge of an experienced operator using a set of production rules in the form of "IF antecedents THEN consequent." The rules are usually obtained from human experts, but can also be extracted automatically from data.

The defuzzification module converts the fuzzy set of modified model output values into a single point-wise value and also performs output denormalization, which maps the point-wise value of the model output onto its physical domain. The design parameter of the defuzzification module is the choice of defuzzification operators. The theory of fuzzy logic is described in detail in many scientific papers and books.[72,77,78,86,88,90–93,98–100,103]

11.4.3.2   Neuro-Fuzzy Systems

In some cases, FKBS developed using the above procedure may not be satisfactory. A useful approach to improve accuracy is incorporating learning ability in the system, e.g., through neural networks (NN).[72,77,78,86,90–93,98,99,103,104] NN and FIS complement each other because NNs are very powerful in making use of data while FIS are capable of capturing human experts' knowledge.

The neuro-fuzzy approach for modeling multivariate nonlinear complex systems mainly includes two phases, structure identification and NN adaptive parameter identification. As the first step, FIS is used and a fuzzy relational model is synthesized. The model is then transformed to an NN structure that can incorporate the basic functions of FIS, while implemented in a connectionist learning structure that has five layers. The first layer consists of neurons representing the input variables, while the last or fifth layer contains neurons corresponding to the output variables. The number of neurons in the second layer is equal to the linguistic values of the input variables. The third layer consists of neurons whose number is equal to the number of the fuzzy rules, while the fourth layer contains neurons whose number is the same as the number of linguistic values of the output variables.

Through training, NN weights are obtained. The back transformation of the NN into the optimized FIS is realized with the help of the training data set and the membership functions of linguistic variables in antecedents part of the rules—tuned by backpropagation gradient descent[90] and the consequent parameters updated by a recursive least-squares method.[90–92] If the adaptive parameter identification is not successful, the weight coefficients of the rules are changed by synaptic weight optimization between the third and fourth layers of the NN. The next step in adaptation of the NN parameters is adaptation of the weight coefficients of the antecedents' implications optimizing synaptic weights between the third and second layers of the NN. The block diagram of the neuro-fuzzy modeling approach is depicted in Figure I.11.13.

11.5   Software Sensors for Yeast Population Morphology and Vitality Prediction

Cell counting and separation by type are the least attractive and exhausting routine operations in industrial and laboratory microbiology. However, they are vital operations because they have significant impact on the cultivation and maintenance of cell cultures, and therefore on the vitality of inocullum and fermentation activity as well as morphological characteristics.

TABLE I.11.11   Plan of Full Factorial Experimental Design of Starter Culture Obtaining

Number Experiment	Temperature T [°C]	Aeration Volume O_2 [l/l.min]	Agitation Speed n [1/min]	Ammonium Nitrogen N_2 [mg/l]
1	1	1	1	1
2	0	1	1	1
3	1	0	1	1
4	0	0	1	1
5	1	1	0	1
6	0	1	0	1
7	1	0	0	1
8	0	0	0	1
9	1	1	1	0
10	0	1	1	0
11	1	0	1	0
12	0	0	1	0
13	1	1	0	0
14	0	1	0	0
15	1	0	0	0
16	0	0	0	0
17–21	0.5	0.5	0.5	0.5

TABLE I.11.12   Factors of Full Factorial Experiment Design and Experimental Levels

Factor	Low Level	High Level	Control Level
Temperature, T [deg C]	13	19	16
Aeration volume, O_2 [l/l.min]	0	1	0.5
Agitation speed, n [1/min]	0	800	400
Ammonium nitrogen, N_2 [mg/l]	0	100	50

Starter yeast cultures play an important role in many biotechnological processes. Basic morphological characteristics of yeast populations, including concentration of living, budding, and dead cells, their relationship, and specific growth rate characterize the activity and the living state of the microbial population. Microbial strains and partial yeasts used in sparkling wine manufacturing show a high degree of specific sensitivity to the cultivation conditions and diversity in their physiological and biochemical properties.[70,97,101,107,108,109,110]

The application described in the following is based on a full factorial experimental design for a batch process of propagation strain *Saccharomyces bayanus 49* (Table I.11.11). The factors are temperature, aeration volume, agitation speed, and the amount of additional ammonium nitrogen on the levels shown in Table I.11.12. During the initial propagation on the morphological characteristics of starter cultures for making red sparkling wine, the budding X_b and dead X_m cells concentration, i.e., % of biomass concentration, are studied. In the final moment of fermentation, the biomass concentration X (million cells/ml) is measured as an additional factor for yeast morphology prediction. The starter culture was cultivated in a microprocessor-controlled laboratory fermentor ABR 01 (Figure I.11.14).

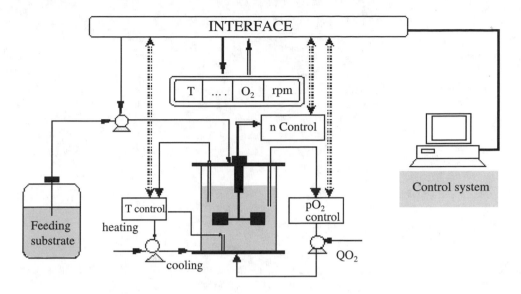

FIGURE I.11.14　Instrumentation of the yeast starter production system with a microprocessor-controlled fermentor ABR 01.

TABLE I.11.13　Randomly Selected Instances of the Training Dataset

T [°C]	O_2 [l/l.min]	n [1/min]	N_2 [mg/l]	X [million cells/ml]	X_b [%]	X_m [%]
19	1	800	100	12	7.1	12
13	1	800	100	13.5	6.9	15
19	0	800	100	14.9	5.06	10
13	0	0	0	10.1	7.4	11.5
16	0,5	400	50	9.5	5.08	10.5
16	0,5	400	50	10.23	2.36	13.5

Experimental data were obtained from 63 experiments with a batch propagation of the studied yeast strain. Table I.11.13 shows some chosen instances from the experimental data set.

The size of the data set is 616. The input partition for five input attributes with two linguistic value leads to thirty-two rules. If we want to stick to a first-order predictive fuzzy model, it results in $(5 + 1)*32 = 192$ linear parameters.

To develop a knowledge-based system that implements all the input variables would lead to a number of practical problems, including unduly large numbers of sensors, many biochemists, extensive computing memory, long computing time, and increased probability of errors. All these lead to high cost.

In order to develop a simple and efficient measurement system, and at the same time address the scarcity problem of data,[90] a neuro-fuzzy modeling system is used.

The approach applied is as follows. In this case, 10 predictive fuzzy models were constructed ($C_2^5 = 10$) for selection of input attributes, i.e., the best predictors. For this reason, the entire database was divided into a data set for training and data set for test; each has 192 and 72 points

TABLE I.11.14   Selection of the Best Predictors

No.	Input Variables	Budding Cells		Dead Cells	
		Minimal Checking Error	Rating	Minimal Checking Error	Rating
1	T, O_2	2.3927	7	4.2047	6
2	T, n	1.9754	2	2.2073	5
3	T, N_2	1.9862	4	1.6544	3
4	T, X	2.3340	6	4.0950	7
5	O_2, n	1.9809	3	2.0854	4
6	O_2, N_2	1.9744	1	1.4279	1
7	O_2, X	2.9665	9	9.0842	8
8	n, N_2	2.0196	5	1.4793	2
9	n, X	2.7493	8	14.5740	9
10	N_2, X	3.6443	10	26.9365	10

TABLE I.11.15   Membership Number Choice for Cell Prediction Based on Best Predictors—O_2 and N_2

No.	Membership Function		Training Epochs	Budding Cells Prediction		Dead Cells Prediction	
	Number	Shape		Minimal Traning Error	Minimal Checking Error	Minimal Traning Error	Minimal Checking Error
1	3	Bellman	5	1.5780	4.9697	9.1367	9.9853
2	3	Bellman	10	1.5780	4.9697	9.0275	9.3678
3	4	Bellman	10	1.5780	4.9275	10.2295	10.5756
4	5	Bellman	5	0.9492	3.8527	10.2295	10.6469
5	5	Bellman	10	1.5780	5.1578	10.2295	10.6469

respectively. The constructed 10 predictive neuro-fuzzy models were compared by the test errors (Table I.11.14).

Human-like perception and expert-knowledge implementation can be seen in the following stage of an intelligent knowledge-based system development. The selection of the best predictors for the industrial process is followed by selection of a suitable structure of the fuzzy inference system (FIS). It involves the selection of linguistic variable presentation with appropriate membership function shapes and number on the one hand, and optimal number of the training epoch determination in relation to the overfitting problem solution on the other hand.[90] According to the experts' experience using two linguistic values, i.e., "low" and "high," represents every input variable. A linear function representation of the output variable showed better model performance. Tables I.11.15 and I.11.16 illustrate the selection of the desired structure of "intelligent microscope" for predicting budding and dead cells on the basis of best predictors, i.e., aeration volume and ammonium nitrogen. It was found that increasing the number of linguistic variables leads to unsatisfactory accuracy of the model, as indicated by Tables I.11.14 and I.11.15. The influence of the shape of the eight different membership functions for cell counting is demonstrated in Table I.11.16. Based on the results, the Bellman and

TABLE I.11.16 Influence of Membership Function Shape on the Budding Cells Best Predictors-Based Model Accuracy

No.	Membership Function Shape	Minimal Checking Error	Rating
		Budding Cells Prediction	
1	Triangular	2.0470	6
2	Trapezoidal	2.0478	7
3	Bellman	1.9744	2
4	Gaussian	1.9723	1
5	Gaussian-2	2.0299	4
6	$\prod$	2.0478	7
7	d-sigmoid	2.0056	3
8	p-sigmoid	2.0412	5

TABLE I.11.17 Budding Cells Best Predictive Models Based on Cell Amount X and a Measurable Physical Variable (T, O_2, or n), 1000 Training Epochs

No.	Membership Function Shape	Model	Minimal Checking Error	Minimal Training Error
1	Gaussian	$Xb = F_1(T, X)$	2.0328	1.1679
2	Gaussian	$Xb = F_2(O_2, X)$	1.7667	1.1378
3	Triangular	$Xb = F_3(n, X)$	1.8700	1.4274

Gauss-shaped membership functions are recommended for the "intelligent microscope" design. The optimal number of training epochs is approximately 1000.

For industrial applications, sometimes it is advantageous to implement additional pairs of predictors. The choice is connected with the fulfillment of two important conditions: easy and secure measurement of the additional predictors.

Although the mathematical experiments have proved that the best predictors are ammonium nitrogen and aeration volume, in industry the variables that are actually measured are such physical variables as cultivation temperature T, aeration volume O_2, and agitation speed n. Additional experiments could be carried out by linking these continuously measured variables with biomass content in order to connect the model design with the input variable.

The tests carried out with the three additional predictor pairs confirmed this assumption. Results in Table I.11.17 satisfy the technological requirements for model accuracy of less than 2% in terms of production error. In comparison with microscope cell counting, the developed knowledge-based system for yeast population morphological characteristics modeling is more accurate and user friendly (see Tables I.11.16 and I.11.17 and Figures I.11.15 and I.11.16).

The model surface with the minimal checking error for budding cell prediction on the basis of the process controlled variables, biomass content X and aeration volume O_2, is presented in Figure I.11.17. The maximum budding cells in the aerobe phase of sparkling wine yeast manufacturing depend on all the biomass and aeration conditions, which assure the desired respiratory conditions for newly formed cells. The optimal model surface for dead cells prediction on the basis of biomass concentration and agitation speed is depicted in Figure I.11.18. The dead cell concentration

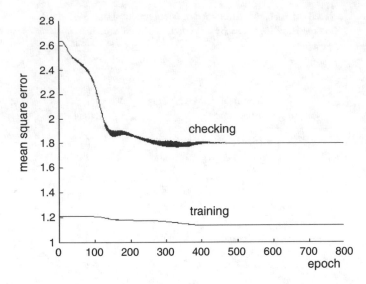

FIGURE I.11.15   Convergence trend of mean square error for budding cells prediction.

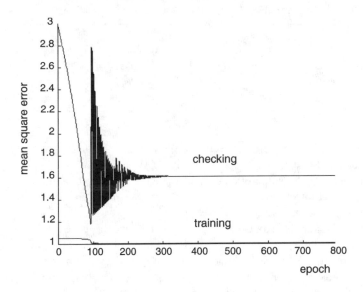

FIGURE I.11.16   Convergence trend of mean square error for dead cells prediction.

increases with increasing biomass volume, decreasing substrate concentration, and also with the lower agitation speed, which deteriorates mass-transfer in the batch bioreactors.

11.6   NN-Model-Based Brewing Yeast Growth Prediction and Optimization

The application described in this section illustrates another method for practical implementation of soft sensors for yeast starter culture prediction and control using a neural network model. The model is developed also for the purpose of optimization of the cultivation conditions in beer manufacturing.

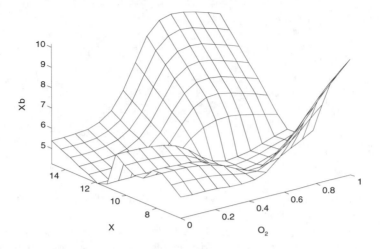

FIGURE I.11.17   Optimal surface for budding cells prediction.

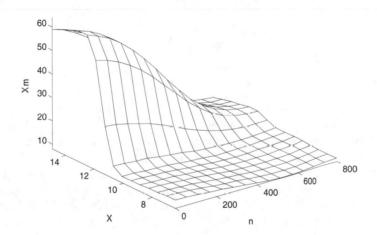

FIGURE I.11.18   Optimal surface for dead cells prediction.

The studied brewing yeast strain *Saccharomyces carlsbergensis 56* is from the collection of the Institute of Brewing and Hop Manufacturing in Sofia, Bulgaria. The focus of the study is on the influence of temperature and pitching rate on morphological composition of the yeast population.

Using the experimental data collected from the process, a neural network model was developed that predicts the dependency of the quantities of living, budding, and dead cells in the population on both the temperature and pitching rate. The neural model was subsequently used to determine the optimal conditions for brewing yeast cultivation. The objective of optimization was to achieve improved fermentation performance, consistent beer quality, and high sensory rating.

Yeast growth involves two phases: the phase of "true" growth when yeast biomass is formed, and the phase of glycogen accumulation. Both phases are related to the different metabolic processes that

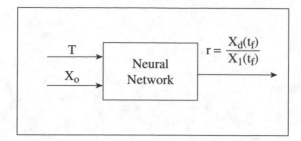

FIGURE I.11.19   Static neural model of the brewing yeast production process.

are most active in earlier hours after pitching. The performance of these processes depends mainly on the temperature and pitching rate.

Yeast belongs to mesophyle microorganisms, therefore the appropriate temperature for its growth is between 15 and 30°C. Sometimes, due to technological considerations in brewing and wine manufacturing the cultivation temperature is maintained at a lower value, between 6 and 10°C.

Other factors affecting the yeast growth, but having less impact, are some medium factors (pH) and the physiological particularities of yeast culture.

Experimental studies on microbial populations show that links exist between morphological composition of population and the operational conditions of the yeast production process. This relation can be expressed in the following general form:

$$X_1 = f_1(X^o, T) \tag{I.11.9}$$

$$X_b = f_2(X^o, T) \tag{I.11.10}$$

$$X_m = f_3(X^o, T) \tag{I.11.11}$$

where the morphological composition of the population is represented by the number of living $X_1(t_f)$ [million cells/ml], budding $X_b(t_f)$ [million cells/ml], and dead cells $X_m(t_f)$ [million cells/ml], which are determined by means of microscope counting. The process conditions that can be varied include temperature T, the pitching rate X_o, and the final time of the process t_f.

Equations (I.11.9), (I.11.10) and (I.11.11) can be used to find the optimal temperature and pitching rate of the yeast production process, in order to minimize the amount of dead cells at the end of the process. In this way, yeast of high viability can be achieved. However, it is difficult to build analytical models since the mechanism is not known. As a result, neural networks were applied to model the process using experimental data.

The neural network structure is shown in Figure I.11.19. The inputs to the model are the temperature and pitching rate, and the output is the ratio of the amount of dead cells to the amount of living cells in the population at the end of the process. This model can be used later to find the optimal values of the inputs so as to minimize the output.

Experiments were carried out to change the temperature in the interval $T \in [6.5°C, 12°C]$ and pitching rate in the interval $X_o \in [10$ mln cells/ml, 30 mln cells/ml].

Each experiment of yeast cultivation lasted 72 hours, and during this period both T and X_o were kept constant. The data for living $X_l(t_f)$ [mln cells/ml], budding $X_b(t_f)$ [mln cells/ml], and dead cells $X_m(t_f)$ were obtained at the end of the fermentation, for $t_f = 72$ hours. The values of the control factors as cultivation temperature T and pitching rate X_o have been used to adjust the static neural model described above.

Based on the neural model developed for the brewing yeast production process, nonlinear programming was used to search for the optimal conditions. The optimization is performed according to the scheme shown in Figure I.11.20.

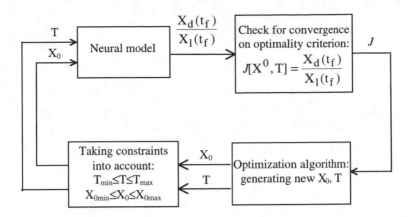

FIGURE I.11.20 Optimization scheme of the brewing yeast production process.

Given the neural model, which predicts the ratio of the amount of dead cells to the amount of living cells in the population at the end of the process using the T and X_o:

$$\frac{X_m(t_f)}{X_1(t_f)} = f_{NN}(X_o, T) \tag{I.11.12}$$

find the values of the temperature T and the pitching rate X_o that will minimise $\mathfrak{J}[X^o, T] = \frac{X_m(t_f)}{X_1(t_f)} \rightarrow$ min subject to the following constraints:

$$6.5[°C] \le T \le 12[°C] \text{ and } 10[m \ln \text{cells/ml}] \le X_o \le 30 \text{ [m ln cells/ml]} \tag{I.11.13}$$

Applying the heuristic algorithm of Box[75] performs the optimization, and the results obtained are:

- Optimal values of the temperature T and the pitching rate X_o:

$$T_{\text{opt}} = 6.5[°C] \tag{I.11.14}$$

$$Xo_{\text{opt}} = 10[m \ln.\text{cells/ml}] \tag{I.11.15}$$

- Minimal value of the optimality criterion:

$$\mathfrak{J}_{\text{opt}} = [Xo_{\text{opt}}, T_{\text{opt}}] = 0.03204 \tag{I.11.16}$$

Therefore, it was recommended to perform the brewing yeast production process from the *S. carlsbergensis 56* strain below a temperature of 6.5°C and a pitching rate of 10.10^6 [mln cells/ml].

It was noticed that the optimal values of the temperature and the pitching rate are within the lowest acceptable ranges. This means that the low values of these factors are favorable for decreasing the ratio between the number of dead and living cells in the population, and thus for producing yeast with high viability. This dependence is confirmed by the experimental data depicted in Figure I.11.21, which are obtained under two cultivation conditions.

11.7 Software Sensors for Prediction of Sparkling Wine Composition and Sensory Rating

Wine manufacturing has been known for centuries, but it is still considered to be an art. Wine composition and sensory rating as final product quality measures are of great importance and are

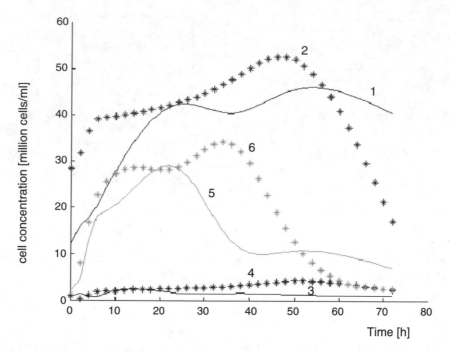

FIGURE I.11.21   Brewing yeast production process—experimental data: 1, 2—number of cells amount; 3, 4—number of dead cells 5, 6—number of pitching cells; 1, 3, 4—a) $T = 6.5°C, X_o = 10$ [mln cells/ml] 2, 4, 6—b) T = 12°C, $X_o = 30$[mln cells/ ml]

chosen as the subject of this section. The purpose is to predict sparkling wine quality using a neuro-fuzzy knowledge-based system.

Sparkling wine manufacturing typically consists of three separated major fermentation processes. The first fermentation process is the production of red base wine, which is also used for the second fermentation process starter culture of *S. bayanus* yeast batch cultivation in conditions similar to that for the third fermentation process, i.e., sparkling wine manufacturing.[70,107,108,109] The final concentration of acetaldehyde, acetoin, and diacetil, as well as products of oxidative metabolism in the starter culture is often analyzed.[87] The third fermentation or pressurization of the red wine medium is carried out by yeast strain *S. bayanus* in bottles, following the primary alcoholic and malolactic fermentation, converting the carbohydrates into ethanol and carbon dioxide, CO_2, which leads to the specific sparkling quality of the wine.[110]

The neuro-fuzzy system for prediction of sparkling wine composition and sensory rating has a two-level hierarchical structure (Figure I.11.22). This is designed to solve the following tasks:

- Prediction of the fermentation activity (FA) of the starter culture on the basis of measured physico-chemical and microbiological variables
- Prediction of base wine composition variables, aldehydes (AL1), acetoin (AC1), and diacetil (D1)
- Prediction of sparkling wine composition variables, aldehydes (AL2), acetoin (AC2), and diacetil (D2)
- Sparkling wine sensory rating (SR) prediction

One hundred and twenty-six industrial fermentation records were used to carry out the prediction of the above-mentioned variables.

The input attributes are the physico-chemical variables, and cultivation temperature T° [degC], aeration volume O_2 [l/l/min] and agitation speed n [rpm], which are continuously measured; the microbiological and biochemical variables, ammonium nitrogen N_2 [mg/l] and final biomass

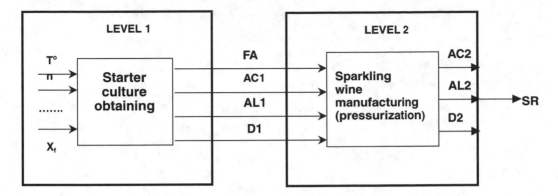

FIGURE I.11.22   Two-level system for sparkling wine composition and quality prediction.

TABLE I.11.18   Randomly Selected Instances Implemented for Prediction of
Fermentation Activity and Red Wine Composition Parameters

			Input Attributes				Output Attributes — Red Wine Medium Composition			
No.	T^o	O_2	n	N_2	X_f	Y_x	AL1	D1	AC1	FA
1	19	1	800	100	240.0	2.38	492.8	23.2	750	0.171
2	13	1	800	100	176.0	1.08	200.2	19.9	470	0.226
3	19	1	0	100	108.0	1.77	248.6	23.3	770	0.216
4	13	1	0	0	125.0	0.72	382.8	25.2	129	0.333
5	19	1	800	0	122.0	1.26	136.4	6.9	288	0.262
6	16	0.5	400	50	102	0.85	255.4	28.6	790	0.353
7	16	0.5	400	50	112	0.93	277.4	30.4	720	0.349
8	16	0.5	400	50	112	0.98	277.4	28.6	760	0.334
9	16	0.5	400	50	113	1.00	297.0	23.0	690	0.326
10	16	0.5	400	50	100	0.85	299.8	26.0	860	0.354

concentration X_f [million cells/ml], which are analyzed off-line; as well as the derived variables such as biomass yield, Y_x [million cells/ml/h], $Y_x = X_f / t$; t[h] is the time of process duration. The output attributes, predicted at the first level, are starter culture fermentation activity, FA [g/g/h] and red wine medium composition parameters such as aldehydes, AL1 [mg/l], acetoin, AC1 [mg/l], and diacetil, D1 [mg/l].

The output attributes, predicted at the second level, are sparkling wine composition variables such as aldehydes, AL2 [mg/l], acetoin, AC2 [mg/l], and diacetil, D2 [mg/l], and sensory rating, SR. The data presented in Table I.11.20 were used for second-level predictions and correspond to the randomly selected records shown in Table I.11.18.

It has already been mentioned that biotechnological processes are complex, nonlinear, and nonstationary, which complicate not only growth modeling and control, but also the evaluation of such derived process variables as fermentation activity or sensory rating of wine composition and quality.

The fermentation activity prediction is a complex problem, as illustrated in Table I.11.19. Experiments confirmed the mathematical results that the fermentation activity of the starter yeast culture

TABLE I.11.19 Best Predictors Selection of the Starter Culture Fermentation Activity

Input Variables				Membership Functions		Minimal Checking Error
				Number	Type	
		T	X_f	2	Bellman	0.0763
		O_2	Y_x	2	Bellman	0.0825
		N_2	Y_x	2	Bellman	0.0847
	T	O_2	N_2	2	Bellman	0.0690
	n	N_2	X_f	2	Bellman	0.0760
	T	O_2	n	2	Bellman	0.0828
	n	N_2	Y_x	3	Trapezoidal	0.0194
	O_2	n	Y_x	3	Gauss	0.0586
	N_2	X_f	Y_x	3	Bellman	0.0655
T	O_2	n	N_2	2	Bellman	0.0072
T	n	N_2	Y_x	2	Bellman	0.0093
T	O_2	n	N_2	3	Bellman	0.0071

TABLE I.11.20 Instances of the Implemented Input Attributes for Sparkling Wine Composition and Sensory Rating Prediction (Numbers Correspond to the Instances in Table I.11.17)

No.	AL2	D2	AC2	SR
1	89.6	1.6	1.7	17.32
2	83.6	1.12	4.8	17.35
3	79.2	0.43	3.6	17.30
4	72.6	0.52	1.2	17.29
5	68.2	0.44	2.1	17.35
6	79.2	0.72	1.3	17.33
7	78.1	0.96	1.8	17.35
8	75.9	0.81	1.8	17.36
9	79.2	0.5	1.8	17.32
10	79.2	0.76	1.8	17.35

is higher when a lower level of aerating volume—approximately 0.4–0.5 l/l/min—and a cultivation temperature of about 15–16°C are implemented. This fact is illustrated in Figure I.11.23, showing the smoothed three-dimensional surface with minimal RMSE = 0.0071 (Figure I.11.24). Figure I.11.25 illustrates the satisfactory results obtained with the developed predictive model.

The biochemical analysis of volatile substances, as well as aldehydes, acetoin, and diacetil, leads to high errors in the observed data, which independently affect model accuracy, on the sophisticated modeling techniques applied. The errors of the implemented biochemical analyses are aldehydes—±6%, diacetil—±25%, acetoin—±20%, and sensory rating—±0.2%. The accuracy is acceptable, proving that these wine constituents are highly reactive and volatile.

The results presented in Table I.11.21 for selection of the best predictors of the red wine medium, the sparkling wine composition parameters, and finally, of sparkling wine sensory rating are the original results. They prove that there is a complex dependency of the final product quality

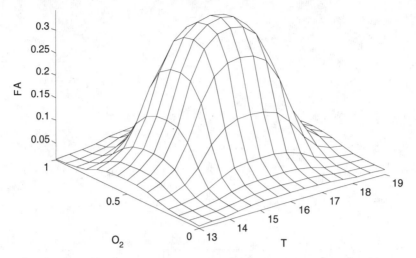

FIGURE I.11.23 Optimal 3-dimensional surface for fermentation activity prediction $FA = f(T, O_2, n, N_2)$.

TABLE I.11.21 Best Predictors of Sparkling Wine Composition and Sensory Rating

Input Variables (Predictors)					Output (predicted) Variable		Model Design		
Continuous Measured			Analyzed	Derived			Membership Functions	Training Epoch	Min RMSE
—	—	n	X_f, N_2	—	AL1	2	Triangular	1000	12.5448
—	—	n	N_2	FA	AL1	3	Bellmann	500	14.5099
—	—	n	N_2	Y_x, FA	D1	2	Triangular	1000	0.9408
—	—	n	X_f, N_2	—	D1	2	Bellmann	1000	1.9279
—	—	n	X_f, N_2	—	AC1	2	Triangular	1000	15.8134
—	—	n	X_f, N_2	—	AL2	2	Bellmann	1000	0.8244
—	—	—	X_f, N_2	FA	AL2	2	Bellmann	300	0.8452
—	—	n	X_f, N_2	—	D2	2	Bellmann	1000	0.1167
—	O_2	n	N_2	Y_x	D2	2	Bellmann	1000	0.1015
—	—	n	X_f, N_2	—	AC2	2	Bellmann	1000	0.3901
—	—	—	AL2, D2, AC2	—	SR	3	Triangular	500	0.0171
—	—	—	X_f	FA	SR	2	Bellmann	1	0.0329
T	O_2	n	—	—	SR	2	Bellmann	1	0.0337

(i.e., the sparkling wine sensory rating) on the technological cultivation conditions and wine medium composition.

The wine quality expressed as a sensory rating is often based on the evaluation by experts on the taste, color, sparkling features, and composition factor. The sensory rating decides the market success of the final product. As demonstrated in the previous described modeling experiments, in our case study (Table I.11.21), the most important factors affecting sensory rating are the medium composition and biochemical variables of the sparkling wine, including aldehydes, acetoin, and diacetil content (Figure I.11.26). This is proved with the satisfactory testing error (Figure I.11.27), but also shows the important role of the implemented yeast starter culture and cultivation conditions.

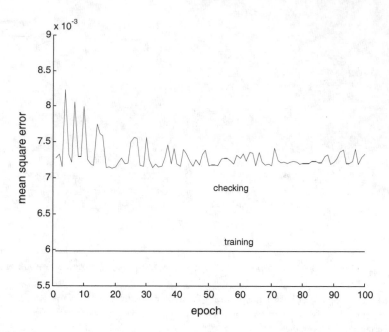

FIGURE I.11.24 Convergence trend of mean square error for fermentation activity prediction with 4 input attributes—T, n, O_2, and N_2.

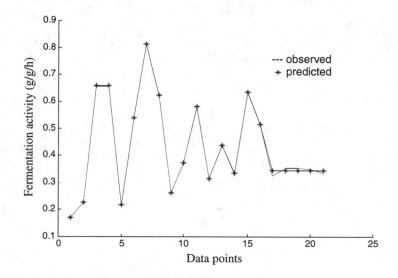

FIGURE I.11.25 Fermentation activity observed and predicated data sets.

11.8 Conclusions and Suggestions for Future Work

In this contribution, five examples of industrial implementations of software sensors using NN and hybrid NN-knowledge-based systems are described. The software sensors were developed to predict:

- Product quality in a refinery fluid catalytic cracking process
- Toxicity of aqueous effluents of a specialized organic manufacturing plant
- Yeast population morphology and vitality

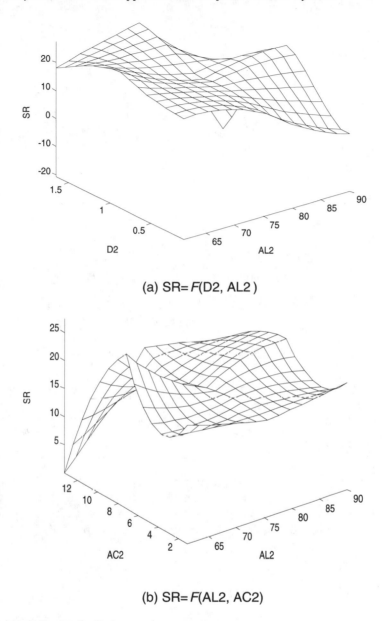

(a) SR= *F*(D2, AL2)

(b) SR= *F*(AL2, AC2)

FIGURE I.11.26   3-D surfaces for sensory rating prediction SR $= F$(AL2, D2, AC2). (a) Sparkling wine sensory rating dependence on diacetil (D2) and aldehydes (AL2). (b) Sparkling wine sensory rating dependence on acetoin (AC2) and aldehydes (AL2).

- Yeast growth and optimization in beer manufacturing
- Sparkling wine composition and sensor rating

By reference to these case studies, some important issues in software sensor development for industrial processes were discussed, including data collection, training and test data selection, input variable dimension reduction, model validation and upgrading, sensitivity study using the software sensor model for knowledge discovery, as well as system architecture. The potential areas in which software sensors can be used in chemical and biological processes are also explored. Due to space limitations a few other issues that can be important to successful software sensor development,

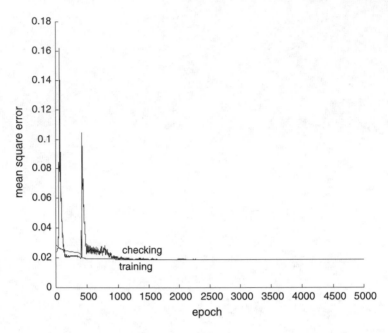

FIGURE I.11.27   Convergence trend of mean square error for SR prediction with 3 input attributes —AL2, D2, and AC2.

e.g., data reconciliation, were not covered in this chapter, because they are relatively well-established technologies and can be found elsewhere.

Acknowledgments

The first author would like to thank the microbiologists from the Institute of Wine Manufacturing and Control and from the Institute of Brewing and Hop Manufacturing, both in Sofia, Bulgaria and Dr. A. Grancharova for help in optimization task solving. The second author would like to thank his current and former colleagues at The University of Leeds, particularly Ms. Bei Yuan and Dr. F.Z. Chen.

References

1. Leonard, J. and Kramer, M.A., Improvement of the backpropagation algorithm for training neural networks, *Comput. Chem. Eng.,* 14, 337–341, 1990.
2. Brent, R.P., Fast training algorithms for multilayer neural nets, *IEEE Trans. Neural Nets,* 2, 346–354, 1991.
3. Chen, S. and Billings, S.A., Neural networks for nonlinear dynamic system modelling and identification, *Int. J. Control,* 56, 319–346, 1992.
4. Peel, C., Willis, M.J., and Tham, M.T., A fast procedure for the training of neural networks, *J. Proc. Cont.,* 2, 205–211, 1992.
5. Powell, M.J.D., Some global convergence properties of a variable metric algorithm for minimisation without exact line searches, in *SIAM-AMS Proc. Symp. on Non-Linear Programming,* Cottle, R. and Lemke, C.E., Eds., 1975, IX: 53–72.
6. Press, W.H. et al., *Numerical Recipes, The Art of Scientific Computing* (Fortran version), Cambridge University Press, Cambridge, 1989.

7. Crowe, E.R. and Vassiliadis, C.A., Artificial intelligence: Starting to realise its practical promise, *Chem. Eng. Progr.,* 91(1), 22–31, 1995.

8. Knight, K., Connectionist ideas and algorithms, *Comm. of the ACM,* 33(11), 59–74, 1990.

9. Chitra, S.P., Use of neural networks for problem solving, *Chem. Eng. Progress,* 89(4), 44–52, 1993.

10. Chen, F., Chen, B., and He, X., Genetic algorithms and artificial neural network. (1) training artificial neural network with extended genetic algorithms, *J. Chem. Ind. Eng.,* 47(3), 280–286, 1996.

11. Kramer, M.A. and Leonard, JA., Diagnosis using backpropagation neural networks—analysis and criticism, *Comput. Chem. Eng.,* 14, 1323–1338, 1990.

12. Shao, R. et al., Confidence bounds for neural network representations, *Comput. Chem. Eng.,* 21, s1173–s1178, 1997.

13. Zhang, J. et al., Inferential estimation of polymer quality using stacked neural networks, *Comput. Chem. Eng.,* 21, s1025–s1030, 1997.

14. Baratti, R., Vacca, G., and Servida, A., Neural network modelling of distillation columns, *Hydrocarbon Processing,* June 35–38, 1995.

15. Riddle, A.L., Bhat, N.V., and Hopper, J.R., Neural networks help optimize solvent extraction, *Hydrocarbon Processing,* Nov., 45–49, 1998.

16. Zhong, W. and Yu, J., Modelling jet fuel endpoint using neural networks with genetic learning, *Hydrocarbon Processing,* Dec., 77–80, 1999.

17. Brambilla, A. and Trivella, F., Estimate product quality with ANNs, *Hydrocarbon Processing,* Sept. 61–66, 1996.

18. Chen, F.Z. and Wang, X.Z., Software sensor design using Bayesian automatic classification and back-propagation neural networks, *Ind. Eng. Chem. Res.,* 37, 3985, 1998.

19. Keaton, M., Keenan, M., and Keenan, J., On-line soft analysers benefit refining, in *Advanced Process Control and Information Systems for the Process Industries,* Kane, L.A., Ed., Gulf Pub. Company, Houston, Texas, 1999, 132–135.

20. Yang, S.II. et al., Soft sensor based predictive control of industrial fluid catalytic cracking processes, *Chem. Eng. Res. Des.—Trans. IChemE,* 76A, 499–508, 1998.

21. McGreavy, C. et al., Characterization of the behaviour and product distribution in fluid catalytic cracking using neural networks, *Chem. Eng. Sci.,* 49, 4717–4724, 1994.

22. He, X.R., Zhao, X.G., and Chen, B.Z., On-line estimation of vapour pressure of stabilized gasoline via ANN's, *Chinese J. Chem. Eng.,* 5(1), 23–28, 1997.

23. Chen, B.Z. and He, X.R., Neural network intelligent system for the on-line optimization in chemical plants, *Chinese J. Chem. Eng.,* 5(1), 57–62, 1997.

24. Malik, S.A., Structuring Knowledge Based Procedures in Process Control, Ph.D. thesis, University of Leeds, 1999.

25. Zhang, J., Inferential estimation of polymer quality using bootstrap aggregated neural networks, *Neural Networks,* 12, 927–938, 1999.

26. Gosden, R.G. et al., Living polymerization reactors: molecular weight distribution control using inverse neural network models, *Polymer Reaction Eng.,* 9(4), in print, 2001.

27. Ogawa, M. et al., Quality inferential control of an industrial high density polyethylene process, *J. Process Control,* 9, 51–59, 1999.

28. Bomberger, J.D., Seborg, D.E., and Ogunnaike, B.A., RBFN identification of an industrial polymerisation reactor model, in *Application of Neural Networks and Other Learning Technologies in Proc. Engineer,* Mujtaba, I.M. and Hussain, M.A., Eds., Imperial College Press, London, 2001, Chap. 2.

29. Mark, W. et al., Application of radial basis function and feedforward artificial neural networks to the *Esherichia coli* fermentation process, *Neurocomputing,* 20, 67, 1998.

30. Ye, K., Jin, S., and Shimizu, K., Fuzzy NN for the control of high cell density cultivation of recombinant *Esherichia coli., J. Fermentation and Bioeng.,* 77, 663, 1994.

31. Latrille, E. et al., Neural network modelling and predictive control of yeast starter production in champagne, in *CD-Proc. ECC'97,* European Control Conference '97, 1997.

32. Acuña, G., Latrille, E., and Corrieu, G. Biomass estimation using neural networks and the extended Kalman filter, in *Preprints of the 6th Intern. Conf. Comp. Appl. Biotech.,* Germany, 209, 1995.

33. Thibault, J., Breusegem, van V., and Cheruy, A., On-line prediction of fermentation variables using neural networks, *Biotech. and Bioeng.,* 36, 1041, 1990.

34. Linko, S., Zhu, Y.-H., and Linko, P., Neural networks in lysine fermentation, in *Preprints of the 6th Int. Conf. Comp. Appl. Biotech.,* Germany, 336, 1995.

35. Ignova, M. et al., Seed data analysis for production fermenter performance estimation, in *Preprints of the 6th Intern. Conf. Comp. Appl. Biotech.,* Germany, 53, 1995.

36. Pfaff, M. et al., Model-aided on-line glucose monitoring for computer-controlled high cell density fermentation, in *Preprints of the 6th Int. Conf. Comp. Appl. Biotech.,* Germany, 6, 1995.

37. Oh, G.S. et al., Neural networks in estimation and control of antibody production using hybridoma cells in fed-batch cultures, in *Preprints of the 6th Intern. Conf. Comp. Appl. Biotech.,* Germany, 183, 1995.

38. Rouzic, Y. Le et al., Soft sensor for adaptive pH control, an industrial application, in *Proc. European Control Conf. ECC'97,* 297, 1997.

39. Najim, K. et al., Constrained long-range predictive control based on artificial neural networks, *Int. J. System Sci.,* 28, 1211, 1997.

40. Rusnak, A. et al., Generalized predictive control based on neural networks, *Neural Processing Letter,* 4, 107, 1996.

41. Kurtanjek, Z., Bioreactor modeling and control by principal component based neural networks, in *Preprints of the 6th Intern. Conf. Comp. Appl. Biotech.,* Germany, 300, 1995.

42. Can, van H.J.L. et al., Neural models in predictive control, in *Preprints of the 6th Int. Conf. Comp. Appl. Biotech.,* Germany, 95, 1995.

43. Himmelblau, D.M., Applications of artificial neural networks in chemical engineering, *Korean J. Chem. Eng.,* 17(4), 373–392, 2000.

44. Baughman, D. and Liu, Y., *Neural Networks in Bioprocessing and Chemical Engineering,* Academic Press Inc., San Diego, 1995.

45. Moody, J. and Darken, C.J., Fast learning in networks of locally-tuned processing units, *Neural Computation,* 1, 28, 1989.

46. Smaragdis, E.I., Developing Neural Network Structures and Using Bayesian Methods for Training Data Analysis, Ph.D. thesis, The University of Leeds, Leeds, UK, 1999.

47. Chen, B.H., Feature extraction and knowledge discovery in process operation analysis, Ph.D. thesis, The University of Leeds, Leeds, UK, 1998.

48. Cheeseman, P. and Stutz, J., Bayesian classification (AutoClass): theory and results, in *Advances in Knowledge Discovery and Data Mining;* Fayyad, U. M., Gregory Piatetsky-Shapiro, P.S., Ramasamy, U., Eds., AAAI Press/MIT Press, 1996.

49. CAS., The latest CAS registry number and substance count, http://www.cas.org/cgi-bin/regreport.pl, 2000.

50. Smith, D., Assessing toxicology quickly and efficiently, *Chem Eng.,* 107, April, 25–128, 2000.

51. Omenn, G., Assessing the risk assessment paradigm, *Toxicology,* 23, 23–28, 1995.

52. Yuan, B., Wang, X.Z., and Morris, T., Software analyser design using data mining technology for toxicity prediction of aqueous effluents, *Waste Management,* 20, 677–686, 2000.

53. Rademan, J.A.M. et al., Neural net based knowledge extraction from the historical data of an industrial leaching process, *Hydrometallurgy,* 43, 95, 1996.

54. Ricotti, M.E. and Zio, E., Neural network approach to sensitivity and uncertainty analysis, *Reliability Engineering and System Safety,* 64, 59, 1999.

55. Gedeon, T.D., Data mining of inputs: analysing magnitude and functional measures, *Int. J. Neural Systems,* 8, 209, 1997.

56. Sung, A.H., Ranking importance of input parameters of neural networks, *Expert Systems with Applications,* 15, 405, 1998.

57. Drechsler, M., Sensitivity analysis of complex models, *Biological Conservation,* 86, 401, 1998.

58. Zhu, J., et al., An on-line wastewater quality prediction system based on a time-delay neural network, *Engineer. Appl. Artificial Intelligence,* 11, 747, 1998.

59. Bartlett, E.B., Self determination of input variable importance using neural networks, *Neural, Parallel and Scientific Computations,* 2, 103, 1994.

60. Kleijnen, J.P.C. and Helton, J.C., Statistical analyses of scatter plots to identify important factors in large-scale simulations, 1: Review and comparison of techniques, *Reliability Engineering and System Safety,* 65, 147, 1999.

61. Garson, G.D., Interpreting neural network connection weights, *AI Expert,* April: 47, 1991.

62. Zurada, J.M., Malinowski, A., and Cloete, I., Sensitivity analysis for minimization of input data dimension for feed forward neural network, *Proc. IEEE Int. Symp. on Circuits and Systems,* IEEE Press, London, 1994.

63. Beck, M.B., Sensitivity analysis, calibration and validation, in *Mathematical Modelling of Water Quality: Streams, Lakes and Reservoirs,* Orlob, G. T. (Ed),. *Int. Series on Applica. System Analysis,* Volume 12, Wiley, Chichester, pp. 425-467, 1983.

64. Okey, R.W. and Martis, M.C., Molecular level studies on the origin of toxicity: Identification of key variables and selection of descriptors, *Chemosphere,* 38, 1419, 1999.

65. Tetko, I.V., Villa, A.E.P., and Livingstone, D.J., Neural network studies. 2. Variable selection, *J. Chem. Inf. Comput. Sci.,* 36, 794, 1996.

66. Lin, V. and Cunningham III, G.A., A new approach to fuzzy-neural system modelling, *IEEE Trans. Fuzzy Systems,* 3, 190, 1995.

67. Acha, V. et al., Model-based estimation of an anaerobic reductive dechlorination process via an ATR-FTIR (attenuated total reflection-Fourier transf. infrared) sensor, *Wat. Sci. Technol.,* 40, 8, 33, 1999.

68. Parulekar, S.J. and Lin, H.C., Agricultural feedstock and waste treatment and engineering, in *Advances in Biochemical Engineering and Biotechnology,* vol. 32, Springer-Verlag, Berlin, 227, 1985.

69. Atanasova, T., Grancharova, A. and Vasileva, S., Adaptive control based on neural network with biotechnological applications, in *Proc. Int. Sci. Conf. Comp. Sci,* Ostrava, Czech, 425, Sept., 1995.

70. Auling, G. et al., Dynamics of growth in batch and continous cultures of *Saccharomyces cerevisiae* during shifts from aerobiosis to anaerobiosis and reverse, *Appl. Microbiol. Biotech.,* 19, 353, 1984.

71. Bakay, A. et al., Declarative programming tools for fermentor control systems, in *Proc. 11th Triennial World Congress IFAC,* vol. 4, Pergamon Press Inc., Elmsford, 293, 1991.

72. Bezdek, J.C., Tsao, E.C.-K., and Pal, N.R., Fuzzy Kohonnen cluster networks, *IEEE Int. Conf. Fuzzy Syst.,* 1035, 1992.

73. Biryukov, V.V., Non-traditional tasks for control of microorganisms' cultivation processes by using computers, in *Theory and Practice of Continuous Micro-Organisms' Cultivation,* Moscow, Nauka (in Russian), 139, 1980.

74. Biryukov, V.V. and Kantere, V.M., *Optimization of Microbial Synthesis Batch Processes,* Moscow, Nauka, 292, 1985.

75. Box, M., A new method of constrained optimization and a comparison with other methods, *Comp. J.,* vol. 8, 1965.

76. Bowerman, C., Readings in intelligent knowledge-based systems, *Comp. Inf. Syst.,* ver.1.0, 1993.

77. Brown, M. and Harris, C. *Neurofuzzy Adaptive Modeling and Control,* Prentice Hall, NY, 1994.

78. Leondes, C., Ed., *Control and Dynamic Systems,* Academic Press, San Diego, vol. 8, 1998.

79. Dochain, D., Design of on-line estimation and adaptive control algorithms for bioreactors, in *Proc. 1st Int. Conf. Modelling and Control of Biotech., Ecol. Biomed. Syst.,* Varna, Bulgaria, 2, 1990.

80. Dochain, D., Dynamical modelling, analysis and monitoring of bioprocesses, in *Proc. Int. Spring School'93 Modelling and Control of Biotech. Envir. Syst.,* Sofia, Bulgaria, 2, 1993.

81. Dochain, D., Software sensors and adaptive control in biotechnology, in *Proc. ACoFoP III,* Paris, p. 12, 1994.

82. Dochain, D. et al., Software sensors: on-line estimation of the specific growth rate in microbial reactors, in *Proc. J. de la Biotech.,* Belgian Society of Industrial Chemistry, 1985.

83. Friedman, J.H., Multivariate adaptive regression splines, *The Analysis of Statistics,* 19, 1, 1, 1991.

84. Garnick, R.L., Solli, N.J., and Papa, P.A., Role of quality control in biotechnology: An analytical perspective, *Anal. Chem.,* 60, 2546, 1988.

85. Grancharova, A., Zaprianov, J., and Vassileva, S., Dynamic modelling of bioprocess with artificial neural network, in *Proc. Int. Sc. Conf. Comp. Sci.,* Ostrava, Czech, 402, 1995.

86. Gupta, M.M. and Rao, D.H., On the principles of fuzzy networks, *Fuzzy Sets and Systems,* 61, 1, 1994.

87. Guymon, J. and Crowell, E., The formation of acetoin and diacetyl during fermentation and the levels found in wine, *Am. J. Enology and Viticulture,* 16, 2, 85, 1965.

88. Hayes-Roth, F. and Jacobstein, N., The state of the knowledge-based systems, *Comm. ACM,* 37, 3, 27, 1994.

89. Heijden, van der, Reinier, T.J.M. et al., State estimators (observers) for the on-line estimation of non-measurable process variables, *Trends Biotech.,* 7, 8, 205, 1989.

90. Jang, J.-S.R., ANFIS: Adaptive network-based fuzzy inference systems, *IEEE Trans. Syst. Man and Cybern.,* 23, 665, 1993.

91. Jang, J.-S.R., Fuzzy modelling using generalized neural networks and Kalman filter algorithm, in *Proc. 9-th Nat. Conf. Artif. Intell. (AAAI'91),* 762, 1991.

92. Jang, J.-S.R. and Sun, C.T., Neuro-fuzzy modeling and control, in *Proc. IEEE,* 83, 3, 379, 1995.

93. Kande, 1 A., Zhang, Y.O., and Miller, T., Hybrid fuzzy NN for fuzzy moves, in *Proc. 3rd World Congress Expert Syst.,* 717, 1994.

94. Karim, M. and Rivera, S., Comparison of feed-forward and recurrent neural networks for bioprocess state estimation, in *Proc. ESCAPE-1,* Elsinore, Denmark, 377, 1992.

95. Kingsbury, D.T., Computational biology for biotechnology, part I: The role of the computational infrastructure, *Trends Biotech.,* 7, 4, 82, 1989.

96. Kleinstreuer, C. and Poweigha, T., Modeling and simulation of bioreactor process dynamics, in *Advanced Biochemical Engineering and Biotechnology,* vol. 30, Springer-Verlag, Berlin, 143, 1984.

97. Latrille, E., Teissier, P., Perret, B., Barille, J.M., and Corrieu, G., Neural network modelling and predictive control of yeast starter production in champagne, in *CD-Proc. ECC'97,* European Control Conference '97, 1997.

98. Lin, C.T. and Lee, C.S.G., Reinforcement structure/parameter learning for neural network based fuzzy logic control systems, *IEEE Trans. Fuzzy Syst.,* 2, 1, 46, 1994.

99. Luo, R.F., Fuzzy-neural-net-based inferential control for a high-purity distillation column, *Control Eng.Practice,* 3, 241, 1995.

100. Mamdani, E.H., Application of fuzzy logic to approximate reasoning using linguistic systems, *Fuzzy Sets and Systems,* 26, 1182, 1977.

101. Monk, P. and Storer, R., The kinetics of yeast growth and sugar utilization in tirage—the influence of different methods of starter culture preparation and inoculum levels, *Am. J. Enology and Viticulture,* 37, 1, 71, 1986.

102. Montellano, R. et al., Knowledge-based system in modeling and control for biotechnological processes, in *Proc. 11th Triennial World Congress IFAC, USSR,* Pergamon, Elmsford, NY, 4, 47, 1991.

103. Monostori, L., Combining neural and fuzzy techniques in monitoring and control of manufacturing processes, in *Proc. 2nd IFAC Symp. Intell. Compon. Instr. Control App.-SICICA'94,* Banyasz, C.S., Ed., Hungary, 199, 1994.

104. Monostori, L., Connectionist and neuro-fuzzy techniques in manufacturing, in *Proc. of Ist World Congr. Intell. Manufact.: Process and Systems,* vol. 2, San Juan, Puerto Rico, 940, 1995.

105. Pons, M.N. et al., Comparison of estimation methods for biotechnological processes, *Chem. Eng. Sci.,* 43, 1909, 1988.

106. Rabinovich, S.G., Efficient calculation for indirect measurements and a new approach to the theory of indirect measurements, in *Proc. Measurement Sci. Conf.,* Newport Beach, Anaheim, CA, 10, 1996.

107. Recueil des methodes internationales d'analyse des vins et des mouts, in *Office International de la Vigne et du Vin,* Paris, 1990.

108. Ribereau-Gayon, J. et al., Trait, d'oenologie sciences et techniques du vin, *Analyse et contr'le des vins,* 1, Paris, Dunod, 1972.

109. Yu, J.J. and Zhou, C.-H., Soft-sensing techniques in process control, *Chinese J. Control Theory Appl.,* 13, 137, 1996.

110. Sarishvili, N.G., Vizelman, B., and Balobina, V.L., Method for yeast culture preparation for sparkling wine manufacture, *Vinodelie i vinogradarstvo,* SSSR, 1, 10, 1978.

111. Solopchenko, G.N., Reznik, L.K., and Johnson, W.C., Fuzzy intervals as a basis for measurement theory, in *Proc. Int. Jnt Conf. NAFIPS-IFIS-NASA 1994. IEE, 94,* 8006, 405, 1994.

112. Swanson, C.H. et al., Bacterial growth as an optimal process, *J. Theoret. Biol.,* 12, 2, 228, 1966.

113. Thibault, J., Breusegem, V. van, and Cheruy, A., On-line prediction of fermentation variables using neural nets, *Biotech.and Bioeng.,* 36, 1041, 1990.

114. Thomas, J.P. and Wei, R.P., Standard error estimates for rates of change from indirect measurements, *Technometrics,* 38, 1, 59, 1996.

115. Yamakawa, T., A neuro-fuzzy neuron and its application to fuzzy system identification and prediction of the system behavior, in *Proc. 2nd Int. Conf. Fuzzy Logic and NN (IIZKA'92),* 477, 1992.

116. Yamane, T. and Shimizu, S., Fed-batch techniques in microbial processes, in *Advanced Biochemical Engineering and Biotechnology,* vol. 30, Springer-Verlag, Berlin 192, 1984.

117. Kresta, J.V., Marlin, T.E., and MacGregor, J.F., Development of inferential process models using PLS, *Comput. Chem. Eng.,* 18, 597–11, 1994.

118. Lightbody, G. and Irwin, G.W., Multi-layer perceptron based modelling of nonlinear systems, *Fuzzy Sets and Systems,* 79, 93–112, 1996.

119. Villar, R.G.D., Thibault, J., and Villar, R.D., Development of a soft sensor for particle size monitoring, *Minerals Eng.,* 9(1), 55–72, 1996.

120. Warnes, M.R. et al., Application of radial basis function and feedforward artificial neural networks to the *Escherichia Coli* fermentation process, *Neurocomputing,* 20, 67, 1998.

121. Bhat, N. and McAvoy, T.J., Use of neural nets for dynamic modelling and control of chemical process systems, *Comput. Chem. Eng.,* 14, 573–583, 1990.

12

From Simple Graphs to Powerful Knowledge Representation: The Conceptual Graph Formalism

Guy W. Mineau
Université Laval

Abstract. The evolution of many computer applications depends on the use of human knowledge as a computable asset. Its integration to computer systems must be planned not only in the development of new applications, but also in the integration of knowledge-based technology to existing applications. The migration of a computer system from an information-based to a knowledge-based system entails a set of constraints that are related to the industrial setting in which these applications are developed and used. Our claim is that these constraints should guide the choice of the knowledge modeling language used in the development of these applications. This chapter presents conceptual graph (CG) formalism as a viable option in light of these constraints. In our opinion, since CG formalism is a simple graphical logic-based representation system, it is quite close to ER diagrams and UML, which should facilitate application developers' ability to learn it, making it a strong candidate to accomplish the task.

12.1 Introduction: The Need for Human Knowledge

The complexity of software applications currently calls for the direct use of human knowledge as a computable asset. By making the applications more knowledgeable about themselves, their users, the world, and the services that they were designed to provide, one aims at improving the quality of service that they would otherwise render. As an example, let us mention *wizards*, which provide automatic task support and user assistance in different software products like word processors. In order to implement an effective wizard that will provide real task support, some modeling of the task and possible user actions is required. A wizard must quickly see what the goal of the task is, analyze the performance of the user with regard to its effectiveness toward reaching it, and propose correct and more efficient courses of action.

As a second example, let us mention *personal digital assistants* (PDA). Since agent technology and mobile computing are currently booming, it is predictable that in the near future PDAs will be in high demand for all sorts of applications. The main purpose of a PDA is to: (1) achieve user profiling, (2) consequently assess the information need of its user, (3) search for the relevant information (and use all the necessary accessible applications in order to produce the required information), and (4) adapt it and present it in the proper format to its user. Using a PDA as a front-end for computer applications transforms them into information providers. Then the interaction between the user and an application is no longer task oriented, but becomes information driven, where information interchange remains close to the user's reality. Again, a PDA must be knowledgeable about the various tasks that its user may want to accomplish, about his needs and preferences, and about the world and its potential resources. Needless to say, this requires the modeling of human knowledge.

As our third and last example, let us mention the *knowledge management problem* that companies are faced with today. With a high turnover of employees, there is an immediate need to exploit the *corporate knowledge* of a company so that the quality of the services rendered and the cost of operations are consistent with what is expected by both customers and management. Furthermore, with strict *time to market* realities that put pressure on the application developers to meet tight delivery deadlines, there is an obvious need for a company to provide its employees with the appropriate knowledge for them to accomplish their tasks efficiently. Knowledge management systems were designed to that effect.

In their most simple form, task-related documents were put on-line and were made accessible through some information retrieval system. Now entering the next era of knowledge management systems, the knowledge contained in these documents is sought to be made accessible to its users so that it can be exploited efficiently at the semantic level. This task is not without much logistic complexity. This corporate knowledge must be acquired, validated, formatted, and provided in a timely manner to the users of the system, and, at best, be integrated to the day-to-day systems that these users must operate in order to achieve their tasks. Such knowledge management systems are really seen as task support systems. Ways to turn corporate documents into knowledge-based systems compose the bulk of the research that goes on today in some scientific research communities: it is still an open research problem. Clearly, the modeling of human knowledge is a big part of that problem.

In all cases above, it is obvious that some human knowledge will need to be acquired and represented in a computable format for these applications to even exist and then evolve. When that happens on a large scale, computer software as we know it today will be seen as middleware serving an application layer where the interactions happen in terms of the information that is needed to answer high-level queries as submitted by human users or other computer applications.

12.2   Knowledge Representation: A Technological Choice

Technological choices should always be based on the assumptions that guide the overall development of an enterprise. In what follows, we describe the assumptions that should be taken into consideration when developing information-related applications.

- Many current computer applications that are actually *data* driven will eventually have to become *knowledge* driven if the kind of services that we describe in Section 12.1 are to be provided.
- There is a very high cost in turning these data driven applications into knowledge based applications, mostly related to the ever present legacy system problem.
- User resistance to these changes, either end-user or application developer, will be high for various reasons, as is always the case.

These assumptions become important as we have to select a knowledge representation formalism that will allow data based systems to *smoothly* evolve into knowledge-based systems. Therefore, they guide our choice of knowledge representation. Based on these assumptions, we consequently developed a list of characteristics that such a knowledge representation language should meet, as described in the following subsections.

12.2.1   Building on the Past

It is expected that actual applications will not be reengineered from scratch because of obvious financial reasons. Therefore, if these applications are to evolve and become knowledge based, that evolution should be done incrementally. So a knowledge representation language that could be seen as a natural extension of the actual description formalism of a system would favor the reuse of the already acquired knowledge about the application, therefore facilitating this evolution. Additional knowledge about the application would then be seen as a new conceptual layer built on top of the previous one. For instance, with relational database systems, some knowledge about the application domain is already available through the description of meta-level information. Providing additional features that would make different inferring procedures available in such systems would certainly be a good way to propose a smooth evolution from relational databases to knowledge-based systems. Consequently, a knowledge modeling language that can be extended to allow its previous syntax (and semantics) to remain valid through this enhancement should be preferred.

In other words, let $S = \{s_i : i \geq 1\}$, the different states of a system undergoing this technology development. Each transition from s_i to s_{i+1} is defined by the addition of some knowledge representation and processing capabilities to the modeling language L_i (used to describe the system in state s_i) which results in L_{i+1}, a more expressive modeling language based on L_i, with which the system now in state s_{i+1} can be described. If a domain theory T_i can be inferred from the system at state s_i, written $s_i \models T_i$, then we should have $s_{i+1} \models T_i$ as well.

12.2.2   Building in Layers

It is not expected that knowledge processing features will be added to some application all at once. Going from database systems where model checking is sufficient, to fuzzy paraconsistent logics would be assuming a capability of technology absorption that is simply unrealistic, and maybe unnecessary. Therefore, we advocate that the evolution of data-driven systems be done in a way so that their additional capabilities are added incrementally, allowing the application of a layer-based development model in order to facilitate technology integration at all levels of the enterprise. So the changes to L_i, written $(L_{i+1} - L_i)$, should be chosen to optimize the cost/benefit ratio of technology integration. Needless to say, such a ratio can be defined in a very large sense, taking into account employee training, database translation, upgrade of task-oriented tools, update of administrative procedures, etc. A modeling language in which the granularity of upgrades $(L_{i+1} - L_i)$ can be determined according to these factors should be preferred.

12.2.3 Expressiveness and Flexibility

In many cases, applications (especially industrial applications) evolve in ways that may be unpredictable. Therefore, the added expressiveness required to L_i to meet these new demands should allow maximal flexibility, i.e., $(L_j - L_i)$ should be determined without the constraint that $j = i + 1$. This flexibility allows the introduction of the added features in layers, so that technology integration can be achieved in an optimized manner each step of the way.

12.2.4 Cognitive Distance

For computer applications where one seeks to implement deterministic information processing procedures, a modeling language must allow a *formal* representation of some reality to be expressed. Furthermore, the soundness and completeness of these procedures depend, among other things, on the valid and complete modeling of that reality. Among different factors that will help achieve this, the application developer must be fluent in that language. As with learning any language, the one closest to our known language, preferably our mother tongue, should prove to be easier to learn and master. Since most computer scientists and programmers today are trained with graphical representations like ER diagrams and UML in order to model information systems, we assume that the modeling languages that will be the easiest for them to acquire should be closely related to them, and should contain a graphical representation of *types* of objects, of objects, and of their relationships. Therefore, we assume that such a language is L_0, usually a modeling language acquired through academic training. Of course, employee training is the lowest if L_1, the modeling language originally used in an industrial setting, is equal to L_0, or in more general terms, if $(L_1 - L_0)$ is minimal.

As a general guideline, $(L_{i+1} - L_i)$ should be minimized in terms of their *cognitive distance*. Here the cognitive distance between two modeling languages is defined empirically as the number of hours required on average for one employee trained in language L_i to master language L_{i+1} sufficiently enough to describe s_{i+1} in a sound and complete manner.

12.2.5 Summary of Requirements

In light of Sections 12.2.1 through 12.2.4, we summarize five principles that should constitute the main characteristics of a modeling language that would ease the development of a knowledge-based system s_j from a data-based system s_i (with $j > i$).

- Expressiveness: there exists a L_j where $j > i$ such that L_j meets the requirements of s_j.
- Manageability: $(L_{i+1} - L_i)$ is an improvement step that can *efficiently* be achieved.
- Flexibility: there exists a set of *manageable* steps from L_i to L_j, i.e., $L_j = (L_{i+n} - L_{i+n-1}) + \cdots + (L_{i+2} - L_{i+1}) + (L_{i+1} - L_i) + L_i$.
- Minimization of cognitive distance: for all $i \geq 1$, $(L_{i+1} - L_i)$ is chosen, so the cognitive distance between L_{i+1} and L_i is minimized.
- Backward compatibility: if $s_i| = T_i$ then $s_{i+1}| = T_i$.

And as explained above, we assume that L_0 is an ER- or UML-like modeling language. Therefore, we propose conceptual graph formalism as L_1. Through the presentation of the conceptual graph formalism in the rest of this chapter, we show how it fits the five principles described above.

12.3 Conceptual Graph Formalism

Conceptual graph (CG) formalism was introduced in and was recently revisited in Sowa.[1,2] In brief, it is a simple graphical notation used to describe knowledge, i.e., object types, objects, and their relationships. It is closely related to first-order logic,[3] but can be extended to higher-order logics.[4]

It is best described as being a graphical interface to logic-based systems. Its graphical nature, close to ER diagrams and UML-like modeling languages (L_0), minimizes its cognitive distance to L_0. Its many possible extensions, abundantly described in the literature,[5-13] prove its expressiveness. We believe that a particular subset of the CG formalism, simple conceptual graphs,[14-22] is particularly relevant to be used as L_1, and we present it in what follows.

12.3.1   The Canon of a CG-Based System

In any formal representation system, there are some fundamental assumptions that constitute the foundation of the language, assumptions that delimit the expressiveness of the language. In a CG-based system, these assumptions are divided into four distinct classes: the existence of a predefined vocabulary (in our case, a type lattice T), the existence of a set of known individuals I, the assignment of individual to types, a conformance relation C between types and individuals (thus defined over $T \times I$, a binary relation), and the set of semantic constraints pertaining to the domain (H). These four categories $\langle T, I, C, H \rangle$ form the *canon* of a CG-based system. All other assertions made in the system must be in accordance with these assumptions. Each category is described in more detail in the following subsections. But first, a brief introduction to the syntax of conceptual graphs is presented in Section 12.3.1.1.

12.3.1.1   The Basics
Conceptual graphs are composed of two types of nodes: concepts and relations. Concepts represent typed individuals, and relations represent their interrelationships. Thus, they are bipartite-oriented graphs. Concepts are graphically represented by boxes in which the type of the individual appears, followed by the name of the individual (or a variable if unknown at the moment). Relations are represented as ovals where the type of the relation appears; they link the different concepts that are in relation with one another. So some order of these concepts, acting as parameters of the relation, is required. The arcs between the relation and the concepts that it links are thus numbered accordingly. By convention, the last number is associated with the only outgoing arc, while all other numbers are associated with the other (incoming) arcs, if any. With binary relations, the numbering system is not necessary since the direction of the arrows is sufficient to disambiguate the role of the various concepts. Figure I.12.1 represents the assertion: *The employee John is assigned to project #584*, a simple conceptual graph.

12.3.1.2   The Vocabulary
Any assertion in any language relies on a set of interrelated terms, called the vocabulary, so that the objects of the language can be given a definition, a meaning. In object-oriented systems, this vocabulary designates *classes* of objects. In CG-based systems, such a vocabulary is called a *type lattice*, and is denoted T. An object o is said to comply with a type t if its definition $d(o)$ *fits* (meets the necessary and sufficient conditions associated with) the definition of type t, $d(t)$, or in other words, if $d(o) \models d(t)$. By finding the definition of t in $d(o)$, o is said to have the properties of type t. In object-oriented systems, classes are created for all objects o such that $d(o) \models d(t)$, for a particular t.

Furthermore, with two distinct types t_1 and t_2, according to the information that their definition carries, either $d(t_1) \models d(t_2)$ (then we could also say that t_1 is more specific than t_2, written

FIGURE I.12.1   An example of a simple conceptual graph.

$t_1 <_g t_2), d(t_2)| = d(t_1), d(t_1) =_g d(t_2)$, or $d(t_1) \neq_g d(t_2)$, where $=_g$, also defined directly between types, represents an equivalence relation, and where $\neq_g$ means that the definitions (or types) are not comparable.

These definitions are sufficient to define a type hierarchy. However, for simplicity reasons pertaining to the graph manipulation operators of CG systems, a restriction is normally imposed on the type hierarchy. Let us define supertype$(t) = \{t' \in T \mid t \leq_g t'\}$ and subtype$(t) = \{t' \in T \mid t' \leq_g t\}$. Then for all pairs of $t_1, t_2 \in T$, we define supset$(t_1, t_2) = $ supertype$(t_1) \cap$ supertype(t_2), and infset$(t_1, t_2) = $ subtype$(t_1) \cap$ subtype(t_2). Let us define sup$(t_1, t_2) = \{t \in$ supset(t_1, t_2) such that $t \leq_g t', \forall t' \in$ supset$(t_1, t_2)\}$, and inf$(t_1, t_2) = \{t \in$ infset(t_1, t_2) such that $t \geq_g t', \forall t' \in$ infset$(t_1, t_2)\}$. The lattice property implies that sup$(t_1, t_2) \neq \emptyset$ and inf$(t_1, t_2) \neq \emptyset$, for any pair t_1, t_2 in T, and that all types in sup(t_1, t_2) (and respectively in inf(t_1, t_2)) are all equivalent. Consequently, in situations where no two type definitions are equivalent, then $|\text{sup}(t_1, t_2)| = |\text{inf}(t_1, t_2)| = 1$. In simpler words, a lattice structure implies that any two types have a *single* supertype and a *single* subtype, provided that equivalent definitions are considered as one. For simplicity reasons, equivalent definitions are rarely allowed to coexist in a logic-based system; they would eventually lead to an explosion of the search space when inference procedures are carried out.

In the CG formalism, one can define the vocabulary, i.e., the type lattice, by providing definitions that specialize previously given definitions. For instance, Figure I.12.2 shows the introduction of type EMPLOYEE in the vocabulary, as a specialization of type PERSON. There is no need to explicitly state the definition between a type and its immediate subtype. For instance, type PERSON may have been defined as in Figure I.12.3, which means that it is a specialization of type ENTITY, but its specificity with regard to type ENTITY is not described at that point. For that reason, contrarily to type EMPLOYEE, type PERSON is seen as a *primitive* type. Primitive types facilitate the implementation of the projection operator (see Section 12.3.7). For that reason we tend to advocate their use.

Finally, since we define the type hierarchy as a lattice structure, we must define a universal supremum, which is T, called the *universal* type, and a universal infimum, which is ⊥, called the *absurd* type. That is, we have: $t \leq_g$ T and $\perp \leq_g t$, $\forall t$ in T.

12.3.1.3 Naming and Typing Objects

Known individuals are given a single name that becomes a unique reference to them. In logic-based systems, these names act as *constants*. The set of constants (I) used to represent the objects of the domain is assumed to be known. The interpretation of each object is done through its association with some type. Figure I.12.4 shows how John, a known individual uniquely referred to by that

type EMPLOYEE(x1) **is**

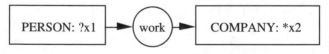

FIGURE I.12.2   The definition of type EMPLOYEE.

type PERSON(x) **is**

```
┌─────────────┐
│  ENTITY: ?x │
└─────────────┘
```

FIGURE I.12.3   The definition of type PERSON.

individual EMPLOYEE(John) **is**

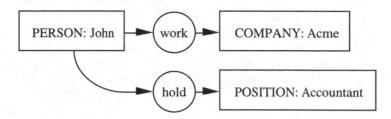

FIGURE I.12.4 Defining the employee John.

type ACCOUNTANT(x1) **is**

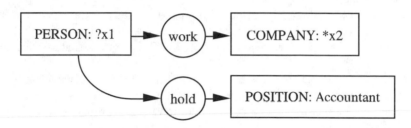

FIGURE I.12.5 The definition of type ACCOUNTANT.

appellation, is associated with type EMPLOYEE. When an individual i conforms to a type t_1, i.e., when $d(i) \mathrel{|=} d(t_1)$, we write $t_1 :: i$. Of course, if $t_1 :: i$ and $t_1 <_g t_2$, then $t_2 :: i$ (Rule 1). When an individual i is compliant with a type t, the concept $[t : i]$ can be formed, where i is used as *referent* to the object in question. Individuals may be known (like in the example of Figure I.12.4 below) or not. In the latter case, they are represented by some (existentially quantified) variable (Figure I.12.5).

As mentioned before, the sole restriction on the definition of John is that $d(\text{John}) \mathrel{|=} d$ (EMPLOYEE). Consequently, the definition of John can carry more information than that of its corresponding type, as illustrated in Figure I.12.4 (in comparison to the definition of Figure I.12.2), either by instantiating some variables of d(EMPLOYEE), or by carrying additional relations to other objects. Again for simplicity reasons pertaining to the join operator, which is a graph derivation operator (see Section 12.3.7), we assume the following restriction:

- There is no $i \in I$ such that $t_1 :: i$ and $t_2 :: i$ and $t_1 \neq_g t_2$ (Rule 2).

This avoids conflicting views on the same data. Let us define $C = \{\langle t, i\rangle \mid t :: i$, and there is no $t' <_g t$ such that $t' :: i$, for $t, t' \in T$ and $i \in I\}$, the set of most specific type assignments for all individuals, either identifiable (by constants) or not (then identified by variables). We call it the conformity set, from which, using Rule 1 above and T, we can deduce all possible conformity relation between types and individuals. In our example, C would contain $\langle$EMPLOYEE, John$\rangle$. But with the definition of ACCOUNTANT (Figure I.12.5), since ACCOUNTANT $<_g$ EMPLOYEE, then C would rather contain $\langle$ACCOUTANT, John$\rangle$ instead of $\langle$EMPLOYEE, John$\rangle$, despite the definition given in Figure I.12.4. From C, T, and Rule 1, all existing objects seen through different interpretation (types) can be represented by a concept $[t : i]$, where i is the referent to (or identifier representing) the object, and t is its type. [ACCOUNTANT: John] and [EMPLOYEE: John] are two examples of valid concepts.

12.3.1.4 Semantic Constraints

All asserted graphs in CG-based systems are derived from the information carried in the canon of the system. The graph derivation operators are specialization operators.[1] In order to ensure that relations will be used according to their associated semantics, Sowa[1] gives a formal mechanism for defining relations. For instance, Figure I.12.6 shows the definition of the relation work as a specialization of the only predefined relation type link, defined between a concept of type AGENT and a concept of type LEGAL_ENTITY. That definition constrains the types of the parameters that the relation may link; it is its signature. The set of all relation signatures is called a canonical basis B. In fact, each signature is a semantic constraint that prevents overgeneralizations from occurring in the data. With our example (Figure I.12.6), no use of the work relation is subsequently possible between concepts whose types do not comply with those of the concepts that the relation links according to its definition. Since work is a (relation) type appearing in T, for any simple conceptual graph g where a relation of type work appears, we must have that $g \models d(\text{work})$, which enforces the *signature* constraint of work on g.

However, Mineau and Missaoui[16] extended the canonical basis of a CG system by defining a simple mechanism, based on *false-unless* clauses, that allows the representation of most semantic constraints found in database literature today. Figure I.12.7 shows the signature constraint of relation type work defined using this mechanism.

The construct of Figure I.12.7 states that no graph should bare the false clause, unless it bares the unless clause. In more general terms, let us define a constraint c_1 as: false g_1 unless g_2 (with $g_2 \models g_1$). Then we have two cases described as follows (for a formal description please refer to Mineau and Missaoui[16] and Mineau[17]):

relation work(x1,x2) **is**

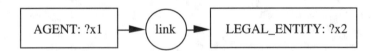

FIGURE I.12.6   The definition of relation type work.

false:

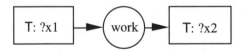

unless:

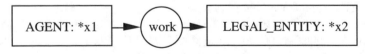

FIGURE I.12.7   Avoiding overgeneralizations of relation type work using *false-unless* clauses.

- If the unless clause is unnecessary and is omitted, then for any graph $g \mid = g_1$ must not hold. Otherwise g bares the information of g_1 and violates the constraint.
- If the unless clause is necessary and expresses an exception to the constraint, then for any graph g, $g \mid = g_1$ must not hold *unless* $g \mid = g_2$ holds. If $g \mid = g_1$ but $g \mid = g_2$ does not hold, then g violates the constraint.

It was proven that this simple mechanism is sufficient to represent most semantic constraints found in database literature.[16] Therefore, we define the final category of fundamental assumptions composing the canon of a CG system as H, the set of semantic constraints of the domain, with $B \subseteq H$. Consequently, the canon of a system $\langle T, I, C, H \rangle$ forms the basic building blocks and constraints that will allow factual knowledge to be thereafter acquired. A graph g is said to conform to the canon of a system, i.e., it is said to be valid, iff:

- All of its concepts are either in C or can be derived from C through the application of Rule 1 on T.
- Graph g does not violate any constraint in H (including the constraints in $B \subseteq H$).

12.3.2 Embedded Graphs

In order to reason about statements represented as conceptual graphs, these CGs must be seen as objects. Therefore, these graphs will need to appear as referents of concepts. So we predefine a concept type ASSERTION, defined as a specialization of another predefined concept type STATEMENT, that will be used in concepts of the form: [ASSERTION: *g], where g is a conceptual graph to be asserted. By asserting [ASSERTION: *g], g is then assumed to be true (since it is asserted as such). For example, Figure I.12.8 shows how the graph of Figure I.12.1 would be asserted.

Conceptual graphs in which some referents are themselves conceptual graphs are called *embedded* graphs. So a conceptual graph-based system can be built in layers, where meta-level information can easily be represented through embedded graphs. The interested reader is referred to Gerbé[18] for a complete treatment of meta-modeling in the CG formalism. In any case, it is obvious that embedded graphs extend the basic representation framework of CGs. In the literature, embedded graphs were used to allow the representation of temporal and modal knowledge using the CG formalism.[19,20]

12.3.3 Asserting Statements

A valid graph g may or may not have a model. Of course, if g was previously asserted, then it is known to be true by definition. For all nonasserted graphs, however, one must be able to decide if g holds, i.e., if it has a model or not. The extensional semantics of CG-based systems are described in Chein and Mugnier,[3] Mineau,[17] and Prediger.[21] and model checking based on relational operators is described in Mineau.[17] A graph is said to have a model if there is at least one true graph g' such that $g' \mid = g$. All asserted graphs are said to be true if they are valid, by definition. From each information $\langle t, i \rangle$ in C, a concept of the form $[t : i]$ is formed, and the graph $[t : i]$ is considered to be asserted ($[t : i]$ can be seen as a single concept graph).

ASSERTION:

FIGURE I.12.8 Asserting a conceptual graph.

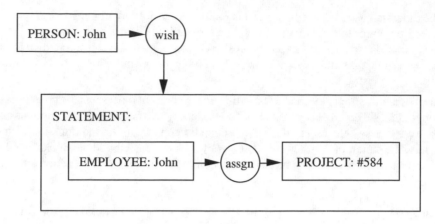

FIGURE I.12.9   An embedded graph.

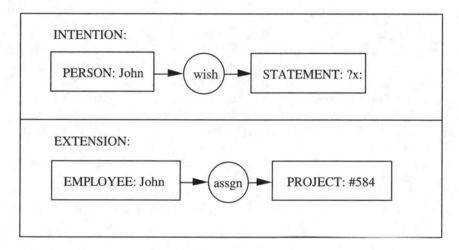

FIGURE I.12.10   A context representing the assertions that hold true in John's wishes.

By default, all asserted graphs are asserted on a main sheet of assertion through the acquisition of a single concept graph of the form: [ASSERTION: *g], as described above. However, it is possible to segment the sheet of assertion and assert a new graph in one of these segments. For instance, the graph of Figure I.12.9 shows an embedded graph that states: *John wishes to be assigned to project #584*. As such, this is a true graph and is asserted in the main sheet of assertion. But the fact that John is assigned to project #584 is only asserted in John's wishes, creating this hypothetical world (sheet) of assertion. So an auxiliary sheet of assertion that would contain John's wishes could be created, as shown in Figure I.12.10. This auxiliary sheet of assertion is identified by both a conceptual graph that constitutes its *intention*: John's wishes, and the conceptual graphs that are asserted under that intention, called its *extension*. Mineau and Gerbé[22] describe this context representation mechanism in detail.

Using such a mechanism allows a complex system to be decomposed into a set of smaller subsystems, called micro-worlds[23] or contexts[1,22] in literature. This reduction in complexity makes more feasible the development of large applications. Mineau[24] shows its usefulness for the representation of agent systems (under the CG formalism).

12.3.4   Negating Statements

Negated knowledge must also be explicitly represented at times. A negated statement g, called a negation, can be explicitly acquired as: [NEGATION: $*g$], with NEGATION $<_g$ STATEMENT. Of course, in order to make sure that g is in fact negated, there must not be any other graph g' in the system such that $g' \models g$; otherwise the truth value of g would be true while the previous negation states that it should be false, leading to an obvious contradiction and the breakdown of the inference procedure. So upon acquiring a valid but negated graph g, consistency with the rest of the system must be checked. If inconsistency is detected, then knowledge revision is required from the knowledge engineer. If not, then this negation is added to H as: false g, thus making sure that further assertions will comply with that negation, since only valid graphs can be asserted. Using negated information as constraints to filter out knowledge allows the fact base of the system to only contain asserted knowledge. Avoiding the explicit use of a negation operator in the inference procedure, which is then applied only on asserted knowledge, is one way to simplify it.

12.3.5   Inference Rules

Now that factual knowledge can be validated and acquired, production rules need to be represented so that an inference engine can support the kind of inference that one would expect from knowledge-based systems. To that effect, *if-then* clauses can be used as shown in Figure I.12.11 where it is stated that *every manager has a salary that is above 60K\$*. Baset, Genest and Mugnier[25] show how colored graphs can be used to express *if-then* clauses in a graphical manner. Figure I.12.12 shows the same example represented as a colored graph, where the darker area of the graph is added to the white background graph as a result of the implication rule. In logic, we would represent the implication rule: if g_1 then g_2, as: $g_1 \Rightarrow g_2$, and in more general terms we would have: $g_1 \wedge g_2 \wedge \cdots \wedge g_{n-1} \Rightarrow g_n$, which is a Horn clause. Because of their simplicity, which has a direct impact on the complexity of the inference procedure, we advocate the use of Horn clauses as much as possible. Other kinds of clausal representation could be considered at a later stage, when the expressiveness of Horn clauses is vitally insufficient for modeling the application domain. Of course, this comment is made in light of the constraints expressed in Section 12.2.

Nevertheless, some information may need to be negated (or retracted if known to be true) as the result of an implication rule. Mineau[26] proposes to represent implication sort rules as processes (Section 12.3.6). For instance, Figure I.12.13 depicts the rule: *every manager has a salary that*

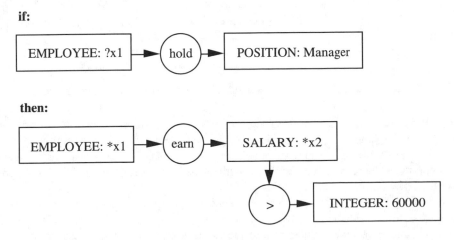

FIGURE I.12.11   An *if-then* clause that encodes some implication rule.

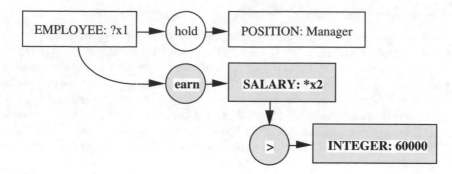

FIGURE I.12.12   The production rule of Figure I.12.11 represented as a colored graph.

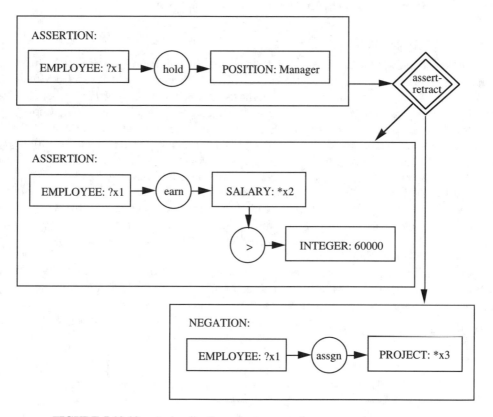

FIGURE I.12.13   An implication rule represented as an assert/retract process.

is above 60K$, and is not assigned to any project. With such rules, negations are first acquired and the resulting updates of *H* are achieved (Section 12.3.4). Then the assertions are acquired (Section 12.3.3). That way, we aim at reducing the complexity issues pertaining to allowing the negation operator in the representation model. With the negation part of Figure I.12.13, the constraint of Figure I.12.14 would be added to *H*.

However, as with any rule-based system, decidability, completeness, and soundness may be troublesome, as the problems related to the order of application of the rules, to the presence of directly or indirectly recursive rules, and to the creation of new concepts, as the result of some production rules, are obviously not solved by a CG-based representation of production rules.

false:

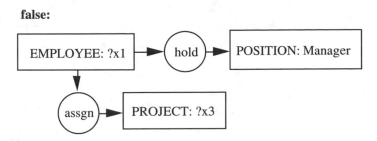

FIGURE I.12.14   A new constraint based on the implication rule of Figure I.12.13.

12.3.6   Complexity Results

Complexity is always a critical issue in knowledge-based systems. Usually the trade-off between expressiveness and tractability leads to compromises. The more expressive the system is, the less guarantee that an inference procedure will terminate in a reasonable amount of time, if it terminates at all. The best analysis of the expressiveness/tractability trade-off in CG-based systems is described in Baget and Mugnier.[14]

In brief, by using no disjunction and by using negations as filters for asserted (or deducted) knowledge, we avoid having to deal with the associated complexity of these two operators in an inference procedure; we seek to keep its associated complexity as minimal as possible.

Additional simplifying assumptions could be made. For instance, the creation of new concepts (of new individuals) in a rule could be forbidden, forcing the rules about add knowledge about already existing individuals. Then, by triggering only the rules that can add new information to a knowledge base, when a finite number of concepts exists, the inference procedure would eventually terminate. The end result would be independent of the order of application of these rules. Again, the trade-off between complexity and expressiveness with regard to the needs of the application domain must be assessed thoroughly by the knowledge engineer. We strongly suggest Baget and Mugnier[14] as a first reading on the subject.

12.3.7   Comparing Graphs: The Projection Operator

The logical consequence operator $g_1 \models g_2$ between two graphs can be evaluated through the *projection* of g_2 onto g_1. If g_2 projects onto g_1, then the logical consequence $g_1 \models g_2$ holds (through the application of generalization operators). The projection operator π is defined in Sowa.[1] In brief, if a *specialization* g' of g_2 exists in g_1 as a subgraph, then g_2 is said to have a projection $s1$ into g_1, written πg_2. A specialization of g_2 is produced: (s1) by replacing (at least) one variable in a concept by an $s2$ constant that conforms to the type of the concept, (s2) by replacing (at least) one concept type by one of its subtypes other than the absurd concept type (as long as the object conforms to this new type), or (3) by joining g_2 with some other graph g_3. The joining of two graphs is defined on *compatible* concepts of both graphs to be joined.[1] That is, if a concept c_1 in g_2 is compatible with a concept c_2 in g_3, then the two graphs can be joined over a common specialization of c_1 and c_2. If the two concepts do not share a common specialization, then they are not compatible. A specialization of a concept is produced only under specialization rules $s1$ and $s2$.

Though some very efficient projection algorithms are described in the literature,[27] in order to simplify even more the computation of projections between graphs, a common variant of the projection operator, the injective projection,[3] here denoted π^*, is sometimes used. In the representation framework for representing semantic constraints described in Mineau and Missaoui,[16] we make use of this injective projection. The injective projection makes sure that no two concepts of g_2 are mapped onto the same concept of g_1. For instance, Figure I.12.15 shows two graphs g_1 and g_2 such that there

graph g₁:

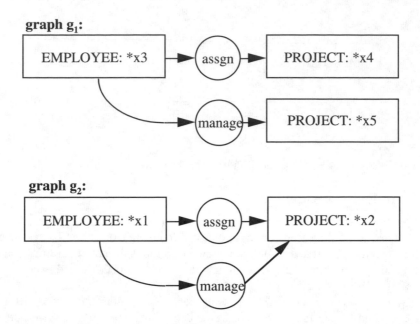

graph g₂:

FIGURE I.12.15   Two graphs g_1 and g_2 that are π-equivalent, but not π^*-equivalent.

is a projection from either one onto the other one. Therefore, under the projection defined in Sowa,[1] $g_1 \models g_2$ and $g_2 \models g_1$. But under an injective projection, only $g_2 \models g_1$, since inequality of distinct concepts is always assumed. So the two concepts of type PROJECT in g_1 are assumed to represent different individuals, and are thus not mappable onto one another. In logical terms, additional axioms about equality and inequality need to be added to a CG system so that π^* is sound; these axioms are listed below. Also, we would like to remind the reader that I, the set of referents (identifiers) to objects of the domain, contain both constants and variables.

- $\forall i \in I, \mathrm{eq}(i, i)$ is true (reflexivity of equality)
- $\forall i, j \in I, \mathrm{eq}(i, j) \Rightarrow \mathrm{eq}(j, i)$ (symmetry of equality)
- $\forall i, j, k \in I, \mathrm{eq}(i, j) \wedge \mathrm{eq}(j, k) \Rightarrow \mathrm{eq}(i, k)$ (transitivity of equality)
- $\forall i, j \in I, \mathrm{not\_}\ \mathrm{eq}(i, j) \Rightarrow \mathrm{not\_}\ \mathrm{eq}(j, i)$ (symmetry of inequality)
- $\forall i, j \in I$, if (i and j appear in the same graph g) $\wedge \neg \mathrm{eq}(i, j)$ (i.e., if we can't prove $\mathrm{eq}(i, j)$) then $\mathrm{not\_eq}(i, j)$
- $\forall i, j \in I, \mathrm{eq}(i, j) \wedge \mathrm{not\_eq}(i, j) \Rightarrow$ *False* (a contradiction)

And in order to precisely define the scope of a graph g, we add the following condition on the system:

- Concepts known to be equal are joined, producing graphs that are said to be in *normal form*.[3]

Together, these axioms restrict the expressiveness of the language, since no ambiguity on the value of the variables in a graph g may subsist with regard to the other concepts in g. However, they enforce π^* and make the implementation of some constraint-checking mechanism possible (as with cardinality constraints important in database applications); and they simplify the applicability of the projection operator.

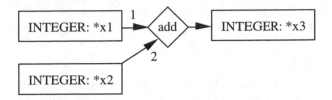

FIGURE I.12.16   The representation of the actor add.

12.3.8   Actors and Processes

Finally, since real-life applications need bridges to other applications, need to communicate with their users, other systems, and the world, and since infinite sets exist and can not be represented exhaustively in clausal form, a way to allow on-the-fly computation is definitely required, particularly for database applications. The CG formalism allows algorithmic procedures to be invoked. For simple computation applied on concepts, Sowa defined actors.[1] Figure I.12.16 gives an example of the arithmetic function add, that instantiates its output concept to the sum of its input concepts. For example, by instantiating its input concepts with 25 and 2, the output concept would be instantiated by the actor, to 27. The instantiation of the input concepts of an actor triggers their processing into its output concept.

Though graphically part of a conceptual graph, the projection operator does not take them into account. Actors are rather seen as graph-transforming operators that change the state of a knowledge base, by specializing some concepts, and therefore the graphs that contain these concepts. Of course, loops could be graphically represented, but the validity of such graphs is questionable; such a use of actor should be formally defined, which is not the case in the literature today. Until then, it would be extremely risky to do so.

One step further, Mineau[26] extended actors and the work of Delugach[28] on demons so that general-purpose procedures could be formally represented under the CG formalism, and could trigger some state change as a result of their invocation. The idea is to have procedures work at the graph level rather than at the concept level as with actors. The integration of this definition framework for processes in a CG-based inference tool is discussed in Benn and Corlett,[29] where a CG-based processing tool is presented; an example of the representation of a process is given in Figure I.12.13. Like actors, processes do not intervene in the computation of projections between graphs.

Since processes are more general than actors, they are sufficient to represent any kind of procedure that would process graphs; they are algorithmic procedures whose input and output arguments are conceptual graphs. Because we may want to describe a process on a declarative level, whether it is activated (waiting for instantiation of its input arguments) or not, we introduce two concept types: PROCESS and ACTIVATED-PROCESS. That way, a process could be described just as any other object in the system, using conceptual graphs, and it could be triggered by specializing its associated concept type in the conformity set C, from PROCESS to ACTIVATED-PROCESS. And the referent of a concept of type PROCESS (or ACTIVATED-PROCESS) could be a URL where a procedure (e.g., applet) could be downloaded to actually carry out the processing.

12.4   Conceptual Graphs as a First Knowledge Modeling Language

With a combined procedural and logical processing system represented under a unified representation model which is closely related to ER diagrams and UML, we feel that the conceptual graph modeling language offers the potential to facilitate the migration of database systems toward knowledge-based systems. Now that Section 12.3 has presented the CG formalism in enough detail for the reader to see what the CG paradigm is, what it offers, and what its complexity is, we now review the requirements summarized in Section 12.2.5, requirements that were identified as being sought-for properties of a modeling language that would assist knowledge-based technology development.

12.4.1   Expressiveness

Through the use of embedded graphs, contexts, a meta-level model, and other extensions, the conceptual graph notation can be extended to meet a variety of representation challenges, as found in the literature.[5–13] For more than 15 years now, researchers have proposed ways to extend the CG formalism to model complex phenomena that require higher-level constructs. Their work and the subsequent variety of extensions covering temporal, modal, fuzzy logic reasoning systems, etc., are proof that the CG modeling language can be extended.

12.4.2   Manageability and Flexibility

The closeness of the extensions found in the CG literature to the original theory as presented in Sowa[1,2] should facilitate manageability of development. As a starting point, Baget and Mugnier[14] propose different subsets of the CG formalism as a representation basis for simpler or more complex types of knowledge-based systems within the framework of first-order logic. With the framework proposed in this article, it should be clear that the integration of knowledge-based technology could be done in steps toward other variants described in Baget and Mugnier[14] and from there to extensions that open the door to more complex logic systems.

12.4.3   Backward Compatibility

By identifying the type of logic-based systems required to closely model the application domain, and considering the modular approach to system development as advocated in this chapter, the use of a meta-level representation of some application domain and of the modeling language itself makes it straightforward to recompile the previously acquired knowledge structures, which guarantees backward compatibility (for the same knowledge base).

Naturally, at some point, nonmonotonicity may be introduced as a result of the added expressiveness of the modeling language. When new statements are acquired, previously inferred knowledge may not hold any more. So monotonicity should be monitored and the end-user should be made aware that the system moved toward nonmonotonic behavior as a result of its development. But in the end, the computation of fixed points, which is often computationally prohibitive, is the only insurance of backward compatibility. In practical terms, there is some point in time where system development may not guarantee backward compatibility in terms of the previously inferred knowledge.

12.4.4   Minimization of Cognitive Distance

In our opinion, this is the strongest property of the CG formalism. Its closeness to ER diagrams and UML should make it easy to learn and use for knowledge modelers who were trained with graphical notations for modeling information systems. As there is a need for database applications to evolve toward knowledge-based systems, in-the-field experiences will reveal if our assumption is correct or not.

12.5   Conclusion

In Section 12.1, we made an argument for the use of human knowledge in software applications. In Section 12.2, we presented five constraints tightly linked to industrial settings where cost factors play a major role in the integration of new technology. Therefore, we feel that they should guide the choice of a knowledge modeling language aimed at building an easy bridge between data-based applications and knowledge-based systems. The conceptual graph notation is a knowledge modeling language that meets the requirements laid out by these constraints, and therefore, we advocate its use. Section 12.3 brushed a quick overview of what the CG representation paradigm is, and how CG-based systems are built. We described the CG formalism with the intent to show that its various features meet the requirements enumerated in Section 12.2. Section 12.4 briefly summarized our point of view on what makes the CG notation suitable for that task of *progressive* integration of knowledge-based technology.

Of course, due to a lack of space and for concision of presentation, we chose to show only the main features of the CG notation, giving references into the vast literature on the subject. The interested reader should be able to find the appropriate material fairly easily. The main objective of this chapter was to show the potential of CGs for the task of representing both data and knowledge-based systems, but also for allowing a layer-by-layer development model to be tailored to the various needs and constraints of an enterprise, hopefully optimizing the cost function normally associated with technology integration. In-depth research on that cost function with regard to CG-based systems is currently under study. Finally, interoperability of different CG-based systems, which is a key factor in modular design and development methods, will be the subject of a forthcoming paper.

References

1. Sowa, J.F., *Conceptual Structures: Information Processing in Mind and Machine*, Addison-Wesley, Reading, MA, 1984.
2. Sowa, J.F., *Knowledge Representation: Logical, Philosophical, and Computational Foundations*, Brookes/Cole, Stamford, CT, 2000.
3. Chein, M. and Mugnier, M.-L., Conceptual graphs: Fundamental notions. *Revue d'Intelligence Artificielle*, 6, 365, 1992.
4. Ghosh, B. and Wuwongse, V., Computational situation theory in the conceptual graph language, in *Lecture Notes in Artificial Intelligence #1115*, Springer-Verlag, Heidelberg, 1996, 188.
5. Mineau, G.W., Moulin, B. and Sowa, J.F., (Eds.), *Lecture Notes in Artificial Intelligence, vol. 699*, Springer-Verlag, Heidelberg, 1993.
6. Tepfenhart, W.M., Dick, J.P., and Sowa, J.F., (Eds.), *Lecture Notes in Artificial Intelligence, vol. 835*, Springer-Verlag, Heidelberg, 1994.
7. Ellis, G., Levinson, R.A., Rich, W., and Sowa, J.F., (Eds.), *Lecture Notes in Artificial Intelligence, vol. 954*, Springer-Verlag, Heidelberg, 1995.
8. Ellis, G. and Eklund, P. (Eds.), *Lecture Notes in Artificial Intelligence, vol. 1115*, Springer-Verlag, Heidelberg, 1996.
9. Lukose, D., Delugach, H., Keeler, M., Searle, L., and Sowa, J.F. (Eds.), *Lecture Notes in Artificial Intelligence, vol. 1257*, Springer-Verlag, Heidelberg, 1997.
10. Mugnier, M.-L. and Chein, M. (Eds.), *Lecture Notes in Artificial Intelligence, vol. 1453*, Springer-Verlag, Heidelberg, 1998.
11. Tepfenhart, W. and Cyre, W. (Eds.), *Lecture Notes in Artificial Intelligence, vol. 1640*, Springer-Verlag, Heidelberg, 1999.
12. Ganter, B. and Mineau, G.W. (Eds.), *Lecture Notes in Artificial Intelligence, vol. 1867*, Springer-Verlag, Heidelberg, 2000.
13. Stumme, G. and Delugach, H. (Eds.), *Lecture Notes in Artificial Intelligence, vol. 2120*, Springer-Verlag, Heidelberg, 2001.

14. Baget, J.-F. and Mugnier, M.-L., The SG family: Extensions of simple conceptual graphs, in *Proc. 17th Int. Joint Conf. on Artificial Intelligence*, 2001, 205.

15. Mineau, G.W., The engineering of a CG-based system: Fundamental issues, in *Lecture Notes in Artificial Intelligence, #1867*, Springer-Verlag, Heidelberg, 2000, 140.

16. Mineau, G.W. and Missaoui, R., The representation of semantic constraints in CG systems, in *Lecture Notes in Artificial Intelligence, #1257*, Springer-Verlag, Heidelberg, 1997, 138.

17. Mineau, G.W., The extensional semantics of the conceptual graph formalism, in *Lecture Notes in Artificial Intelligence, #1867*, Springer-Verlag, Heidelberg, 2000, 221.

18. Gerbé, O., Un modèle uniforme pour modélisation et métamodélisation d'une mémoire d'entreprise. Université de Montréal: Ph.D. thesis, 2000.

19. Moulin, B., The representation of linguistic information in an approach used for modeling temporal knowledge in discourses, in *Lecture Notes in Artificial Intelligence, #699*, Springer-Verlag, Heidelberg, 1993, 182.

20. Moulin, B. and Dumas, S., The temporal structure of a discourse and verb tense determination, in *Lecture Notes in Artificial Intelligence, #835*, Springer-Verlag, Heidelberg, 1994, 45.

21. Prediger, S., Simple concept graphs: A logic approach, in *Lecture Notes in Artificial Intelligence, #1453*, Springer-Verlag, Heidelberg, 1998, 225.

22. Mineau, G.W. and Gerbé, O., Contexts: A formal definition of worlds of assertions, in *Lecture Notes in Artificial Intelligence, #1257*, Springer-Verlag, Heidelberg, 1997, 80.

23. Lenat, D.B. and Guha, R.V., *Building Large Knowledge-Based Systems, Representation and Inference in the CYC Project*, Addison-Wesley, Reading, MA, 1990.

24. Mineau, G.W., Constraints on processes: Essential elements for the validation and execution of processes, in *Lecture Notes in Artificial Intelligence, #1640*, Springer-Verlag, Heidelberg, 1999, 66.

25. Baget, J.F., Genest, D., and Mugnier, M.L., A pure graph-based solution to the SCG-1 initiative, in *Lecture Notes in Artificial Intelligence, #1640*, Springer-Verlag, Heidelberg, 1999, 335.

26. Mineau, G.W., From actors to processes: The representation of dynamic knowledge using conceptual graphs, in *Lecture Notes in Artificial Intelligence, #1453*, Springer-Verlag, Heidelberg, 1998, 65.

27. Levinson, R.A. and Ellis, G., Multi-level hierarchical retrieval, *Knowledge-Based Systems*, vol. 5(3), 1992, 233.

28. Delugach, H.S., Dynamic assertion and retraction of conceptual graphs, in *Proc. 6th Ann. Workshop on Conceptual Structures*, E. Way, Ed., Binghamton, NY: SUNY at Binghamton, 1991, 15.

29. Benn, D. and Corbett, D., An application of the process mechanism to a room allocation problem using the pCG language, in *Lecture Notes in Artificial Intelligence, #2120*, Springer-Verlag, Heidelberg, 2001, 360.

13

Autonomous Mental Development by Robots: Vision, Audition, and Behaviors

Juyang Weng
Michigan State University

Wey-Shiuan Hwang
Michigan State University

Yilu Zhang
Michigan State University

Abstract. Machine learning techniques are useful to alleviate hand programming for vision, audition, and autonomous robots, especially in partially unknown or complex environments. However, traditional machine learning approaches are task specific in nature. We discuss some fundamental problems of the current task-specific paradigm to building complex systems and contrast it with some recent studies in neuroscience that indicate the power of developmental mechanisms in animals that enables autonomous development of cognitive and behavioral capabilities in humans and animals. What do we mean by autonomous development? Does it lead us to a more tractable and more systematic approach to vision, audition, robotics, and beyond? Motivated by human mental development from infancy to adulthood, the work presented here aims to enable robots to develop their mental skills autonomously, through online, real-time interactions with their environment. The SAIL developmental robot built at MSU is an early prototype of such a new kind of robot. Our experiments indicated that it appears feasible to develop vision, audition, and other cognitive capabilities as well as various cognition-based behaviors through online interactions by an autonomous robot.

13.1   Introduction

In order to understand the motive of the work, we need to first examine the established engineering paradigm. The new approach requires that we rethink the paradigm with which we all are familiar.

13.1.1   The Traditional Manual Development Paradigm

The process for developing an artificial system (e.g., an image analysis system) is not automatic—the human designer is in the loop. It follows a traditional, well-established paradigm for creating a manmade device:

- Start with a task. Given a task to be executed by a machine, it is the human engineer who understands the task (not the machine).
- Design a task-specific representation. The human engineer translates his understanding into a representation (e.g., develops some symbols or rules that represent particular concepts for the task and the correspondence between the symbols and physical concepts). The representation reflects how the human engineer understands the task.
- Programming for the specific task. The human engineer then writes a program (or designs a mechanism) that controls the machine to perform the task using the representation.

Run the program on the machine. If machine learning is used, sensory data are then used to modify the parameters of the task-specific representation. However, since the representation is designed for the specific task only, the machine cannot do anything beyond the predesigned representation. In fact, it does not even know what it is doing. All it does is run the program.

Although the above manual developmental paradigm is very effective for clean tasks, it has encountered tremendous difficulties for tasks that cannot be clearly formulated and thus include a large number of unknowns in task specification. If the task performer is a human adult, these task specification unknowns are dealt with through the cognitive and behavioral capabilities developed since infancy. However, the situation is very different with a machine that is programmed following the traditional paradigm. The machine is not able to automatically generate new representation for environments or tasks that its programmer has not considered at the programming stage.

13.1.2   Is the Human Vision System Totally Genetically Predetermined?

One may think that the human brain has an innate representation for the tasks that humans generally do. For example, one may believe that the human vision system and audition system are very much determined by human genes. However, recent studies of brain plasticity have shown that the human brain is not as task specific as commonly believed. Rich studies of brain plasticity in neuroscience exist including varying the extent of sensory input, redirecting input, transplanting cortex, lesion studies, and sensitive periods. Redirecting input seems illuminating in explaining how task-specific our brain really is. For example, Mriganka Sur and his coworkers rewired visual input to primate auditory cortex early in life. The target tissue in the auditory cortex, which is supposed to take auditory representation, was found to take on *visual* representation instead.[11] Furthermore, they have successfully trained animals to form visual tasks using the rewired auditory cortex.[9] Why are the self-organization schemes that guide development in the human brain so general that they can deal with either speech or vision, depending on what input it takes through development? Why do vision systems, audition systems, and robots that are programmed using human designed, task-specific representation not do well in complex, changing, partially unknown environments? What is the fundamental limitation of programming a single-modality system (e.g., vision or speech) without developing a multimodal agent?* What are the self-organization schemes that robots can use to automatically develop mental skills through interactions with the environment? Is it more advantageous to enable robots to automatically develop their mental skills than to program robots using human-specified, task-specific representation? Therefore, it is useful to rethink the traditional engineering paradigm.

13.1.3   The New Autonomous Development Paradigm

To overcome fundamental difficulties that face computer vision researchers, we have been investigating a new paradigm—the autonomous development paradigm, which is motivated by human mental development from infancy to adulthood. The new paradigm is as follows:

- Design body. According to the general ecological condition in which the robot will work (e.g., on land or underwater), human designers determine the sensors, the effectors, and the computational resources that the robot needs and then design a sensor-rich robot body.
- Design a developmental program. A human programmer designs a developmental program for the robot.
- Birth. A human operator turns on the robot whose computer then runs the developmental program.
- Develop mind. Humans mentally "raise" the developmental robot by interacting with it. The robot develops its cognitive skills through real-time, online interactions with the environment which includes humans (e.g., let them attend special lessons). Human trainers teach robots through verbal, gestural, or written commands in much the same way as parents teach their children. New skills and concepts are autonomously learned by the robot every day. The software (brain) can be downloaded from robots of different mental ages to be run by millions of other computers, e.g., desktop computers.

A robot that runs a developmental program is called a developmental robot. Such a robot is not simply an incremental learning system that can grow from small to large in terms of its occupied memory size. Such systems have already existed (e.g., some systems that use neural network techniques). Traditional machine learning systems still operate in the manual development mode outlined above, but cognitive development requires the new autonomous development mode.

What is the most basic difference between a traditional learning algorithm and a developmental algorithm? Autonomous development does require the ability to learn, but it requires something more fundamental. A developmental algorithm must be able to learn tasks that its programmer does not know or cannot even predict. This is because a developmental algorithm, being designed before the robot's "birth," must be able to learn new tasks and skills without requiring re-programming. The representation of a traditional learning algorithm is designed by a human for a given task, but one for a developmental algorithm must be automatically generated based on its own experience. This basic capability enables humans to learn more and more new tasks and skills using the same developmental program in the human genes.

---

*By definition, an agent is something that senses and acts.

13.1.4   The Developmental Approach

Since 1996,[14] we have been working on a robotic project called SAIL (short for Self-organizing, Autonomous, Incremental Learner); SHOSLIF is its predecessor.[15] The goal of the SAIL project is to *automate* the process of mental *development* for robots by following the new autonomous development paradigm.

An important issue with a developmental robot is what should be programmed and what should be learned. The nervous system of a primate may operate at several levels:

1. Knowledge level (e.g., symbolic skills, thinking skills, general understanding of the world around us, learned part of emotions, and rich consciousness).
2. Inborn behavior level (e.g., sucking, breathing, pain avoidance, and some primitive emotions in neonates). In neurons, they are related to synapses at birth.
3. Representation level (e.g., how neurons grow based on sensory stimuli).
4. Architecture level (corresponding to anatomy of an organism, e.g., a cortex area is prepared for eyes, if everything develops normally).
5. Timing level (the time schedule of neural growth for each area of the nervous system during development).

Studies in neuroscience seem to show that all of the above five levels are experience dependent.* In fact, experience can to a very great extent shape all these levels. But it seems that our genes have specified a lot for levels 2 through 5. Level 1 is made possible by levels 2 through 5 plus experience, but level 1 is not wired in. Thus, levels 2 through 5 seem to be what a programmer for a developmental algorithm may want to design—but not rigidly, they should be experience dependent. The designer of a developmental robot may have some information about the ecological condition of the environment in which the robots will operate, very much in a way that we know the ecological condition of a typical human environment. Such known ecological conditions are very useful for designing a robot body. However, the designer does not known what particular tasks that the robot will end up learning.

According to the above view, our SAIL developmental algorithm has some innate reflexive behaviors built-in. At the "birth" of the SAIL robot, its developmental algorithm starts to run. It runs in real time, through the entire "life span" of the robot. In other words, the design of the developmental program cannot be changed once the robot is "born," no matter what tasks it ends up learning. The robot learns while performing simultaneously. Its innate reflexive behaviors enable it to explore the environment while improving its skills. The human trainer teaches the robot by interacting with it, very much like the way a human parent interacts with his/her infant, letting it see around, demonstrating how to reach objects, teaching commands with the required responses, delivering reward or punishment (pressing "good" or "bad" buttons on the robot), etc. The SAIL developmental algorithm updates the robot memory in real time according to what was received by the sensors, what the robot did, and what it got as feedback from the human trainer.

13.1.5   Comparison of Approaches

The new developmental approach is fundamentally different from all the existing approaches. Table I.13.1 outlines the major characteristics of existing approaches to constructing an artificial system and the new developmental approach.

The developmental approach relieves humans from explicit design of (a) any task-specific representation and knowledge and (b) task-specific behavior representation, behavior modules, and their interactions. Some innate behaviors are programmed into a developmental program, but they are not

---

*The literature about this subject is very rich. A good start is *Rethinking Innateness* (pp. 270–314).[4]

TABLE I.13.1   Comparison of Approaches

Approach	Species Architecture	World Knowledge	System Behavior	Task Specific
Knowledge-Based	Programming	Manual modeling	Manual modeling	Yes
Behavior-Based	Programming	Avoid modeling	Manual modeling	Yes
Learning-Based	Programming	Model with parameters	Model with parameters	Yes
Evolutionary	Genetic search	Model with parameters	Model with parameters	Yes
Developmental	Programming	Avoid modeling	Avoid modeling	No

task specific. In other words, they are generally applicable and can be overridden by new, learned behaviors. As indicated by Table I.13.1, the developmental approach is the first approach that is not task specific.*

13.1.6   More Tractable

Is it true that the developmental approach makes system development more difficult? Not really, if the tasks to be executed by the system are very muddy. Task-nonspecific nature of a developmental program is a blessing. It relieves the human programmer from the daunting tasks of programming task-specific visual recognition, speech recognition, autonomous navigation, object manipulation, etc. for unknown environments. The programming task for a developmental algorithm concentrates on self-organization schemes, which are more manageable by human programmers than the above task-specific programming tasks.

Although the concept of developmental program for a robot is very new,[14] a lot of well-known self-organization tools can be used in designing a developmental program. In this chapter, we informally describe the theory, method, and experimental results of our SAIL-2 developmental algorithm tested on the SAIL robot. In the experiments presented here, our SAIL-2 developmental algorithm was able to automatically develop low-level vision and touch-guided motor behaviors.

13.2   The SAIL-2 Developmental Program

13.2.1   Mode of Operation: AA-Learning

A robot agent M may have several sensors. By definition, the *extroceptive, proprioceptive,* and *interoceptive* sensors are, respectively, those that sense stimuli from an external environment (e.g., visual), relative position of internal control (e.g., arm position), and internal events (e.g., internal clock).

The operational mode of automated development can be termed AA-learning (named after *a*utomated, *a*nimal-like learning without claiming to be complete) for a robot agent.

Definition   A robot agent M conducts AA-learning at discrete time instances, $t = 0, 1, 2, \ldots$, if the following conditions are met: (i) M has a number of sensors, whose signal at time t is collectively denoted by $x(t)$. (ii) M has a number of effectors, whose control signals at time t are collectively denoted by $a(t)$. (iii) M has a "brain" denoted by $b(t)$ at time t. (iv) At each time t, the time-varying state-update function f_t updates the brain based on sensory input $x(t)$ and the current brain $b(t)$:

$$b(t + 1) = f_t(x(t), b(t)) \tag{I.13.1}$$

---

*In the engineering application of the evolutionary approach, the representation of chromosomes is tasks specific.

and the action-generation function g_t generates the effector control signal based on the updated brain $b(t + 1)$:

$$a(t + 1) = g_t(b(t + 1)) \qquad (I.13.2)$$

where $a(t + 1)$ can be a part of the next sensory input $x(t + 1)$. The brain of M is closed in that after the birth (the first operation); $b(t)$ cannot be altered directly by humans for teaching purposes. It can only be updated according to Equation I.13.1.

As can be seen, AA-learning requires that a system not have two separate phases for learning and performance. An AA-learning agent learns while performing.

13.2.2   SAIL-2 Developmental Architecture

Figure I.13.1 gives a schematic illustration of the implemented architecture of SAIL-2 robot.

The current implementation of the SAIL-2 system includes extroceptive sensors and proprioceptive sensors. In the SAIL robot, the color stereo images come from two CCD cameras with wide-angle lens. The robot is equipped with 32 touch/switch sensors. Each eye can pan and tilt independently, and the neck can turn. A six-joint robot arm is the robot's manipulator.

13.2.3   Sensory Vector Representation

A developmental program may preprocess the sensory signal but the human programmer should not directly program feature detectors, since these predefined features are not sufficient to deal with an unknown environment. Thus, we must use a very general vector representation that keeps almost all the essential information in the raw sensory signal.

A digital image with r pixel rows and c pixel columns can be represented by a vector in a (rc)-dimensional space S without loss of any information. For example, the set of image pixels $\{I(i, j) | 0 \le i < r, 0 \le j < c\}$ can be written as a vector $X = (x_1, x_2, \ldots, x_d)^t$ where $x_{ri+j+1} = I(i, j)$ and $d = rc$. The actual mapping from the two-dimensional position of every pixel to a component in the d-dimensional vector X is not essential but is fixed once it is selected. Since the pixels of all the practical images can only take values in a finite range, we can view S as bounded. If we consider X as a random vector in S, the cross-pixel covariance is represented by the corresponding element

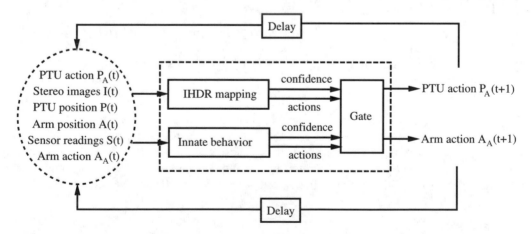

FIGURE I.13.1   A schematic illustration of the architecture of SAIL-2 robot. The sensory inputs of the current implementation include stereo images, position of the pan-tilt unit (PTU) for each camera, touch/ switch sensors, and the position of arm joints, as well as the action of every effector. The gate is to select an appropriate action from either cognitive mapping (learned) or innate behaviors (programmed in) according to the confidence values.

of the covariance matrix Σ_x of the random vector X. This representation, used early by Kirby and Sirovich[8] and Turk and Pentland,[13] have been widely used by what are now called *appearance-based* methods in the computer vision literature. Using this new representation, the correlation between any two pixels is considered in the covariance matrix Σ_x, not just between neighboring pixels.

13.2.4   Working Memory and Long-Term Memory

In the SAIL-2 system, the brain contains a working memory called state $w(t)$ and long-term memory implemented as a regression tree, which is discussed in detail in Section 13.3. The state keeps information about the previous actions (context). If $x(t)$ is the vector of all sensory inputs and action outputs at time t, the state is a long vector $w(t) = (x(t-1), x(t-2), \ldots x(t-k))$, where k is the temporal extent of the state. Typically, to save space, we make k small for sensory input, but large for action so that action retains more context. This gives a way of updating the working memory of the brain by function f_t. The updating of long-term memory (part of f_t) as well as the generation of action (what g_t does) are realized by the IHDR mapping in Figure I.13.1. The IHDR mapping accepts $(x(t), w(t))$ as input and generates $a(t+1)$ as the output, as well as updating the long-term memory of $b(t+1)$ for each time t. The IHDR is a general mapping approximator and is discussed in the following section.

13.2.5   Innate and Learned Behaviors

Innate behavior is programmed before the machine is born. The currently implemented built-in innate behavior is the motion detection and tracking mechanism for vision. When an object is moving in the scene, the absolute difference of each pixel between two consecutive image frames gives another image called the intensity-change image, which is directly mapped to the PTU control of each eye, also using the IHDR mapping technique but this mapping was generated in a prenatal offline learning process. In other words, this offline learning generates innate behaviors in the newborns. Our experience indicates that it is computationally faster and more reliable to generate innate behavior this way than explicitly finding the regions of moved objects through explicit programming.

The online learned IHDR mapping and the innate behavior may generate PTU motion signals at the same time. The resolution of such conflict is performed by the gate system. In the current implementation, the gate system performs subsumption, namely, the learned behavior takes the higher priority. Only when the learned behavior does not produce actions can the innate behavior be executed. A more resilient way of conducting subsumption is to use the confidence of each action source, but this subject is beyond the scope of this chapter.

13.3   The Mapping Engine: IHDR

One of the most challenging components of a developmental program is the mapping engine, which maps from sensory input and state (for context) to the effector control signal. Existing neural networks are not applicable due to the following reasons:

- The mapping engine must perform one-instance learning. An event represented by only one input sensory frame needs to be learned and recalled. Thus, iterative learning methods such as backpropagation learning or iterative pulling in self-organizing maps are not applicable.
- It must adapt to increasing complexity dynamically. It cannot have a fixed number of parameters, like a traditional neural network, since a developmental program must dynamically create system parameters to adapt to regions where increased complexity behaviors are needed due to, e.g., increased practice for some tasks.
- It must deal with the local minima problem. In methods that use traditional feed-forward neural networks with backpropagation learning, typically many instances of neural networks are

created in the development stage, each with a different random initial guess. The best performing network is chosen as the final system. However, the mapping engine of the developmental program cannot use this kind of method due to the real-time requirement. The system must perform on the fly in real time and thus does not allow a separate offline system evaluation stage. Further, the system that performs the best now may not necessarily perform best later. We use a coarse-to-fine local fitting scheme.

- It must be incremental. The input must be discarded as soon as it is used for updating the memory. It is not possible to keep all the training samples during open-ended incremental development.

- It must be able to retain information from some old memory. The effect of old samples used to train an artificial network is lost if these old samples do not appear repeatedly in the stream of training samples.

- It must have a very low time complexity so that response time is within a fraction of a second even if memory size has grown very large. Thus, any slow learning algorithm is not applicable here. Although the entire developmental process of a robot can extend to a long time period, the response time for each sensory input must be very short, e.g., a fraction of a second.

These considerations have been taken into account in our IHDR mapping engine described below.

13.3.1   Regression

Therefore, a major technical challenge is to incrementally generate the IHDR mapping. In the work reported here, online training is done by supplying desired action at the right time. When action is not supplied, the system generates its own actions using the updated IHDR mapping. In other words, the robot runs in real time. When the trainer wants to teach the robot, he/she pushes the effector through the corresponding touch sensor that directly drives the corresponding motor. Otherwise, the robot runs on its own.

Thus, the major problem is to approximate a mapping $h : \chi \mapsto \gamma$ from a set of training samples $\{(x_i, y_i)|x_i \in \chi, y_i \in \gamma, i = 1, 2, \ldots, n)\}$ that arrives one pair (x_i, y_i) at a time, where $y_i = *$ if y_i is not given (in this case, the approximator will produce estimated y_i corresponding to x_i). The mapping must be updated for each (x_i, y_i). If y_i was a class label, we could use linear discriminant analysis (LDA)[5] since the within-class scatter and between-class scatter matrices are all defined. However, if y_i is a numerical output that can take any value for each input component, it is a challenge to figure out an effective discriminant analysis procedure that can disregard input components that are either irrelevant to output or contribute little to the output.

We introduce a new hierarchical statistical modeling method. Consider the mapping $h : \chi \mapsto \gamma$, which is to be approximated by a regression tree called the incremental hierarchical discriminating regression (IHDR) tree, for the high-dimensional space χ. Our goal is to automatically derive discriminating features although no class label is available (other than the numerical vectors in space γ). In addition, for a real-time requirement, we must process each sample (x_i, y_i) to update the IHDR tree using only a minimal amount of computation (e.g., 0.05 sec).

13.3.2   Clustering in Both Input and Output Space

Two types of clusters are incrementally updated at each node of the IHDR tree, y-clusters and x-clusters, as shown in Figure I.13.2. The y-clusters are clusters in the output space γ and x-clusters are those in the input space χ. There are a maximum of q (e.g., $q = 10$) clusters of each type at each node. The q y-clusters determine the virtual class label of each arriving sample (x, y) based on its y part. Each x-cluster approximates the sample population in χ space for the samples that belong to it. It may spawn a child node from the current node if a finer approximation is required. At each node, y in (x, y) finds the nearest y-cluster in Euclidean distance and updates (pulling) the

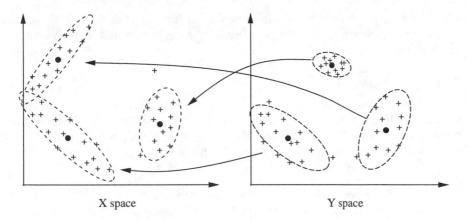

X space　　　　　　Y space

FIGURE I.13.2 y-clusters in space γ and the corresponding x-clusters in space χ. The first and the second order statistics are updated for each cluster.

center of the y-cluster. This y-cluster indicates to which corresponding x-cluster the input (x, y) belongs. Then, the x part of (x, y) is used to update the statistics of the x-cluster (the mean vector and the covariance matrix). These statistics of every x-cluster are used to estimate the probability of the current sample (x, y) belonging to the x-cluster, whose probability distribution is modeled as a multidimensional Gaussian at this level. In other words, each node models a region of the input space χ using q Gaussians. Each Gaussian will be modeled by more small Gaussians in the next tree level if the current node is not a leaf node. Each x-cluster in the leaf node is linked with the corresponding y-cluster.

Moreover, the center of these x-clusters provides essential information for discriminating subspace, since these x-clusters are formed according to virtual labels in γ space. We define a discriminating subspace as the linear space that passes through the centers of these x-clusters. A total of q centers of the q x-clusters give $q - 1$ discriminating features that span $(q - 1)$-dimensional discriminating space. A probability-based distance called size-dependent negative-log-likelihood (SNLL)[7] is computed from x to each of the q x-clusters to determine which x-cluster should be further searched. If the probability is high enough, the sample (x, y) should further search the corresponding child (maybe more than one but with an upper bound k) recursively, until the corresponding terminal nodes are found.

The algorithm incrementally builds an IHDR tree from a sequence of training samples. The deeper a node is in the tree, the smaller the variances of its x-clusters are. When the number of samples in a node is too small to give a good estimate of the statistics of q x-clusters, this node is a leaf node. If y is not given in the input, the x part is used to search the tree, until the nearest x-cluster in a leaf node is found. The center of the corresponding y-cluster is the produced estimated y output.*

Why do we use a tree? Two major reasons: (1) automatically derived features (instead of human defined features) and (2) fast search. The number of x-clusters in the tree is a very large number. The y-clusters allow the search to disregard input components that are not related to the output. For example, a humanoid robot has both visual and auditory sensors. When the robot is expected to separate a red ball from a blue one, visual information is enough. In this context, the corresponding node in the IHDR mapping spans only the $q - 1$-dimensional discriminating subspace within the visual portion of the whole sensory input. Consequently, the information from the auditory sensors is disregarded automatically. This subspace is the automatically derived feature space for the samples

---

*In each leaf node, we allow more than q clusters to fully use the samples available at each leaf node.

in the subtree. Further, the tree allows a large portion of faraway clusters to be disregarded. This results in the well-known logarithmic time complex for tree retrieval: $O(\log m)$ where m is the number of leaf nodes in the tree.

13.3.3 IHDR Procedure

The algorithm incrementally builds a tree from a sequence of training samples. The deeper a node is in the tree, the smaller the variances of its x-clusters are. When the number of samples in a node is too small to give a good estimate of the statistics of q x-clusters, this node is a leaf node. The following outlines the incremental algorithm for tree building (also tree retrieval when y is not given).

Procedure 1: Update-Node

Update-node: Given a node N and (x, y) where y is either given or not given, update the node N using (x, y) recursively. Output: Top matched terminal nodes. The parameters include k, which specifies the upper bound in the width of parallel tree search; δx, the sensitivity of the IHDR in χ space as a threshold to further explore a branch; and c, representing if a node is on the central search path. Each returned node has a flag c. If $c = 1$, the node is a central cluster and $c = 0$ otherwise.

1. Find the top matched x-cluster in the following way. If $c = 0$ skip to step (2). If y is given, do (a) and (b); otherwise do (b).

   (a) Update the mean of the y-cluster nearest y in Euclidean distance by using amnesic averages. Update the mean and the covariance matrix of the x-cluster corresponding to the y-cluster by using the amnesic average.

   (b) Find the x-cluster nearest x according to the probability-based distances. The central x-cluster is this x-cluster. Update the central x-cluster if it has not been updated in (a). Mark this central x-cluster as active.

2. For all the x-clusters of the node N, compute the probability-based distances for x to belong to each x-cluster.

3. Rank the distances in increasing order.

4. In addition to the central x-cluster, choose peripheral x-clusters according to increasing distances until the distance is larger than δx or a total of k x-clusters has been chosen.

5. Return the chosen x-clusters as active clusters.

From the above procedure, we can observe the following points. (a) When y is given, the corresponding x-cluster is updated, although this x-cluster is not necessarily the one on the central path from which the tree is explored. Thus, we may update two x-clusters, one corresponding to the given y, the other being the one used for tree exploration. The update for the former is an attempt to pull it to the right location. The update for the latter is an attempt to record the fact that the central x-cluster has hit this x-cluster once. (b) No matter if y is given or not, the x-cluster along the central path is always updated. (c) Only the x-clusters along the central path are updated, other peripheral x-clusters are not. We would like to avoid, as much as possible, storing the same sample in different brother nodes.

Procedure 2: Update-Tree

Update-tree: Given the root of the tree and sample (x, y), update the tree using (x, y). If y is not given, estimate y and the corresponding confidence. The parameters include k, which specifies the upper bound in the width of parallel tree search.

1. From the root of the tree, update the node by calling Update-node using (x, y).

2. For every active cluster received, check if it points to a child node. If it does, mark it inactive and explore the child node by calling Update-node. At most, q^2 active x-clusters can be returned this way if each node has at most q children.

3. The new central x-cluster is marked as active.

4. Mark additional active x-clusters according to the smallest probability-based distance d, up to k total if there are that many x-clusters with $d \leq \delta x$.
5. Do steps 2 through 4 recursively until all the resulting active x-clusters are terminal.
6. Each leaf node keeps samples (or sample means) $(\hat{x}_i, \hat{y}_i)$ that belong to it. If y is not given, the output is $\hat{y}_i$ if $\hat{x}_i$ is the nearest neighbor among these samples. If y is given, do the following: If $\|y - \hat{y}_i\|$ is smaller than an error tolerance, (x, y) updates $(\hat{x}_i, \hat{y}_i)$ through the amnesic average discussed below. Otherwise, (x, y) is a new sample to keep in the leaf.
7. If the current situation satisfies the spawn rule, i.e., the number of samples exceeds the number required for estimating statistics in a new child, the top-matched x-cluster in the leaf node along the central path spawns a child that has q new x-clusters. All the internal nodes are fixed in that their clusters do not update further using future samples, so their children do not get temporarily inconsistent assignment of samples.

The above incrementally constructed tree gives a coarse-to-fine probability model. If we use Gaussian distribution to model each x-cluster, this is a *hierarchical* version of the well-known mixture of Gaussian distribution models: the deeper the tree is, the more Gaussians are used and the finer these Gaussians are. At shallow levels, the sample distribution is approximated by a mixture of large Gaussians (with large variances). At deep levels, the sample distribution is approximated by a mixture of many small Gaussians (with small variances). The multiple search paths guided by probability allow a sample x that falls between two or more Gaussians at each shallow level to explore the tree branches that contain its neighboring x-clusters. Those x-clusters to which the sample (x, y) has little chance to belong are excluded for further exploration. This results in the well-known logarithmic time complex for tree retrieval: $O(\log m)$ where m is the number of leaf nodes in the tree, assuming that the number of samples in each leaf node is bounded above by a constant.

13.3.4 Amnesic Average

In incremental learning, the initial centers of each state clusters are largely determined by early input data. When more data are available, these centers move to more appropriate locations. If these new locations of the cluster centers are used to judge the boundary of each cluster, the initial input data were typically incorrectly classified. In other words, the center of each cluster contains some earlier data that do not belong to this cluster. To reduce the effect of these earlier data, the amnesic average can be used to compute the center of each cluster. The amnesic average can also track dynamic change of the input environment better than a conventional average.

The average of n input data $x_1, x_2, \ldots, x_{n+1}$ can be recursively computed from the current input data x_{n+1} and the previous average $\bar{x}^n$ by

$$\bar{x}^{(n+1)} = \frac{n\bar{x}^{(n)} + x_{n+1}}{n+1} = \frac{n}{n+1}\bar{x}^{(n)} + \frac{1}{n+1}x_{n+1}. \tag{I.13.3}$$

In other words, the previous average $\bar{x}^{(n)}$ gets a weight $n/(n+1)$ and the new input x_{n+1} gets a weight $1/(n+1)$. These two weights sum to one. The recursive Equation I.13.3 gives an equally weighted average. In amnesic average, the new input gets more weight than old inputs as given in the following expression: $\bar{x}^{(n+1)} = \frac{n-l}{n+1}\bar{x}^{(n)} + \frac{l+1}{n+1}x_{n+1}$, where l is a parameter.

The amnesic average can also be applied to the recursive computation of a covariance matrix Γ_x from incrementally arriving samples: $x_1, x_2, \ldots, x_n$ where x_i is a column vector for $i = 1, 2, \ldots$. The unbiased estimate of the covariance matrix from these n samples $x_1, x_2, \ldots, x_n$ is given in a batch form as

$$\frac{1}{n-1}\sum_{i=1}^{n}(x_i - \bar{x})(x_i - \bar{x})^T \tag{I.13.4}$$

with $n > 1$, where $\bar{x}$ is the mean vector of the n samples. Using the amnesic average, $\bar{x}^{(n+1)}$, up to the $(n+1)$-th sample, we can compute the amnesic covariance matrix up to the $(n+1)$-th sample as

$$\Gamma_x^{(n+1)} = \frac{n-1-l}{n}\Gamma_x^{(n)} + \frac{1+l}{n}(x_{n+1} - \bar{x}^{(n+1)})(x_{n+1} - \bar{x}^{(n+1)})^T \qquad (I.13.5)$$

for $n > l + 1$. When $n \leq l + 1$, we may use the batch version as in expression Equation I.13.4. Even with a single sample x_i, the corresponding covariance matrix should not be estimated as a zero vector, since x_i is never exact if it is measured from a physical event. For example, the initial variance matrix $\Gamma_x^{(1)}$ can be estimated as $\sigma^2 I$, where σ^2 is the expected digitization noise in each component and I is the identity matrix of the appropriate dimensionality.

13.3.5 Discriminating Subspace

Due to a very high input dimensionality (typically at least a few thousands), for computational efficiency, we should not represent data in the original input space χ. Further, for better generalization characteristics, we should use discriminating subspaces in which input components that are irrelevant to output are disregarded.

We first consider x-clusters. Each x-cluster is represented by its mean as its center and the covariance matrix as its size. However, since the dimensionality of the space χ is typically very high, it is not practical to keep the covariance matrix. If the dimensionality of χ is 3000, for example, each covariance matrix requires $3000 \times 3000 = 9,000,000$ numbers! We adopt a more efficient method that uses subspace representation.

As explained in Section 13.3.1, each internal node keeps up to q x-clusters. The centers of these q x-clusters are denoted by

$$C = \{c_1, c_2, \ldots, c_q \mid c_i \in \chi, i = 1, 2, \ldots, q\} \qquad (I.13.6)$$

The q center locations tell us the subspace $\mathfrak{R}$ in which these q centers lie. $\mathfrak{R}$ is a discriminating space since the clusters are formed based on the clusters in output space γ.

The discriminating subspace $\mathfrak{R}$ can be computed as follows. Suppose that the number of samples in cluster i is n_i and thus the grand total of samples is $n = \sum_{i=1}^{q} n_i$. Let $\overline{C}$ be the mean of all the q x-cluster centers,

$$\overline{C} = \frac{1}{n} \sum_{i=1}^{a} n_i c_i.$$

The set of scatter vectors from their center then can be defined as $s_i = c_i - \overline{C}, i = 1, 2, \ldots, q$. These q scatter vectors are not linearly independent because their sum is equal to a zero vector. Let S be the set that contains these scatter vectors $S = \{s_i \mid i = 1, 2, \ldots, q\}$. The subspace spanned by S, denoted by span (S), consists of all the possible linear combinations from the vectors in S, as shown in Figure I.13.3.

The orthonormal basis $a_1, a_2, \ldots, a_{(q-1)}$ of the subspace span (S) can be constructed from the radial vectors $s_1, s_2, \ldots, s_q$ using the *Gram-Schmidt Orthonormalization* (GSO) procedure. The number of basis vectors that can be computed by the GSO procedure is the number of linearly independent radial vectors in S.

Given a vector $x \in \chi$, we can compute its scatter part $s = x - \overline{C}$. Then we compute the projection of x onto the linear manifold by $f = M^T s$, where $M = [a_1, a_2, \ldots, a_{q-1}]$. We call the vector f the discriminating features of x in the linear manifold S. The mean and the covariance of the clusters then are computed on the discriminating subspace.

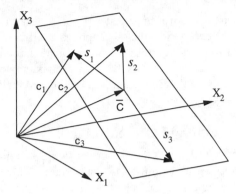

FIGURE I.13.3 The linear manifold represented by $\overline{C} + span(S)$, the spanned space from scatter vectors translated by the center vector $\overline{C}$.

13.3.6 The Probability-Based Metric

Let us consider the negative-log-likelihood (NLL) defined from Gaussian density of dimensionality $q - 1$:

$$G(x, c_i) = \frac{1}{2}(x - c_i)^T \Gamma_i^{-1}(x - c_i) + \frac{q-1}{2} \ln(2\pi) + \frac{1}{2} \ln(|\Gamma_i|) \qquad (\text{I.13.7})$$

We call it the Gaussian NLL for x to belong to the cluster i. c_i and Γ_i are the cluster sample mean and sample covariance matrix, respectively, computed using the amnesic average in Section 13.3.4. Similarly, we define the Mahalanobis NLL and Euclidean NLL as:

$$M(x, c_i) = \frac{1}{2}(x - c_i)^T \Gamma^{-1}(x - c_i) + \frac{q-1}{2} \ln(2\pi) + \frac{1}{2} \ln(|\Gamma|) \qquad (\text{I.13.8})$$

$$E(x, c_i) = \frac{1}{2}(x - c_i)^T \rho^2 I^{-1}(x - c_i) + \frac{q-1}{2} \ln(2\pi) + \frac{1}{2} \ln(|\rho^2 I|) \qquad (\text{I.13.9})$$

where Γ is the within-class scatter matrix of each node—the average of covariance matrices of q clusters:

$$\Gamma = \frac{1}{q-1} \sum_{i=1}^{q-1} \Gamma_i \qquad (\text{I.13.10})$$

computed using the same technique of the amnesic average.

Suppose that the input space is χ and the discriminating subspace for an internal node is $\mathfrak{R}$. The Euclidean NLL treats all the dimensions in the discriminating subspace $\mathfrak{R}$ the same way, although some dimensionalities can be more important than others. It has only one parameter ρ to estimate. Thus it is the least demanding among the three NLLs in the observation richness required. When very few samples are available for all the clusters, the Euclidean likelihood is the suited likelihood.

The Mahalanobis NLL uses within-class scatter matrix Γ computed from all the samples in all the q x-clusters. Using the Mahalanobis NLL as the weight for subspace $\mathfrak{R}$ is equivalent to using the Euclidean NLL in the basis computed from Fisher's LDA procedure.[5,12] It decorrelates all dimensions and weights, each dimension using a different weight. The number of parameters in Γ is $q(q-1)/2$, and thus, the Mahalanobis NLL requires more samples than the Euclidean NLL.

The Mahalanobis NLL does not treat different x-clusters differently because it uses a single within-class scatter matrix Γ for all the q x-clusters in each internal node. For the Gaussian NLL,

$L(x, c_i)$ in Equation I.13.7 uses the covariance matrix Γ_i of x-cluster i. In other words, the Gaussian NLL not only decorrelates the correlations but also applies a different weight at different locations along each rotated basis. However, it requires that each x-cluster has enough samples to estimate the $(q-1) \times (q-1)$ covariance matrix. It thus is the most demanding on the number of observations. Note that the decision boundary of the Euclidean NLL and the Mahalanobis NLL is linear but that by the Gaussian NLL is quadratic.

13.3.7 The Transition among Different Likelihoods

We prefer to use the Euclidean NLL when the number of samples in the node is small. Gradually, as the number of samples increases, the within-class scatter matrix of q x-clusters is better estimated. Then, we would use the Mahalanobis NLL. When a cluster has very rich observations, we use the full Gaussian NLL for it. We make an automatic transition when the number of samples increases. We define the number of samples n_i as the measurement of maturity for each cluster i; $n = \sum_{i=1}^{q} n_i$ is the total number of samples in a node.

For the three types of NLLs, we have three matrices, $\rho^2 I$, Γ, and Γ_i. Since the reliability of estimates are well indicated by the number of samples, we consider the number of scales received to estimate each parameter, called the number of scales per parameter (NSPP) in the matrices. The NSPP for $\rho^2 I$ is $(n-1)(q-1)$, because the first sample does not give any estimate of the variance and each independent vector contains $q - 1$ scales. For the Mahalanobis NLL, there are $(q-1)q/2$ parameters to be estimated in the (symmetric) matrix Γ. The number of independent vectors received is $n - q$ because each of the q x-clusters requires a vector to form its mean vector. Thus, there are $(n-q)(q-1)$ independent scalars. The NSPP for the matrix Γ is

$$\frac{(n-q)(q-1)}{(q-1)q/2} = \frac{2(n-q)}{q}.$$

To avoid a negative value when $n < q$, we take NSPP for Γ to be

$$\max \left\{ \frac{2(n-q)}{q}, 0 \right\}.$$

Similarly, the NSPP for Γ_i for the Gaussian NLL is

$$\frac{1}{q} \sum_{i=1}^{q} \frac{2(n_i - 1)}{q} = \frac{2(n-q)}{q^2}.$$

Table I.13.2 summarizes the results of the NSPP values of the above derivation.

A bounded NSPP is defined to limit the growth of NSPP so that other matrices that contain more scalars can take over when there are a sufficient number of samples for them. Thus, the bounded NSPP for $\rho^2 I$ is $b_e = \min\{(n-1)(q-1), n_s\}$, where n_s denotes the soft switch point for the next more complete matrix to take over. To estimate n_s, we consider a series of random variables drawn independently from a distribution with a variance σ^2, the expected sample mean of n random variables has a expected variance $\sigma^2/(n-1)$. We can choose a switch confidence value α for $1/(n-1)$. When $1/(n-1) = \alpha$, we consider that the estimate can take about a 50% weight. Thus,

TABLE I.13.2   Characteristics of Three Types of Scatter Matrices

Type	Euclidean $\rho^2 I$	Mahalanobis Γ	Gaussian Γ_i
NSPP	$(n-1)(q-1)$	$2\dfrac{(n-q)}{q}$	$2\dfrac{(n-q)}{q^2}$

$n = 1/\alpha + 1$. As an example, let $\alpha = 0.05$ meaning that we trust the estimate with 50% weight when the expected variance of the estimate is reduced to about 5% of that of a single random variable. This is like a confidence value in hypothesis testing except that we do not need an absolute confidence, a relative one suffices. We then get $n = 21$, which leads to $n_s = 21$.

The same principle applies to the Mahalanobis NLL and its bounded NSPP for Γ is

$$b_m = \min\left\{\max\left\{\frac{2(n-q)}{q}, 0\right\}, n_s\right\}.$$

It is worth noting that the NSPP for the Gaussian NLL does not need to be bounded, since among our models it is the best estimate when the number of samples increases further. Thus, the bounded NSPP for Gaussian NLL is $b_g = 2\frac{(n-q)}{q^2}$.

How do we realize automatic transition? We define a size-dependent scatter matrix (SDSM) W_i as a weighted sum of three matrices:

$$W_i = \omega_e \rho^2 I + \omega_m \Gamma + \omega_g \Gamma_i \tag{I.13.11}$$

where $\omega_e = b_e/b$, $\omega_m = b_m/b$, $\omega_g = b_g/b$ and b are normalization factors so that these three weights sum to 1: $b = b_e + b_m + b_g$. Using this size-dependent scatter matrix W_i, the size-dependent negative log likelihood (SDNLL) for x to belong to the x-cluster with center C_i is defined as

$$L(x, c_i) = \frac{1}{2}(x - c_i)^T W_i^{-1}(x - c_i) + \frac{q-1}{2}\ln(2\pi) + \frac{1}{2}\ln(|W_i|). \tag{I.13.12}$$

With b_e, b_m, and b_g changed automatically, $L(x, c_i)$ transit smoothly through the three NLLs. It is worth noting the relation between the LDA and SDNLL metric. LDA in space $\mathfrak{R}$ with original basis η gives a basis ε for a subspace $\mathfrak{R}' \subseteq \mathfrak{R}$. This basis ε is a properly oriented and scaled version for $\mathfrak{R}$ so that the within-cluster scatter in $\mathfrak{R}'$ is a unit matrix (Sections 13.2.3 and 13.10.2).[5] In other words, all the basis vectors in ε for $\mathfrak{R}'$ are already weighted according to the within-cluster scatter matrix Γ of $\mathfrak{R}$. If $\mathfrak{R}'$ has the same dimensionality as $\mathfrak{R}$, the Euclidean distance in $\mathfrak{R}'$ on ε is equivalent to the Mahalanobis distance in $\mathfrak{R}$ on η, up to a global scale factor. However, if the covariance matrices are very different across different x-clusters and each of them has enough samples to allow a good estimate of individual covariance matrix, LDA in space $\mathfrak{R}$ is not as good as Gaussian likelihood because covariance matrices of all x-clusters are treated as the same in LDA while Gaussian likelihood takes into account such differences. The SDNLL in Equation I.13.12 allows automatic and smooth transition between three different types of likelihood: Euclidean, Mahalanobis, and Gaussian, according to the predicted effectiveness of each likelihood.

13.3.8   Computational Considerations

The matrix-weighted squared distance from a vector $x \in \chi$ to each x-cluster with center c_i is defined by

$$d^2(x, c_i) = (x - c_i)^T W_i^{-1}(x - c_i) \tag{I.13.13}$$

which is the first term of Equation I.13.12.

This distance is computed only in $(q-1)$-dimensional space using the basis M. The SDSM W_i for each x-cluster in then only a $(q-1) \times (q-1)$ square symmetric matrix, of which only $q(q-1)/2$ parameters need to be estimated. When $q = 6$, for example, this number is 15.

Given a column vector v represented in the discriminating subspace with an orthonormal basis whose vectors are the columns of matrix M, the representation of v in the original space χ is $x = Mv$.

To compute the matrix-weighted squared distance in Equation I.13.13, we use a numerically efficient method, Cholesky factorization.[6] The Cholesky decomposition algorithm computes a lower triangular matrix L from W so that W is represented by $W = LL^T$. With the lower triangular matrix L, we first compute the difference vector from the input vector x and each x-cluster center c_i: $v = x - c_i$. The matrix-weighted squared distance is given by

$$d^2(x, c_i) = v^T W_i^{-1} v = v^T (LL^T)^{-1} v = (L^{-1}v)^T (L^{-1}v) \qquad (I.13.14)$$

We solve for y in the linear equation $Ly = v$ and then $y = L^{-1}v$ and $d^2(x, c_i) = (L^{-1}v)^T (L^{-1}v) \|y\|^2$. Since L is a lower triangular matrix, the solution for y in $Ly = v$ is trivial since we simply use the backsubstitution method as described in Press et al. (p. 42).[10]

13.4   Experiments

13.4.1   SAIL Robot

A human-size robot called SAIL was assembled at MSU, as shown in Figure I.13.4. SAIL robot's "neck" can turn. Each of its two "eyes" is controlled by a fast pan-tilt head. Its torso has four pressure sensors to sense push actions and force. It has 28 touch sensors on its arm, neck, head, and bumper to allow humans to teach it how to act by direct touch. Its drive-base is adapted from a wheelchair and thus the SAIL robot can operate both indoor and outdoor. Its main computer is a high-end dual-processor dual-bus PC workstation with 512MB RAM and an internal 27GB three-drive disk array for real-time sensory information processing, and real-time memory recall and update as well as

FIGURE I.13.4   The SAIL robot built at the Pattern Recognition and Image Processing Laboratory at Michigan State University.

real-time effector controls. This platform is being used to test the architecture and the developmental algorithm outlined here.

13.4.2 Autonomous Navigation

At each time instance, the vision-based navigation system accepts a pair of stereo images, updates the states which contain past sensory inputs and actions, and then outputs the control signal C to update the direction the vehicle is heading. In the current implementation, the state transition function f_i in Equation I.13.1 is programmed so that the current state includes a vector that contains the sensory input and past heading direction of the last T cycles. The key issue then is to approximate the action generation function g_t in Equation I.13.2. This is a very challenging approximation task since the function to be approximated is for a very high-dimensional input space and the real application requires the navigator to perform in real time.

We applied our IHDR algorithm to this challenging problem. Some of the example input images are shown in Figures I.13.5 and I.13.6. We first applied the IHDR algorithm to simulate the actual vision-based navigation problem. A totally of 2106 color stereo images with corresponding heading directions were used for training. The resolution of each image was 30 by 40. The input dimensionality of the IHDR algorithm was $30 \times 40 \times 3 \times 2 = 7200$, where 3 represents the red, green, and blue color bands. We used the other 2313 stereo images to test the performance of the trained system. Figure I.13.7 shows the error rate vs. the number of training epochs, where each epoch corresponds to a one-time feeding of the entire training sensory sequence. As shown even after the first epoch, the performance of the IHDR tree is already reasonably good. With the increased number of epochs, we observed improvements of the error rate. The error rate for the test set was 9.4% after 16 epochs.

The IHDR algorithm then was applied on the real training/testing experiment. The SAIL robot was trained interactively by a human trainer using the force sensors equipped on the body of the robot. The forces sensed by the sensors are translated to the robot's heading direction and speed. The training is online in real time. The trainer pushed just two force sensors to guide the robot to navigate a corridor of about 3.5 wide in the engineering building of Michigan State University. The navigation site includes a turn, and two straight sections that include a corridor door. Trips were found sufficient to reach reliable behavior. During the training, the IHDR algorithm receives the color stereo images as input and the heading direction as output. It rejects samples (not used for learning) if the input images are too similar to samples already learned. We tested performance by letting the robot go through the corridor 10 times. All the tests were successful. The closest distance between the SAIL robot and the wall was about 40 cm among the 10 tests. The test showed that

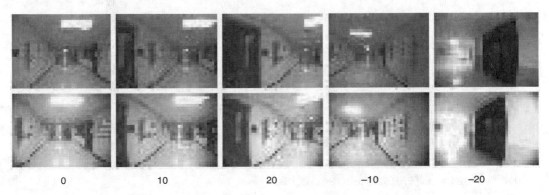

| 0 | 10 | 20 | −10 | −20 |

FIGURE I.13.5 A subset of images used in autonomous navigation. The number right below the image shows the needed heading direction (in degrees) associated with that image. The first row shows the images from the right camera while the second row shows those from the left camera.

FIGURE I.13.6 A subset of images that were inputs to guide the robot's turn. Rows one and three show the images from the left camera. The second and fourth rows show the images taken from the right camera.

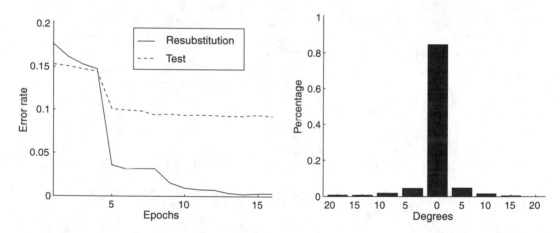

FIGURE I.13.7 The performance for autonomous navigation. (a) The plot for the error rates vs. epochs. The solid line rsepresents the error rates for the resubstitution test. The dashed line represents the error rates for the testing set; (b) the error histogram of the testing set after 16 epochs.

the SAIL robot can successfully navigate in an indoor environment after the interactive training, as indicated in Figure I.13.8. We plan to extend the area of navigation in future work.

13.4.3 Visual Attention Using Motion

The SAIL robot has embedded some innate behaviors, behaviors either programmed in or learned offline. For this behavior we used offline supervised learning. One such behavior is vision attention driven by motion. Its goal is to move the eyes so that a moving object of interest is moved to the

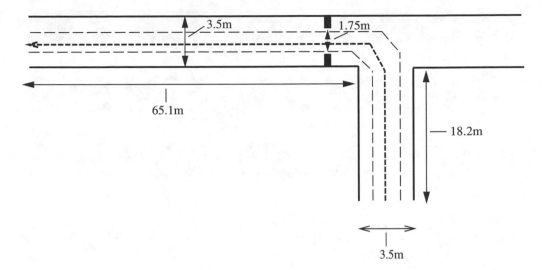

FIGURE I.13.8 The navigation path. The thick line is the desired navigation path and the thin lines are the navigation bounds. During the test, the SAIL robot navigated within the boundaries.

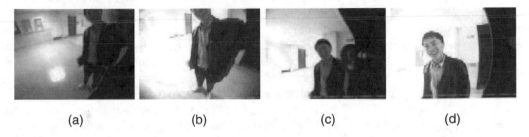

(a)   (b)   (c)   (d)

FIGURE I.13.9 An example of motion tracking, or motion-guided visual attention. (a) and (b) are the left and right images when an object moves in, (c) and (d) are the images after desired pan and tilt of the eyes.

"fovea," the center of the image. With this mechanism, perception and measurement are performed mainly for the fovea, while the periphery of an image frame is used only to find the object of interest.

To implement this mechanism, we first collect a sequence of images with moving objects. The input to the IHDR mapping is an image in which each pixel is the absolute difference of pixels in consecutive images. For training, we acquire the center of the moving object and the amount of motion that the pan-tilt unit must perform to bring the position to the image center. We used the IHDR algorithm to build the mapping between the motion (image difference) and pan-tilt control signals. For each training sample point $i, i = 1, 2, \ldots n$, we have an image difference as input and pan-and-tilt angle increments as output. Some example images are shown in Figure I.13.9.

13.4.4   Test for the Developmental Algorithm SAIL-2

We ran the developmental algorithm on the SAIL robot. Since tracking objects and reaching objects are sensorimotor behaviors first developed in early infants, we trained our SAIL robot for two tasks. In the first, called the finding-ball task, we trained the SAIL robot to find a nearby ball and then turn its eyes toward it so that the ball was located on the center of the sensed image. In the second, called pre-reaching task, we trained the SAIL robot to reach for the object once it has been located and the eyes fixate on it.

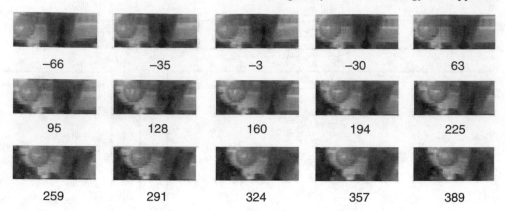

−66	−35	−3	−30	63
95	128	160	194	225
259	291	324	357	389

FIGURE I.13.10 A subset of images used in the tracking problem. The number right below the image shows the PTU position associated with that image. From left to right, one image sequence of ball-tracking is shown.

Existing studies on visual attention selection are typically based on low-level saliency measures, such as edges and texture.[1] In Birnbaum's work,[2] the visual attention is based on the need to explore geometrical structure in the scene. In our case, the visual attention selection is a result of past learning experiences. Thus, we do not need to define any task-specific saliency features. It is the SAIL robot that automatically derives the most discriminating features for the tasks being learned. At the time of learning, the ball was presented in the region of interest (ROI) inside the stereo images. The human trainer interactively pulled the robot's eyes toward the ball (through the touch sensors for the pan-tilt heads) so that the ball was located on the center of the ROI (fixating the eyes on the ball).* The inputs to the developmental algorithm are the continuous sequence of stereo images and the sequence of the pan-tilt head control signal. Three actions are defined for the pan-tilt head in pan direction: 0 (stop), 1 (move to the left), or −1 (move to the right). The size of the ROI we chose for this experiment is defined as 120×320. In the mind of the trainer, the ROI is divided into five regions so that each region is 120×64. The goal of the finding-ball task is to turn the pan-tilt head so that the ball is at the center region. Figure I.13.10 shows some example images for the tracking task.

The transitions during the training session are described below:

1. The task input is initiated by pushing a pressure sensor of the robot (or typing in a letter via keyboard) before imposing action to pan the camera. The action of the pan is zero at this time since no action is imposed.
2. The action of the pan is imposed at time t. The initialization flag is on at the same time. The main program issues a control signal to pan the camera.
3. The PTU starts to pan. The pan position as well as the image changes. Note that at time $t + 1$ the previous pan action is zero. When the ball is at the view fixation at time T, we stop the imposition of the pan action, and the initialization flag is off.
4. At time $T + 1$, the PTU stops moving and the image does not change any more. It is worth noting that the pan action is all zero after time $T - 1$.

---

*This is not typically done with human infants, since we cannot pull an infant's eye. However, this makes the robot learn much faster than a human baby can. This is, in fact, an advantage of robot over humans in that the robot can be built to facilitate training.

Similarly, the testing session can be explained as follows:

1. The human tester pushes a pressure sensor to simulate a task command and the initialization flag is on at time t.
2. The action of the pan is automatically generated by the IHDR tree. A nonzero action is expected according to the training process.
3. The PTU starts to move automatically and the image changes.
4. When the ball is at the fixation of the view at time T, the query result of the IHDR is a zero action. This zero action (stop) is sent to the PTU and the initialization flag is off.
5. At time $T + 1$, the PTU stops moving and the image does not change any more.

Why is the state important here? If the state that keeps the previous pan action is not used as input to the IHDR tree, the image and the pan position will be very similar at the point where the action should stop. This will make the PTU stop and go in a random fashion at this boundary point. The context (direction from which the arm is from) resolves the ambiguity.

The online training and testing were performed successfully and the robot performed the finding-ball and pre-reaching tasks successfully after interactive training, although the developmental algorithm was not written particularly for these two tasks.

To quantitatively evaluate the online learning and performance, we recorded the sensory data and studied the performance offline. Since the developmental algorithm runs indefinitely, does its memory grow without bound? Figure I.13.11(a) shows the program's memory usage. In the first stage, the tree grows because the samples are accumulated in the shallow nodes. When the performance of the updated tree is consistent with the desired action, the tree does not grow and thus the memory curve becomes flat. Recall step 6 of Procedure II in Section 13.3.3. In cases when the imposed action is significantly different from that of the tree, a new sample will be kept, which contributes to memory growth. Otherwise, the new samples (input–output pairs) only participate in the average of the nearest x- and y-clusters, simulating sensorimotor refinement of repeated practice, and the size of the tree does not grow. This is a kind of forgetting – not remembering every detail of a repeated practice. How fast does the developmental algorithm learn? Figure I.13.11(b) shows the accuracy of the PTU action in terms of the percentage of the view field. After the third epoch (repeated training), the systems can reliably move the eye so that the ball is at the center of ROI. Does the developmental algorithm slow down when it has learned more? Figure I.13.11(c) gives the plot of the average CPU time for each sensory action update. The average CPU time for update is within 100 msec, meaning that the system runs at about 10 Hz, 10 refresh of sensory input and 10 updated actions per second. Since the IHDR tree is dynamically updated, all the updating and forgetting are performed in each cycle. This relatively stable time profile is due to the use of the tree structure. The depth of the tree is stable.

13.4.5   Speech Recognition—A Simulation on Spoken Digit Recognition

Speech recognition has achieved significant progress in the past 10 years. It still faces, however, many difficulties, one of which is the training mode. Before training any acoustic models, such as hidden Markov model (HMM), the human trainer must do data transcription, a procedure for translating a speech waveform into a string of symbols representing the acoustic unit, like phonemes. In other words, the training data must be organized manually according to acoustic characteristics. This procedure requires some linguistic expertise and is very labor intensive. Moreover, inherently, this training can only be done offline, making online experience learning not possible. We used our SAIL-2 developmental algorithm to realize online learning instead, without a need for data transcription. The association between acoustic stream and the desired action is established through real-time experience.

In the experiment presented here, we trained the robot to act correctly according to real-time online auditory inputs. The effector of the system is represented by a 10-D action vector. Ten desired behaviors are defined, each for one of the ten digits ("one" to "ten"). Behaviors are identified by the component of the action vector with the maximum value. For example, if the first component of the action vector has the maximum value, it is identified as action 1.

The training procedure consists of two phases, supervised learning followed by reinforcement learning. In supervised learning, the imposed action is given to the system by the end of each utterance. In reinforcement learning, each reward is decided as follows. If the system makes an action at the end of an utterance, and the action is correct, the robot receives reward 1. If the action is wrong, the system gets reward -1. If no action is made within the time window, the system gets a reward -1. In all other cases including the silence period, the system gets reward -0.001.

The auditory data were collected as follows: 63 persons of various nationalities, including American, Chinese, French, India, Malaysian, and Spanish, and ages, from 18 to 50, participated in our speech data collection. Each person made five utterances for each of the ten digits. There is a silence of a length of about 0.5 sec between two consecutive utterances. This way, we got a speech data set with a total of 3150 isolated utterances.

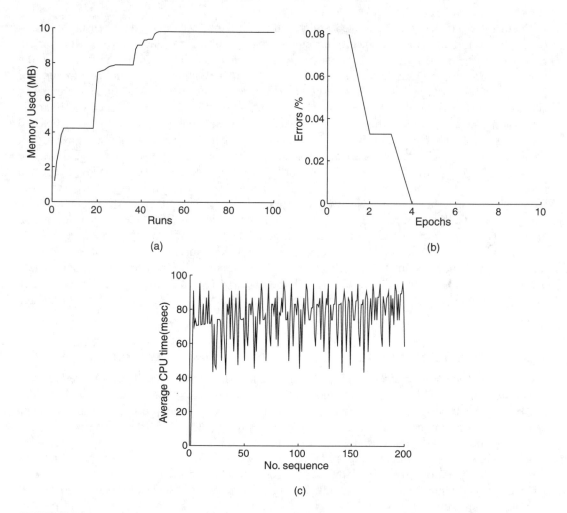

FIGURE I.13.11 (a) Memory usage for the offline simulation of the finding-ball task. (b) The accuracy of the finding-ball task vs. the number of training cases. (c) The CPU time for each update.

TABLE I.13.3   Results of Digit Recognition

	1	2	3	4	5	6	7	8	9	10	Ave
C.R.[a] (%)	98.4	95.2	93.7	96.8	95.2	93.7	96.8	92.1	93.7	93.7	94.9
E.R.[b] (%)	1.6	3.2	3.2	3.2	4.8	3.2	3.2	3.2	3.2	3.2	3.2
R.R.[c] (%)	0	1.6	3.2	0	0	3.2	0	4.8	3.2	3.2	1.9

[a]Correction rate
[b]Error rate
[c]Rejection rate

Performance was evaluated as follows. Within a short period before and after the end of an utterance, if there is one incorrect action or if the system does not take any action, we marked it as an error. If the system reacted correctly once or more within that time window, we marked it as correct. The test was done using five-fold leave-one-out cross validation. The results are summarized in Table I.13.3. The robot learned reliable responses through this challenging online interactive learning mode.

13.4.6   Speech Recognition—Grounded Speech Learning

In the experiments reported here, the SAIL robot was taught to follow verbal commands by moving its body, arm, and eyes. The verbal commands include "go left," "go right," "forward," "backward," "freeze," "arm left," "arm right," "arm up," "arm down," "hand open," "hand close," "see left," "see right," "see up," and "see down."

The training process* was conducted online in real time through physical interactions between a trainer and the SAIL robot. Once SAIL starts running, the microphone keeps collecting environmental sound. A SoundBlast card digitizes the signals from the microphone at 10 kHz. For every segment of 256 speech data points, which is roughly 25 msec of data, 13-order Mel-frequency Cepstral Coefficients (MFCCs)[3] are computed and serve as features. At the same time, the readings from the touch and pressure sensors are collected as feedback from the environment. Similar to the experiment described in Section 13.4.5, the training procedure consisted of two phases, supervised learning followed by reinforcement learning.

After training for 15 min, the SAIL robot could follow commands with about a 90% correct rate. Table I.13.4 summarizes the response performance of the SAIL robot when it was guided by the verbal commands navigating it through the corridors of the Engineering Building at MSU (see Figure I.13.12). The arm and eye commands were issued ten times each at different locations. We have been using learned verbal commands to teach the SAIL robot to perform vision-guided navigation. The navigation result of such an interactive verbal teaching process will be reported in detail elsewhere.

To further test the capability of dealing with speaker variation, we have conducted a corresponding multi-trainer experiment. Eleven persons participated in training. They spoke each of 15 commands five times, which resulted in 825 utterances. The speech data (4 out of 5 utterances of each commands) were fed into the SAIL robot offline, appended with appropriated actions at the end of each utterance. The SAIL robot with a partially trained brain started to run in real time. Then, the 12th trainer taught the SAIL robot using physical interactions four times for each command. In this way, we

---

*For more detailed discussion on the training procedure, the reader is referred to Zhang and Weng.[17]

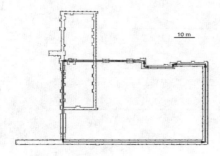

FIGURE I.13.12   Engineering Building, 2nd floor at MSU.

TABLE I.13.4   Performance of the SAIL Robot in the One-Trainer Case

Commands	Go left	Go right	Forward	Backward	Freeze
Correct rate (%)	97.1	91.3	93.8	100	80
Commands	Arm left	Arm right	Arm up	Arm down	Hand open
Correct rate (%)	100	90	100	100	90
Commands	Hand close	See left	See right	See up	See down
Correct rate (%)	90	100	100	100	100

TABLE I.13.5   Performance of the SAIL Robot Following the 12th Trainer's Command in the Multi-Trainer Case

Commands	Go left	Go right	Forward	Backward	Freeze
Correct rate (%)	94.5	89.9	92.7	100	100
Commands	Arm left	Arm right	Arm up	Arm down	Hand open
Correct rate (%)	100	90.9	96.3	92.7	89.9
Commands	Hand close	See left	See right	See up	See down
Correct rate (%)	89.9	90.9	92.7	100	100

simulated the situation in which a partially developed SAIL robot continuously developed its audio-driven behaviors.

After training, the 12th trainer tested the SAIL robot by guiding it through the second floor of the Engineering Building, just as was done in the one trainer case. Performance is summarized in Table I.13.5. More trainers introduced more variance in speech. The results show that the SAIL robot's performance in multi-trainer case degraded a little when compared to the one-trainer case, but it was still reasonable. Performance for other trainers was evaluated using the leftout utterances to test it offline. Performance is summarized in Table I.13.6.

This is the first work, as far as we know, about speech online learning without a predesigned model, with the number of words and speakers totally open, and learning directly from untranscribed real-time sensory data streams. Although the performance described above does not match those of traditional SR systems in terms of vocabulary, this work has made solid progress in this very difficult new learning mode.

TABLE I.13.6 Performance of the SAIL Robot on offline Test Data in the Multi-Trainer Case

Commands	Go left	Go right	Forward	Backward	Freeze
Correct rate (%)	88.9	89.3	92.8	87.5	88.9

Commands	Arm left	Arm right	Arm up	Arm down	Hand open
Correct rate (%)	90	90	100	100	90

Commands	Hand close	See left	See right	See up	See down
Correct rate (%)	80	100	100	100	100

13.5   Conclusions

We have introduced here a new kind of robot: robots that can develop their mental skills autonomously through real-time interactions with the environment. Representation of the system is automatically generated through online interaction between the developmental program and the experience. This new type of robot opens an array of new research problems, from computer vision to speech recognition to robotics. From the perspective of mental development, the work here raised the need for rethinking traditional static ways of programming and teaching a system, either for vision, speech, or an autonomous robot.

A technical challenge for the developmental algorithm is that the mapping engine must be scalable—keeping real-time speed and stable performance for a very large number of high-dimensional sensory and effector data. In our IHDR mapping engine, the developmental algorithm operates in real time. The SAIL-2 developmental algorithm has successfully run on the SAIL robot for real-time interactive training and real-time testing for two sensorimotor tasks: finding a ball and reaching the centered ball, two early tasks that infants learn to perform. These two tasks do not seem very difficult as judged by a human layperson, but they mark a significant technical advance because the program has little to do with the task. First, the same developmental program can be continuously used to train other tasks. This marks a significant paradigm change. Second, if a task-specific program was used for the two tasks that the SAIL robot infant learned, it cannot run in real time without special image process hardware, due to the extensive computation required for image analysis. Apart from appearance-based methods, almost no other image analysis methods can run in real time without special-purpose image processing hardware. Third, detecting an arbitrary object from an arbitrary background is one of the most challenging tasks for a robot. The main reason that our developmental algorithm can learn to do this is that it does not rely on humans to predefine representation. The same is true for our autonomous navigation experiment—the amount of scene variation along the hallways of our engineering building is beyond hand programming.

Automatically generated representation is able to use context very intimately. Every action is tightly dependent on the rich information available in the sensory input and the state. In other words, every action is context dependent. The complexity of the rules of such context dependence is beyond human programming. A human-defined representation is not be able to keep such rich information, without making the hand-designed representation too complicated to create any effective rules.

Since the developmental algorithm is not task specific, we plan to train the SAIL robot for other tasks to study the limitation of the current SAIL-2 developmental algorithm as well as the SAIL robot design. Future research directions include using longer context, attention selection, incorporating more sophisticated reinforcement learning mechanisms, and the value system. As pointed out by a recent article,[16] computational studies of mental development may be a common ground for understanding both machine and human intelligence.

Acknowledgments

The work is supported in part by National Science Foundation under grant No. IIS 9815191, DARPA ETO under contract No. DAAN02-98-C-4025, DARPA ITO under grant No. DABT63-99-1-0014, and research gifts from Microsoft Research, Siemens Corporate Research, and Zyvex.

References

1. Bichsel, M., *Strategies of Robust Object Recognition for the Automatic Identification of Human Faces*, Swiss Federal Institute of Technology, Zurich, Switzerland, 1991.
2. Birnbaum, L., Brand, M., and Cooper, P., Looking for trouble: Using causal semantics to direct focus of attention, in *Proc, IEEE Int'l Conf. Computer Vision*, IEEE Computer Press. Berlin, Germany, May 1993, pp. 49–56.
3. Deller, J.R., Proakis, J.G., and Hansen, J.H.L., *Discrete-Time Processing of Speech Signals*, Macmillan, New York, NY, 1993.
4. Elman, J.L. et al., *Rethinking Innateness: A Connectionist Perspective on Development*, MIT Press, Cambridge, MA, 1997.
5. Fukunaga, K., *Introduction to Statistical Pattern Recognition*, second edition, Academic Press, New York, NY, 1990.
6. Golub, G.H. and van Loan, C.F., *Matrix Computations*, The Johns Hopkins University Press, Baltimore, MD, 1989.
7. Hwang, W. et al., A fast image retrieval algorithm with automatically extracted discriminant features, in *Proc. IEEE Workshop Content-Based Access of Image and Video Libraries*, Fort Collins, Colorado, June 1999, pp. 8–15.
8. Kirby, M. and Sirovich, L., Application of the Karhunen-loeve procedure for the characterization of human faces. *IEEE Trans. Pattern Analysis and Machine Intelligence*, 12(1):103–108, Jan. 1990.
9. Pallas, S.L., von Melchner, L., and Sur, M., Visual behavior mediated by retinal projections directed to the auditory pathway, *Nature*, 404:871–876, 2000.
10. Press, W.H. et al., *Numerical Recipes*. Cambridge University Press, New York, 1986.
11. Sur, M., Angelucci, A., and Sharm, J., Rewiring cortex: The role of patterned activity in development and plasticity of neocortical circuits. *J. Neurobiol*, 41:33–43, 1999.
12. Swets, D.L. and Weng, J., Hierarchical discriminant analysis for image retrieval. *IEEE Trans. Pattern Anal. Machine Intel.*, 21(5):386–401, 1999.
13. Turk, M. and Pentland, A., Eigenfaces for recognition, *J. Cog. Neurosci.*, 3(1):71–86, 1991.
14. Weng, J., Learning in computer vision and beyond: Development, in C. W. Chen, and Y. Q. Zhang (Eds.), *Visual Communication and Image Processing*, Marcel Dekker, New York, NY, 1999. (Technical Report CPS96-60, Department of Computer Science, Michigan State University, East Lansing, MI, Dec. 1996.)
15. Weng, J. and Chen, S., Vision-guided navigation using SHOSLIF. *Neural Networks*, 11:1511–1529, 1998.
16. Weng, J. et al., Autonomous mental development by robots and animals, *Science*, 291:599–600, 2001.
17. Zhang, Y. and Weng, J., Grounded auditory development by a developmental robot, in *Proc. of INNS/IEEE Internat. Joint Conf. Neural Networks 2001 (IJCNN 2001)*, Washington, D.C., July 14–19, 2001, pp. 1059–1064.

Index

C